Lecture Notes in Computer S

Edited by G. Goos, J. Hartmanis, and J.

T0250561

Springer
Berlin
Heidelberg
New York
Barcelona
Hong Kong
London
Milan
Paris
Tokyo

Stefano Cagnoni   Jens Gottlieb   Emma Hart
Martin Middendorf   Günther R. Raidl (Eds.)

Applications of Evolutionary Computing

EvoWorkshops 2002:
EvoCOP, EvoIASP, EvoSTIM/EvoPLAN
Kinsale, Ireland, April 3-4, 2002
Proceedings

 Springer

Series Editors

Gerhard Goos, Karlsruhe University, Germany
Juris Hartmanis, Cornell University, NY, USA
Jan van Leeuwen, Utrecht University, The Netherlands

Main Volume Editor

Stefano Cagnoni
University of Parma, Dept. of Computer Engineering
Parco Area delle Scienze 181/a, 43100 Parma, Italy
E-mail: cagnoni@ce.unipr.it

The logo entitled "Evolving Human Figure Geometry" appearing on the cover is the
work of Matthew Lewis (http://www.cgrg.ohio-state.edu/˜mlewis/).

Cataloging-in-Publication Data applied for

Die Deutsche Bibliothek - CIP-Einheitsaufnahme

Applications of evolutionary computing : proceedings / EvoWorkshops 2002:
EvoCOP ... Kinsale, Ireland, April 3 - 4, 2002. Stefano Cagnoni .. (ed.). -
Berlin ; Heidelberg ; New York ; Barcelona ; Hong Kong ; London ; Milan ;
Paris ; Tokyo : Springer, 2002
  (Lecture notes in computer science ; Vol. 2279)
  ISBN 3-540-43432-1

CR Subject Classification (1998): D.1, F.2, I.2, G.2.1, I.4, I.5, G.1.6

ISSN 0302-9743
ISBN 3-540-43432-1 Springer-Verlag Berlin Heidelberg New York

Springer-Verlag Berlin Heidelberg New York
a member of BertelsmannSpringer Science+Business Media GmbH

http://www.springer.de

© Springer-Verlag Berlin Heidelberg 2002
Printed in Germany

Typesetting: Camera-ready by author, data conversion by Steingräber Satztechnik GmbH, Heidelberg
Printed on acid-free paper   SPIN: 10846254   06/3142   5 4 3 2 1 0

Volume Editors

Stefano Cagnoni
Dept. of Computer Engineering
University of Parma
Parco Area delle Scienze 181/a
43100 Parma, Italy
Email: cagnoni@ce.unipr.it

Jens Gottlieb
SAP AG
Neurottstrasse 16
69190 Walldorf, Germany
Email: jens.gottlieb@sap.com

Emma Hart
Napier University
School of Computing
219 Colinton Road
Edinburgh EH14 1DJ, UK
Email: emmah@dcs.napier.ac.uk

Martin Middendorf
Computer Science Group
Catholic University of Eichstätt–Ingolstadt
Ostenstr. 28
D-85072 Eichstätt, Germany
Email: martin.middendorf@ku-eichstaett.de

Günther R. Raidl
Algorithms and Data Structures Group
Institute of Computer Graphics
Vienna University of Technology
Favoritenstrasse 9-11/186
A-1040 Vienna, Austria
Email: raidl@ads.tuwien.ac.at

Preface

Evolutionary Computation (EC), which involves the study of problem solving, optimization, and machine learning techniques inspired by genetics and natural selection, has been able to draw the attention of an increasing number of researchers in several different fields. The number of applications and different disciplines that benefit from EC techniques is probably the most immediate proof of EC's high flexibility and potential. In recent years, a number of studies and results have been reported in the literature which have shown the capability of EC techniques to solve difficult problems in several domains.

When EvoNet organized its first events in 1998, the chosen format was a collection of workshops that dealt with both theoretical and application-oriented aspects of EC. While EuroGP soon became the main European event dedicated to Genetic Programming (GP), the number of workshops dedicated to specific EC applications increased. This led, in 2000, to the re-organization of EvoNet events into two main co-located independent sections: EuroGP became a single-track conference (the First European Conference on Genetic Programming), while the application-oriented workshops were merged in a multi-track event: EvoWorkshops 2000.

This volume contains the proceedings of EvoWorkshops 2002, which included the Second European Workshop on Evolutionary Computation in Combinatorial Optimization (EvoCOP), the Fourth European Workshop on Evolutionary Computation in Image Analysis and Signal Processing (EvoIASP), and the Third European Workshop on Evolutionary Scheduling and Timetabling (EvoSTIM). These workshops were held in Kinsale, Ireland, on 3-4 April 2002, jointly with EuroGP 2002, the Fifth European Conference on Genetic Programming.

EvoCOP is concerned with a variety of combinatorial optimization problems of academic and industrial interest. The workshop covers problem analyses, studies of algorithmic techniques used within evolutionary algorithms and related heuristics, and performance comparisons of different approaches.

EvoIASP was the first event ever specifically dedicated to the applications of EC to image analysis and signal processing. After four consecutive workshops, held between 1999 and 2002, it has become a traditional appointment for European and non-European researchers in those fields.

EvoSTIM/EvoPLAN is a merger of the third EvoSTIM workshop – a European event specifically dedicated to the applications of evolutionary computation to scheduling and timetabling – and the second EvoPLAN workshop – a forum specialized on evolutionary methods in the field of AI Planning. The workshop presents the latest results on evolutionary techniques in the fields of scheduling, timetabling, and AI planning. These fields are amongst the most successful applications for evolutionary techniques and are challenging for the developement of new methods.

EvoWorkshops 2002 has confirmed its tradition in providing researchers in these fields, as well as people from industry, an opportunity to present their latest research, discuss current developments and applications, besides fostering closer future interaction between members of all scientific communities that may benefit from the application of EC techniques.

The workshops were sponsored by EvoNet, the European Network of Excellence in Evolutionary Computation. Their organization was made possible thanks to the active participation of members of EvoIASP and EvoSTIM, the EvoNet working groups on Evolutionary Image Analysis and Signal Processing and on Evolutionary Scheduling and Timetabling, and of several other EvoNet members.

A total of 33 papers were accepted for publication out of more than 50 papers submitted to the 3 workshops. The rather low acceptance rate of EvoWorkshops 2002, along with the significant number of submissions, guarantees for the quality of the papers presented at the workshops, while showing the liveliness of the scientific movement in the corresponding fields. We would like to give credit to all members of the program committee, to whom we are very grateful for their quick and thorough work.

April 2002

Stefano Cagnoni
Jens Gottlieb
Emma Hart
Martin Middendorf
Günther R. Raidl

Organization

EvoWorkshops 2002 were organized by EvoNet jointly with EuroGP 2002.

Organizing Committee

EvoCOP co-chair:	Jens Gottlieb, SAP AG, Germany
EvoCOP co-chair:	Günther R. Raidl, Vienna University of Technology, Austria
EvoIASP chair:	Stefano Cagnoni, University of Parma, Italy
EvoSTIM/EvoPLAN co-chair:	Emma Hart, Napier University, Edinburgh, UK
EvoSTIM/EvoPLAN co-chair:	Martin Middendorf, Catholic University of Eichstätt–Ingolstadt, Germany
EvoWorkshops chair:	Stefano Cagnoni, University of Parma, Italy
EuroGP co-chair:	James Foster, University of Idaho, USA
EuroGP co-chair:	Evelyne Lutton, INRIA, France
Local co-chair:	Conor Ryan, University of Limerick, Ireland

Program Committee

Giovanni Adorni, University of Genoa, Italy
Wolfgang Banzhaf, University of Dortmund, Germany
Daniel Borrajo, Universidad Carlos III de Madrid, Spain
Alberto Broggi, University of Pavia, Italy
Edmund Burke, University of Nottingham, UK
Stefano Cagnoni, University of Parma, Italy
Jie Cheng, J. D. Power & Associates, MI, USA
Ela Claridge, The University of Birmingham, UK
David Corne, University of Reading, UK
Carlos Cotta-Porras, University of Malaga, Spain
Peter Cowling, University of Nottingham, UK
Marco Dorigo, Université Libre de Bruxelles, Belgium
Agoston E. Eiben, Free University Amsterdam, The Netherlands
Terry Fogarty, Napier University, UK
David Fogel, Natural Selection Inc., CA, USA
Jens Gottlieb, SAP AG, Germany
Jin-Kao Hao, University of Angers, France
Emma Hart, Napier University, UK
Daniel Howard, DERA, UK
Michiel de Jong, CWI, The Netherlands
Bryant Julstrom, St. Cloud State University, MN, USA
Dimitri Knjazew, SAP AG, Germany

Joshua Knowles, Université Libre de Bruxelles, Belgium
Gabriele Kodydek, Vienna University of Technology, Austria
Mario Köppen, Fraunhofer IPK Berlin, Germany
Jozef Kratica, Serbian Academy of Sciences and Arts, Yugoslavia
John Levine, University of Edinburgh, UK
Yu Li, University of Picardie, France
Ivana Ljubic, Vienna University of Technology, Austria
Evelyne Lutton, INRIA, France
Elena Marchiori, Free University Amsterdam, The Netherlands
Dirk Mattfeld, University of Bremen, Germany
Helmut Mayer, University of Salzburg, Austria
Daniel Merkle, University of Karlsruhe, Germany
Martin Middendorf, Catholic University of Eichstätt-Ingolstadt, Germany
Peter Nordin, Chalmers University of Technology, Sweden
Ben Paechter, Napier University, UK
Georgios I. Papadimitriou, Aristotle University, Greece
Riccardo Poli, University of Essex, UK
Günther Raidl, Vienna University of Technology, Austria
Marc Reimann, University of Vienna, Austria
Colin Reeves, Coventry University, UK
Peter Ross, Napier University, UK
Claudio Rossi, Ca' Foscari University of Venice, Italy
Franz Rothlauf, University of Bayreuth, Germany
Conor Ryan, University of Limerick, Ireland
Marc Schoenauer, INRIA, France
Jim Smith, The University of Western England, UK
Giovanni Squillero, Polytechnic of Turin, Italy
Thomas Stützle, Darmstadt University of Technology, Germany
Peter Swann, Rolls Royce plc, UK
El-ghazali Talbi, Laboratoire d'Informatique Fundamentale de Lille, France
Andrea G. B. Tettamanzi, Genetica srl, Italy
Andy Tyrrell, University of York, UK
Christine Valenzuela, Cardiff University, UK
Hans-Michael Voigt, GFaI - Center for Applied Computer Science, Germany

Sponsoring Institution

EvoNet, the Network of Excellence on Evolutionary Computing.

Table of Contents

EvoCOP Talks

EvoIASP Talks

EvoSTIM/EvoPLAN Talks

Hyperheuristics: A Tool for Rapid Prototyping in Scheduling and Optimisation

Peter Cowling, Graham Kendall, and Eric Soubeiga*

Automated Scheduling, optimisAtion and Planning (ASAP) Research Group
School of Computer Science and Information Technology (CSIT)
The University of Nottingham
Jubilee Campus, Wollaton Road, Nottingham NG8 1BB, England, United Kingdom

Abstract. The term *hyperheuristic* was introduced by the authors as a high-level heuristic that adaptively controls several low-level knowledge-poor heuristics so that while using only cheap, easy-to-implement low-level heuristics, we may achieve solution quality approaching that of an expensive knowledge-rich approach. For certain classes of problems, this allows us to rapidly produce effective solutions, in a fraction of the time needed for other approaches, and using a level of expertise common among non-academic IT professionals. Hyperheuristics have been successfully applied by the authors to a real-world problem of personnel scheduling. In this paper, the authors report another successful application of hyperheuristics to a rather different real-world problem of personnel scheduling occuring at a UK academic institution. Not only did the hyperheuristics produce results of a quality much superior to that of a manual solution but also these results were produced within a period of only three weeks due to the savings resulting from using the existing hyperheuristic software framework.

Key words: Hyperheuristic, Heuristic, Rapid prototyping, Personnel Scheduling.

1 Introduction

Personnel scheduling involves the allocation of timeslots and possible locations to people. This subject has been the focus of research since the 1970's [1,2,3]. As for other combinatorial optimisation problems, the resulting NP-hard problem is often solved using heuristic techniques [4,5,6,7,8], many of which use sophisticated metaheuristic methods and problem-specific information to arrive at a good solution. For instance Levine [9] used a hybrid genetic algorithm to solve an airline crew scheduling problem. To obtain better results, the genetic algorithm is hybridised with a local search heuristic which tries to repair infeasibilities present in the rows of the constraint matrix. Experiments also compared the hybrid genetic algorithm with branch-and-cut and branch-and-bound algorithms and these latter algorithms both outperformed the hybrid genetic algorithm.

---

* Corresponding author

S. Cagnoni et al. (Eds.): EvoWorkshops 2002, LNCS 2279, pp. 1–10, 2002.

Dowsland [6] used tabu search combined with strategic oscillation to schedule nurses. Using a variety of sophisticated local search neighbourhoods, the search is allowed to make some moves into infeasible regions in the hope to quickly reach a good solution beyond. The result is a robust and effective method capable of producing solutions of similar quality to those of a human expert. The same problem was solved in [5] using a co-evolutionary strategy based on co-operating subpopulations and problem-specific considerations.

However, as noted in [10,11] heuristic solution methods are often tailored specifically for the problem they are being applied to, so that it is unlikely that they may be successfully applied to a different problem. Heuristic and meta-heuristic approaches tend to be knowledge-rich, requiring substantial expertise in both the problem domain and appropriate heuristic techniques [5]. It is in this context that we proposed a *hyperheuristic* approach [10] as a heuristic that operates at a higher level of abstraction than current metaheuristic approaches. The hyperheuristic manages a set of simple, knowledge-poor, low-level heuristics (for example swap, add and drop moves). At any given decision point the hyperheuristic must choose which low-level heuristic to apply, without access to domain-knowledge. Hence we may use hyperheuristics in cases where little domain-knowledge is available (for instance when dealing with a new, poorly understood or unusual problem) or when a solution must be produced quickly (for example for prototyping). A hyperheuristic could be regarded as an "off-the-peg" method as opposed to a "made-to-measure" metaheuristic. A hyperheuristic is therefore a generic and fast method, which should produce solutions of acceptable quality, based on a set of easy-to-implement low-level heuristics. In order for a hyperheuristic to be applicable to a given problem, all that is needed is a set of low-level heuristics and one or more measures for evaluating solution quality. In [11] we described various ways of choosing the low-level heuristic to apply at each decision point and reported successful applications to a real-world problem of scheduling a sales summit. In this paper, we use our hyperheuristic techniques to solve another real-world problem of personnel scheduling occuring at a UK university. We aim to demonstrate that hyperheuristics are not only readily applicable to a wide range of problems of scheduling and other combinatorial optimisation problems, but also are capable of generating good quality solutions given very little development time.

In the remainder of the paper, sections 2, 3, 4 and 5 are devoted respectively to the application problem (Project Presentation Scheduling), the hyperheuristic approaches that we have used, an experimental study and conclusions.

2   The Project Presentation Scheduling Problem

Every academic year the School of Computer Science and Information Technology of the University of Nottingham is faced with the problem of scheduling final year BSc students' project presentations during a period of up to 4 weeks. As part of their course requirements, final year BSc students have to give a 15-minute presentation of their project. Each student works on a chosen project topic and is assigned a member of academic staff to supervise the project. Project presentations are then organised and each student must present his/her project before a panel of three members of academic staff who will mark the student's presentation: The Chair or First Marker, the Second Marker and the Observer. Ideally, the project's supervisor should be involved in the presentation (as Chair or Observer) but this is rarely the case in practice. Once every student has been assigned to a supervisor for his/her project, the problem is to schedule all individual presentations, that is, determine a first marker, a second marker and an observer for each individual presentation, and allocate both a room and a timeslot to the resulting quadruple (student, 1st marker, 2nd marker, observer). The presentations are organised in sessions, each containing up to six presentations. Typically the same markers and observers will see all of the presentations in a particular session. So the problem can be seen as that of determining (student, 1st marker, 2nd marker, observer, room, timeslot) tuples, that respect the following constraints: (1) Each presentation must be scheduled exactly once; (2) No more than six presentations for each room and for each session; (3) No member of staff (whether as 1st marker or as 2nd marker or as observer) can be scheduled to 2 different rooms within the same session. In addition presentations can only be scheduled in a given session when both the academic members of staff and the room assigned to those presentations are available during that session. There are four objectives to be achieved: (A) Fair distribution of the total number of presentations per staff member; (B) Fair distribution of the total number of sessions per staff member; (C) Fair distribution of the number of "bad" sessions per staff member, i.e. sessions at bad times (before 10:00 am, after 4:00 pm); (D) Optimise the match between staff research interest and project themes, and try to ensure that a supervisor attends presentations for projects which they supervise. To formulate the problem, we denote by $\mathbf{I}$ the set of students, $\mathbf{S}$ the set of academic staff members, $\mathbf{Q}$ the set of sessions and $\mathbf{R}$ the set of seminar rooms. Our decision variables are denoted by x_{ijklqr} $(i \in \mathbf{I}, j, k, l \in \mathbf{S}, j \neq k, j \neq l, k \neq l, q \in \mathbf{Q}, r \in \mathbf{R})$, where x_{ijklqr} is 1 if presentation of student i is assigned to 1st marker j, 2nd marker k, observer l and allocated to session q in seminar room r, otherwise x_{ijklqr} is 0; and y_{jqr} $(j \in \mathbf{S}, q \in \mathbf{Q}, r \in \mathbf{R})$ where y_{jqr} is 1 if staff j is in room r during session q, otherwise y_{jqr} is 0. We may then formulate the problem as follows:

Minimise $E(x) = 0.5A + B + 0.3C - D$

s.t.

$$\sum_{j,k,l \in \mathbf{S}} \sum_{q \in \mathbf{Q}} \sum_{r \in \mathbf{R}} x_{ijklqr} = 1, \quad (i \in \mathbf{I}) \tag{1}$$

$$\sum_{i \in \mathbf{I}} \sum_{j,k,l \in \mathbf{S}} x_{ijklqr} \leq 6, \quad (q \in \mathbf{Q}, r \in \mathbf{R}) \tag{2}$$

$$\sum_{r \in \mathbf{R}} y_{jqr} \leq 1, \quad (j \in \mathbf{S}, q \in \mathbf{Q}) \tag{3}$$

$$\sum_{i \in \mathbf{I}} \sum_{k,l \in \mathbf{S}} (x_{ijklqr} + x_{ikjlqr} + x_{ikljqr}) \leq M y_{jqr}, \quad (j \in \mathbf{S}, q \in \mathbf{Q}, r \in \mathbf{R}) \tag{4}$$

$$x_{ijklqr}, y_{jqr} \in \{0, 1\}, \quad i \in \mathbf{I}, j, k, l \in \mathbf{S}, j \neq k \neq l, q \in \mathbf{Q}, r \in \mathbf{R} \qquad (5)$$

where $A = \sum_{j \in \mathbf{S}} \left(\sum_{q \in \mathbf{Q}} \sum_{r \in \mathbf{R}} \sum_{i \in \mathbf{I}} \sum_{k,l \in \mathbf{S}} (x_{ijklqr} + x_{ikjlqr} + x_{ikljqr}) - K \right)^2$,

$B = \sum_{j \in \mathbf{S}} \left(\sum_{q \in \mathbf{Q}} \sum_{r \in \mathbf{R}} y_{jqr} - K_1 \right)^2$, $C = \sum_{j \in \mathbf{S}} \left(\sum_{q \in \mathbf{Q}_{bad}} \sum_{r \in \mathbf{R}} y_{jqr} - K_2 \right)^2$,

$D = \sum_{j \in \mathbf{S}} \left(\sum_{q \in \mathbf{Q}} \sum_{r \in \mathbf{R}} \sum_{i \in \mathbf{I}} \sum_{k,l \in \mathbf{S}} (p_{ij} + 10Sup_{ij})(x_{ijklqr} + x_{ikjlqr} + x_{ikljqr}) \right)$.

Equations (1), (2), (3) express constraints (1), (2), (3) respectively. Equation (4) links variables x_{ijklqr} with y_{jqr}, where M is a large number. $K = \frac{3|I|}{|S|}$, $K_1 = \frac{6P_1}{|S|}$ and $K_2 = \frac{6P_2}{|S|}$ where K, (K_1/K_2) is the average number of presentations (sessions/"bad" sessions) per member of staff, with P_1 (P_2) the total number of (bad) sessions used in the solution and Q_{bad} a subset of Q containing early sessions (before 10:00 am) and late sessions (after 4:00pm). In objective D, p_{ij} is an integer value associated with the level of matching between the topic of presentation i and the research interest of staff member j if he/she is involved in presentation i. The higher p_{ij}, the better the matching. Sup_{ij} is an indicator of whether staff member j is the supervisor of presentation i ($Sup_{ij} = 1$) or not ($Sup_{ij} = 0$). The different coefficients in the objective function were set so as to reflect the relative importance of each objective. The problem admits a feasible solution if there is enough room-time to allocate each presentation to, hence if $6|R||Q| > |I|$. In this instance of the problem, $|R| = 2$, $|Q| = 80$, $|I| = 151$, $|S| = 26$ and therefore the problem admits a feasible solution. At present the problem is solved manually. We developed a constructive heuristic that produces an initial solution better than the manual one. The constructive heuristic iteratively chooses a triple of staff members and assigns them to as many as 6 presentations in the first available session and room. Priority is given to non-bad sessions (to optimise objective C), to presentations whose supervisor is among the three staff and whose project topic is most related to the concerned staff research interest (to optimise objective D) , and the staff members are chosen on a cyclic basis (to optimise objectives A and B). The solution of the constructive heuristic is used as starting solution for all our algorithms presented in this paper. In the next section we present the different hyperheuristics used for this work.

3   Hyperheuristic Solution Techniques

We used two types of hyperheuristics, simple ones and choice function-based ones as described in [11]. The first type of hyperheuristic comprises simple multiple-neighbourhood search techniques which choose the low-level heuristics cyclically or at random. We used the following simple hyperheuristic algorithms. *SimpleRandom*: This algorithm repeatedly chooses one low-level heuristic at random and applies it once, until some stopping criterion is met. *RandomDescent*: This algorithm repeatedly chooses one low-level heuristic at random and applies it until no further improvement is possible, until some stopping criterion is met. *RandomPerm*: This algorithm repeatedly chooses a random sequence of all the low-level heuristics and applies each low-level heuristic once in the sequence order until some stopping criterion is met. It cycles from the last low-level heuristic in the sequence order back to the first one. *RandomPermDescent*: This algorithm does the same thing as *RandomPerm* but each low-level heuristic is applied in a steepest descent fashion.

The second type of hyperheuristic is based on a *Choice-Function* which provides guidance regarding which low-level heuristic to choose. The choice function adaptively ranks the low-level heuristics. In its original version presented in [10], the choice function is determined based upon information regarding recent improvement of each low-level heuristic (first order improvement) denoted by f_1, recent improvement for consecutive pairs of heuristics (second order improvement) denoted by f_2 and the amount of time elapsed since the heuristic was last called denoted by f_3. Thus we have $f_1(N_j) = \sum_n \alpha^{n-1}(\frac{I_n(N_j)}{T_n(N_j)})$ and $f_2(N_j, N_k) = \sum_n \beta^{n-1}(\frac{I_n(N_j, N_k)}{T_n(N_j, N_k)})$ where $I_n(N_j)/I_n(N_j, N_k)$ (respectively $T_n(N_j)/T_n(N_j, N_k)$) is the change in the objective function (respectively the number of CPU seconds) the nth last time heuristic N_j was called/called immediately after heuristic N_k. Both α and β are parameters between 0 and 1, which reflects the greater importance attached to recent performance. f_1 and f_2 aim at intensifying the search. The idea behind the expressions of f_1 and f_2 is analogous to the exponential smoothing forecast of their performance [12]. f_3 provides an element of diversification, by favouring those low-level heuristics that have not been called recently. Then we have $f_3(N_j) = \tau(N_j)$ where $\tau(N_j)$ is the number of CPU seconds which have elapsed since low-level heuristic N_j was last called. If the low-level heuristic just called is N_j then for any low-level heuristic N_k, the choice function f of N_k is defined as

$$f(N_k) = \alpha f_1(N_k) + \beta f_2(N_j, N_k) + \delta f_3(N_k) \tag{6}$$

In the above expression, the choice function attempts to predict the overall performance of each low-level heuristic. In [11] we presented a choice function which separately predicts the performance of each low-level heuristic with respect to each criterion of the objective function instead (i.e. A, B, C and D). The choice function f is then decomposed into

$$f(N_k) = \sum_{l \in \mathbf{L}} f_l(N_k) = \sum_{l \in \mathbf{L}} \left[\alpha_l f_{1l}(N_k) + \beta_l f_{2l}(N_j, N_k) + \frac{\delta}{|L|} f_3(N_k) \right] \tag{7}$$

where $\mathbf{L} = \{A, B, C, D\}$ is the set of the objective function criteria, and $f_{1l}(N_k)$ (respectively $f_{2l}(N_j, N_k)$) is obtained by replacing $I_n(N_k)$ (respectively $I_n(N_j, N_k)$) with $I_{ln}(N_k)$ (respectively $I_{ln}(N_j, N_k)$) in the expression of $f_1(N_j)$ (respectively $f_2(N_j, N_k)$) above. $I_{ln}(N_k)$ (respectively $I_{ln}(N_j, N_k)$) is the first (respectively second) order improvement with respect to criterion $l \in \mathbf{L}$. We will consider both variants of our choice function in the experiments. In [10], parameters of the original choice function were tuned manually. Instead, we presented a procedure in [11] that monitors the choice function parameters α, β and δ. The procedure can be applied to both choice function expressions (6) and (7).

The procedure adaptively adjusts the values of the different parameters so that, in light of the observed historical performance of each low-level heuristic, the weighting assigned to each factor is modified. More precisely it rewards/penalises the choice function intensification (f_1 and f_2) and diversification (f_3) factors by increasing/decreasing the values of the corresponding parameters α, β, δ. Overall the hyperheuristic works as follows

    Do

        For choice function (7) only: Choose a search criterion l
          - Select the low-level heuristic that maximises f (f_l for (7)) and apply it.
          - Update choice function f (f_l for (7))'s parameters using the adaptive procedure
        Until Stopping condition is met.

We would like to emphasize the fact that the implementation of the hyperheuristic techniques was quite fast. In effect all hyperheuristics presented here are "standard" approaches which worked well for another real-world problem [10,11]. Indeed all that was needed was a set of low-level heuristics to be input to the hyperheuristic black box. The way the hyperheuristic works is independent of both the nature of the low-level heuristics and the problem to be solved except for the objective function's value and CPU time which are passed from the low-level heuristics to the hyperheuristic. Whilst producing the hyperheuristic framework has taken over 18 months, using this framework took us only the equivalent of 101 hours of work (or two and a half weeks at 40 hours work per week) from understanding the problem to obtaining good hyperheuristic solutions. In the next section we report experiments carried out on all hyperheuristics when applied to the CSIT third year problem of scheduling project presentations.

4   Experiments

All algorithms were coded in Micosoft Visual C++ version 6 and all experiments were run on a PC Pentium III 1000MHz with 128MB RAM running under Microsoft Windows 2000 version 5. In all experiments the stopping condition was 600 seconds of CPU time. All experimental results were averaged over 10 runs. For each algorithm we distinguished the case where all moves (AM) are accepted and the case where only improving moves (OI) are accepted. We used three types of low-level heuristics based on "Replacing" one staff member in a session with a diferent one, "Moving" a presentation from one session to another, and "Swapping" two staff members, one from each presentation. The "Replace" and "Move" type have three variants, and the "Swap" type two variants. Overall we used the following $n = 8$ low-level heuristics.

1. *Replace one staff member in a session (N_1):* This heuristic chooses a random staff member, say j_1, chooses a random session, say q during which staff j_1 is scheduled for presentations and replaces j_1 with another random staff member, say j_2, in all presentations involving staff j_1 during session q. Staff

j_2 must not be involved in any presentations during session q prior to the substitution.

2. *Replace one staff member in a session (N_2) Version 2*: Same as previous heuristic but staff j_1 has the largest number of scheduled sessions.
3. *Replace one staff member in a session (N_3) Version 3*: Same as N_2 but session q is the one where staff j_1 has the smallest number of presentations. Also staff j_2 may be involved in presentations during session q prior to the substitution.
4. *Move a presentation from one session to another (N_4)*: This heuristic chooses a random presentation, removes it from its current session and reschedules it in another random session and a random room.
5. *Move a presentation from one session to another (N_5) Version 2*: Same as previous heuristic but the chosen presentation is that for which the sum of presentations involving all three staff (i.e. 1st marker, 2nd marker, observer) is smallest of all sessions.
6. *Move a presentation from one session to another (N_6)*: Same as N_5 but the new session is one where at least one of the staff members (i.e. 1st marker, 2nd marker, observer) is already scheduled for presentations.
7. *Swap 2nd marker of one presentation with observer of another (N_7)*: This heuristic chooses two random presentations and swaps the 2nd marker of the first presentation with the observer of the second presentation. The swap cannot involve the removal of a supervisor.
8. *Swap 1st marker of one presenation with 2nd marker of another (N_8)*: This heuristic chooses two random presentations and swaps the 1st marker of the first presentation with the 2nd marker of the second presentation. The swap cannot involved the removal of a supervisor.

For each of "Replace" and "Move" types of low-level heuristic the third version generally yields solutions of better quality than the two others. We shall see later on that the choice function hyperheuristic is capable of detecting this behaviour.

In Table 1, we present results for the simple hyperheuristics, the original choice function hyperheuristic (*OriginalHyperheuristic*), which needs manual tuning of its choice function parameters and both choice function (6) and (7) hyperheuristics (*Hyperheuristic6* and *Hyperheuristic7*). The choice of the search criterion l for *Hyperheuristic7* is based on a probability distribution which assigns the probability with which a criterion is chosen depending on the relative weight of that criterion in the objective function [11]. In this case we choose A with probability $p_a = \frac{0.5}{0.5+1+0.3+1}$, B with probability $p_b = \frac{1}{0.5+1+0.3+1}$, C with probability $p_c = \frac{0.3}{0.5+1+0.3+1}$, and D with probability $p_d = \frac{1}{0.5+1+0.3+1}$. Table 1 presents the results for all our algorithms. We show the objective value of both the manual and constructive heuristic solutions.

We see that all algorithms produced results much better than the manual solution and the constructive heuristic solution. We note that among the simple hyperheuristics, the best results are from *RandomPermDescent* and *RandomDescent*, which apply the low-level heuristics in a descent fashion. This was also

Table 1. All algorithms start from constructive heuristic solution

Algorithm	A	B	C	D	**E**
Manual Solution	455	40	18	-363	-90.1
Constructive Heuristic	1313	51	0	-1616	-908.5
RandomPerm-AM	962.80	66.00	18.10	-1616.80	-1063.97
RandomPerm-OI	790.40	20.75	4.7	-1614.8	-1197.59
RandomPermDescent-AM	624.60	17.40	3.50	-1618.00	-1287.25
RandomPermDescent-OI	631.20	16.10	3.50	-1617.50	-1284.75
SmpleRandom-AM	837.40	76.50	17.30	-1621.50	-1121.11
SimpleRandom-OI	796.00	22.10	2.70	-1614.10	-1193.19
RandomDescent-AM	645.40	20.10	3.50	-1618.40	-1274.55
RandomDescent-OI	583.00	18.30	4.40	-1614.50	-1303.38
Hyperheuristic6-AM	344.40	14.60	17.70	-1637.10	**-1444.99**
Hyperheuristic6-OI	592.40	15.20	4.80	-1629.40	-1316.56
Hyperheuristic7-AM	671.80	20.60	4.90	-1628.40	-1270.43
Hyperheuristic7-OI	671.40	18.70	2.60	-1620.70	-1265.52
OriginalHyperheuristic-AM	545.60	17.00	3.20	-1617.10	-1326.34
OriginalHyperheuristic-OI	665.20	19.00	5.70	-1621.00	-1267.69

the case in [10,11] when these algorithms were applied to a different scheduling problem. While *Hyperheuristic7* gave results comparable to those of the simple hyperheuristics, *Hyperheuristic6* produced results much better than those of the simple hyperheuristics. It should be noted that as the search goes on and as solution quality improves, finding a better solution becomes very challenging. In light of this, it would appear that *Hyperheuristic6* performs a very effective search. We note that the original hyperheuristic whose parameters were manually set to $\alpha = 0.1, \beta = 0.1, \delta = 2.5$ for the AM case and $\alpha = 0.1, \beta = 0.1, \delta = 1.5$ for the OI case, produced good results too, as was the case in [10] when applied to another real-world problem. The fact that *Hyperheuristic6* outperforms the original hyperheuristic confirms the net advantage of having an adaptive procedure for setting parameters as noted in [11]. Beyond the obvious advantage of the reduction in intervention from automatically setting parameters the procedure in some sense learns the interplay between all factors of the choice function and adjusts values of α, β, δ accordingly.

In Table 2, we compare the frequency of call of each low-level heuristic (the number of times that each low-level heuristic has been called) in the case of *Hyperheuristic6* and *Hyperheuristic7*. We can see that these two hyperheuristics do not treat the low-level heuristics in the same way. More precisely, we see that although low-level heuristic N_3 is the most important for both hyperheuristics, low-level heuristics N_1, N_4, N_5, N_7 are given more importance by *Hyperheuristic7* than by *Hyperheuristic6*, while low-level heuristic N_6 is treated the same way by both hyperheuristics. It seems that the three most important heuristics for *Hyperheuristic6* are N_3, N_2 and N_6 whereas heuristics N_3, N_1 and N_8 are the top three from *Hyperheuristic7*'s perspective. As mentioned earlier, a comparison of the proportion of calls of the low-level heuristics within each of "Replace" and

Table 2. Comparison between *Hyperheuristic6* and *Hyperheuristic7* of the frequency of calls of the low-level heuristics

Algorithm	E	N_1	N_2	N_3	N_4	N_5	N_6	N_7	N_8
Hyperheuristic12-AM	-1402.4	4	42	76	3	16	20	5	15
Proportion	-	0.02	0.23	0.42	0.02	0.09	0.11	0.03	0.08
Hyperheuristic13-AM	-1314	25	19	35	10	21	20	17	22
Proportion	-	0.15	0.11	0.21	0.06	0.12	0.12	0.10	0.13

"Move" types shows that the third version for each type is called more often than each of the other two versions of that type. Thus N_3 is called by both choice-function based hyperheuristics more often than N_1 and N_2. Similarly, N_6 is called more often by *Hyperheuristic6* than N_4 and N_5 while *Hyperheuristic7* slightly favours N_5 over N_6. Overall the choice function hyperheuristic appears capable of detecting good low-level heuristics [11].

The superiority of the choice function-based hyperheuristics was also noticeable when we ran each of our algorithms starting from the manual solution (instead of that produced by the constructive heuristic). Again all algorithms produced results better than the manual solution and *Hyperheuristic6* produced the best results of all. However the results are very poor when compared to the solution obtained by the constructive heuristic. It seems that the search starts from a region of such poor quality, since the manual scheduler was unable to handle the objective of matching research interests, that it is difficult to move to a good area. *OriginalHyperheuristic* produced results of comparable quality to those of the simple hyperheuristics. Whilst *Hyperheuristic6* produces better results than *Hyperheuristic7* for the project presentation scheduling problem, the opposite happened with the sales summit scheduling problem [11]. Effectively in [11], *Hyperheuristic7*, which decomposed the objective value into three criteria, gave better results than *Hyperheuristic6*, which did not. The reason why *Hyperheuristic7* does not outperform *Hyperheuristic6* here is probably due to the fact that it would have to deal with more individual objectives than in [11]. The more individual objectives there are (i.e. the bigger $|L|$), the more parameters α_l, β_l ($l \in \mathbf{L}$) need to be managed. Thus convergence to a good solution for the hyperheuristic search could slow down when in presence of a substantial number of individual objectives in E. It is also worth noting that more time was spent "tuning" model parameters in [11] than has been undertaken for the project presentation problem.

5 Conclusions

We have applied various hyperheuristics to a real-world problem of scheduling project presentations in a UK university. Prior to our intervention the problem was solved manually. Our hyperheuristics produced solutions dramatically better than the manual one when starting from both a constructive heuristic solution and (even) the original manual solution. Comparing the performance of our hyperheuristics for two real-world problems it appears that the choice function hyperheuristic produces the best result of all hyperheuristics. This type of hyperheuristic is based on a choice function which adaptively ranks the low-level heuristics and thus provides effective guidance to the hyperheuristics. In this paper we have added evidence to our claim that hyperheuristic approaches are easy to implement which deliver solutions of acceptable quality, providing a useful tool in rapid prototyping of optimisation systems. Ongoing research will investigate other types of hyperheuristics applied to a wider range of real-world problems.

References

1. K. Baker. Workforce allocation in cyclical scheduling problems: A survey. *Operational Research Quarterly*, 27(1):155–167, 1976.
2. J. M. Tien and A. Kamiyama. On manpower scheduling algorithms. *SIAM Review*, 24(3):275–287, July 1982.
3. D. J. Bradley and J. B. Martin. Continuous personnel scheduling algorithms: a literature review. *Journal Of The Society For Health Systems*, 2(2):8–23, 1990.
4. G. M. Thompson. A simulated-annealing heuristic for shift scheduling using non-continuously available employees. *Computers and Operations Research*, 23(3):275–288, 1996.
5. U. Aickelin and K. A. Dowsland. Exploiting problem structure in a genetic algorithm approach to a nurse rostering problem. *Journal of Scheduling*, 3:139–153, 2000.
6. K. A. Dowsland. Nurse scheduling with tabu search and strategic oscillation. *European Journal of Operational Research*, 106:393–407, 1998.
7. B. Dodin, A. A. Elimam, and E. Rolland. Tabu search in audit scheduling. *European Journal of Operational Research*, 106:373–392, 1998.
8. J. E. Beasley and B. Cao. A tree search algorithm for the crew scheduling problem. *European Journal of Operational Research*, 94:517–526, 1996.
9. D. Levine. Application of a hybrid genetic algorithm to airline crew scheduling. *Computers and operations research*, 23(6):547–558, 1996.
10. P. Cowling, G. Kendall, and E. Soubeiga. A hyperheuristic approach to scheduling a sales summit. In E. Burke and W. Erben, editors, *Selected Papers of the Third International Conference on the Practice And Theory of Automated Timetabling PATAT'2000*, Springer Lecture Notes in Computer Science, 176-190, 2001.
11. P. Cowling, G. Kendall, and E. Soubeiga. A parameter-free hyperheuristic for scheduling a sales summit. Proceedings of the 4th Metaheuristic International Conference, MIC 2001, 127-131.
12. S. C. Wheelwright and S. Makridakis. *Forecasting methods for management*. John Wiley & Sons Inc, 1973.

SavingsAnts for the Vehicle Routing Problem

Karl Doerner, Manfred Gronalt, Richard F. Hartl, Marc Reimann,
Christine Strauss, and Michael Stummer

Institute of Management Science, University of Vienna, Brünnerstrasse 72,
A-1210 Vienna, Austria
{karl.doerner, manfred.gronalt, richard.hartl, marc.reimann,
christine.strauss}@univie.ac.at, a9705488@unet.univie.ac.at
http://www.bwl.univie.ac.at/bwl/prod/index.html

Abstract. In this paper we propose a hybrid approach for solving vehicle routing problems. The main idea is to combine an Ant System (AS) with a problem specific constructive heuristic, namely the well known Savings algorithm. This differs from previous approaches, where the subordinate heuristic was the Nearest Neighbor algorithm initially proposed for the TSP. We compare our approach with some other classic, powerful meta-heuristics and show that our results are competitive.

1 Introduction

The European situation in freight transportation reflects the need for improved efficiency, as the traffic volume increases much faster than the street network grows. Thus, given the current efficiency, this will eventually lead to a breakdown of the system. However, with rapidly increasing computational power intelligent optimization methods can be developed and used to increase the efficiency in freight transportation and circumvent the above mentioned problem.

The VRP involves the design of a set of minimum cost delivery routes, originating and terminating at a depot, which services a set of customers. Each customer must be supplied exactly once by one vehicle route. The total demand of any route must not exceed the vehicle capacity. The total length of any route must not exceed a pre-specified bound. This problem is known to be NP-hard (cf. [1]), such that exact methods like Dynamic Programming or Branch & Bound work only for relatively small problems in reasonable time. Thus, a large number of approximation methods have been proposed. Most of the recent approaches are based on meta-heuristics like Tabu Search, Simulated Annealing and Ant Systems.

The Ant System approach, belonging to a class of methods called Ant Colony Optimization, is based on the behavior of real ants searching for food. Real ants communicate with each other using an aromatic essence called pheromone, which they leave on the paths they traverse. If ants sense pheromone in their vicinity, they are likely to follow that pheromone, thus reinforcing this path. The pheromone trails reflect the 'memory' of the ant population. The quantity of the pheromone deposited on paths depends on both, the length of the paths as well as the quality of the food source found.

S. Cagnoni et al. (Eds.): EvoWorkshops 2002, LNCS 2279, pp. 11–20, 2002.

The observation of this behavior has led to the development of Ant Systems by Colorni et al. (see e.g. [2]) in the early nineties. In artificial terms the optimization method uses the trail following behavior described above in the following way. Ants construct solutions by making a number of decisions probabilistically. In the beginning there is no collective memory, and the ants can only follow some local information. As some ants have constructed solutions, pheromone information is built. In particular, the quantity of pheromone deposited by the artificial ants depends on the solution quality found by the ants. This pheromone information guides other ants in their decision making, i.e. paths with high pheromone concentration will attract more ants than paths with low pheromone concentration. On the other hand, the pheromone deposited is not permanent, but rather evaporates over time. Thus, over time, paths that are not used will become less and less attractive, while those used frequently will attract ever more ants.

This approach has been applied to a number of combinatorial optimization problems, such as the Graph Coloring Problem [3], the Quadratic Assignment Problem (e.g. [4]), the Travelling Salesman Problem (e.g. [5], [6]), the Vehicle Routing Problem ([7], [8]) and the Vehicle Routing Problem with Time Windows ([9]). Recently, a covergence proof for a generalized Ant System has been developed by Gutjahr ([10]).

In the previous approaches for the VRP ([7], [8]) the construction of solutions was based on a sequential tour building approach, which utilized a parametrized savings criterion. The main idea of our new approach is to transfer the simultaneous tour construction mechanism proposed in [11] into a rank based Ant System. Our computational findings show that a considerable improvement is achieved through this new approach.

The remainder of this paper is organized as follows. In the next section we briefly describe the Savings algorithm before we propose our new approach, which we will refer to as SavingsAnts. After that we will report on computational results with our SavingsAnts. In Section 4 we conclude with a discussion of our findings.

2   Savings and Ant System Algorithms for VRPs

In this section we describe the Savings algorithm and the Ant Systems algorithm. The basic structure of our Ant System algorithm is identical to the one proposed in [8]. Thus, we will focus on our improvements to the original algorithm.

2.1   The Savings Algorithm

The Savings algorithm, proposed in [11], is the basis of most commercial software tools for solving VRPs in industrial applications. It is initalized with the assignment of each customer to a separate tour.

After that for each pair of customers i and j the following savings measure is calculated:

$$s(i,j) = d(i,0) + d(0,j) - d(i,j), \tag{1}$$

where $d(i,j)$ denotes the distance between locations i and j and the index 0 denotes the depot. Thus, the values $s(i,j)$ contain the savings of combining two customers i and j on one tour as opposed to serving them on two different tours.

In the iterative phase, customers or partial tours are combined according to these savings, starting with the largest savings, until no more combinations are feasible. A combination is infeasible if it violates either the capacity or the tourlength constraints.

The result of this algorithm is a (sub-)optimal set of tours through all customers.

2.2   The Ant System Algorithm for the VRP

Bullnheimer et al. ([7], [8]) have first applied the Ant System to the VRP. Their approach centers on the similarity of VRPs with TSPs, namely the fact that for a given clustering of customers the problem reduces to several TSPs. Thus, their Ant System is strongly influenced by the Ant System algorithms applied to the TSP. On the contrary, our approach, to our best knowledge, is the first combination of a heuristic algorithm for the VRP with an Ant System.

The Ant System algorithm mainly consists of the iteration of three steps:

- Generation of solutions by ants according to private and pheromone information
- Application of a local search to the ants' solutions
- Update of the pheromone information

In addition to that our approach features a fourth step, namely:

- Augmentation of the Attractiveness list, which stores the desirability of all feasible combinations.

The implementation of these four steps is described below.

Generation of Solutions. As stated above, the solution generation technique we implemented is the main contribution of this paper. So far, in Ant Systems solutions for the VRP have been built using a Nearest Neighbor heuristic (see e.g. [7], [8]). As opposed to that we use the Savings algorithm described above to generate solutions. To that end we need to modify the deterministic version of the algorithm. This modification is done in the following way.

Initially, we generate a sorted list of attractiveness values ξ_{ij} in decreasing order. These attractiveness values feature both the savings values as well as the pheromone information.

Thus the list consists of the following values

$$\xi_{ij} = [s(i,j)]^{\beta}[\tau_{ij}]^{\alpha} \tag{2}$$

where τ_{ij} denotes the pheromone concentration on the arc connecting customers i and j, and α and β bias the relative influence of the pheromone trails and the savings values, respectively. The pheromone concentration τ_{ij} contains information about how good the combination of two customers i and j was in previous iterations.

In each decision step of an ant, we consider the k best combinations still available, where k is a parameter of the algorithm which we will refer to as 'neighborhood' below.

Let Ω_k denote the set of k neighbors, i.e. the k feasible combinations (i, j) yielding the largest savings, considered in a given decision step, then the decision rule is given by equation 3, where $\mathcal{P}_{ij}$ is the probability of choosing to combine customers i and j on one tour.

$$
\mathcal{P}_{ij} = \begin{cases} \dfrac{\xi_{ij}}{\sum_{(h,l) \in \Omega_k} \xi_{hl}} & \text{if } \xi_{ij} \in \Omega_k \\ \\ 0 & \text{otherwise.} \end{cases} \tag{3}
$$

The construction process is stopped when no more feasible combinations are possible.

Local Search. Following [8] we apply the 2-opt algorithm (c.f. [12]) to all vehicle routes built by the ants, before we update the pheromone information. The 2-opt algorithm was developed for the traveling salesman problem and iteratively exchanges two edges with 2 new edges until no further improvements are possible.

Pheromone Update. After all ants have constructed their solutions, the pheromone trails are updated on the basis of the solutions found by the ants. According to the rank based scheme proposed in [6] and [8], the pheromone update is as follows

$$
\tau_{ij}^{new} = \rho \tau_{ij}^{old} + \sum_{\mu=1}^{\sigma-1} \Delta \tau_{ij}^{\mu} + \sigma \Delta \tau_{ij}^{*} \tag{4}
$$

where $0 \le \rho \le 1$ is the trail persistance and σ is the number of elitists. Using this scheme two kinds of trails are laid. First, the best solution found during the process is updated as if σ ants had traversed it. The amount of pheromone laid by the elitists is $\Delta \tau_{ij}^{*} = 1/L^{*}$, where L^{*} is the objective value of the best solution found so far. Second, the $\sigma - 1$ best ants of the iteration are allowed to lay pheromone on the arcs they traversed. The quantity laid by these ants depends on their rank μ as well as their solution quality L^{μ}, such that the μ-th best ant lays $\Delta \tau_{ij}^{\mu} = (\sigma - \mu)/L^{\mu}$. Arcs belonging to neither of those solutions just lose pheromone at the rate $(1 - \rho)$, which constitutes the trail evaporation.

Augmentation of the Savings List. After the pheromone information has been updated the attractiveness values ξ_{ij} are augmented with the new pheromone information as in equation 2.

After the augmentation the attractiveness values are again sorted in decreasing order. This mechanism is the second important contribution of our new approach. In the beginning the attractiveness values are sorted according to the savings values, as the pheromone is equal on all arcs. As learning occurs, and some arcs are reinforced through the update of the pheromone information, the attractiveness values ξ_{ij} change, as they become more and more biased by the pheromone information. Thus, values that were initially high but turned out not to be in good solutions will decrease, while combinations with initially low values that appeared in good solutions will become more attractive. As the attractiveness values are re-sorted after each iteration, this leads to dynamic effects. In particular, 'good' arcs are reinforced twice. First, they receive more pheromone than others, and second as their attractiveness increases they are considered earlier in the constructive process.

3   Numerical Analysis

In this section we will present numerical results for our new approach and compare them with results from previous Ant System approaches as well as different meta-heuristics.

3.1   The Benchmark Problem Instances

The numerical analysis was performed on a set of benchmark problems described in [13]. The set of benchmark problems consists of 14 instances containing between 50 and 199 customers and a depot. The first ten instances were generated with the customers being randomly distributed in the plane, while instances 11-14 feature clusters of customer locations. All instances are capacity constrained. In addition to that, the instances 6-10 and 13-14 are also restricted with respect to tour length. In these instances, all customers have identical service times δ. Apart from the additional time constraints, instances 1-5 and 6-10 are identical. The same is true for instances 11-12 and 13-14. Table 1 contains the data for the 14 problem instances [1].

3.2   Parameter Settings

In order to keep the results comparable and to isolate the effects of our new approach, we chose to basically use the same parameter values as proposed in [8]. Thus, we used n artificial ants, $\alpha = \beta = 5$ and $\sigma = 6$ elitist ants. We only found that for our approach an evaporation rate $\rho = 0.95$ is preferable to $\rho = 0.75$ as proposed in earlier works.

Apart from that we varied the number of iterations in order to be able to estimate the performance of our algorithm for different run times. More specifically we will provide results for $\lfloor n/2 \rfloor$, n and $2 \cdot n$ iterations together with the corresponding computation times.[2]

---

[1] The instances can be found at http://www.ms.ic.ac.uk/jeb/orlib/vrpinfo.html
[2] The algorithms were implemented in Borland C and run on a Pentium 3 with 900MHz.

Table 1. Characteristics of the benchmark problem instances

Random Problems					
Instance	n	Q	L	δ	best publ.
C1	50	160	∞	0	524.61 [14]
C2	75	140	∞	0	835.26 [14]
C3	100	200	∞	0	826.14 [14]
C4	150	200	∞	0	1028.42 [14]
C5	199	200	∞	0	1291.45 [15]
C6	50	160	200	10	555.43 [14]
C7	75	140	160	10	909.68 [14]
C8	100	200	230	10	865.94 [14]
C9	150	200	200	10	1162.55 [14]
C10	199	200	200	10	1395.85 [15]

Clustered Problems					
Instance	n	Q	L	δ	best publ.
C11	120	200	∞	0	1042.11 [14]
C12	100	200	∞	0	819.56 [14]
C13	120	200	720	50	1541.14 [14]
C14	100	200	1040	90	866.37 [14]

n ... number of customers
Q ... vehicle capacity
L ... maximum tour length
δ ... service time
best publ. ... best published solution

Apart from that, given our dynamic savings list, we performed runs with different sizes of the neighborhood, i.e. with different numbers of alternatives in each decision step of an ant. We provide results for neighborhood sizes of $k = \lfloor n/5 \rfloor$, $k = \lfloor n/4 \rfloor$, $k = \lfloor n/2 \rfloor$ and $k = n$.

3.3 Experiments with the Size of the Neighborhood

In this section let us first provide some results we obtained using the different neighborhood sizes. In Table 2 we show for all neighborhood sizes the deviations (RPD) of both our best (best) and average (avg.) solutions (over 10 runs of all 14 problem instances) from the best known solutions after $\lfloor n/2 \rfloor$, n and $2 \cdot n$ iterations. In addition to that we report the corresponding computation times in seconds.

Table 2 mainly shows two different effects. First, we see that increasing the size of the neighborhood generally increases the computation times. This is clear

Table 2. Effects of neighborhood sizes and numbers of iterations on solution quality and computation times.

neighborhood size k	$\lfloor n/2 \rfloor$			n			$2 \cdot n$		
	RPD (in %)		CPU	RPD (in %)		CPU	RPD (in %)		CPU
	best	avg.	sec.	best	avg.	sec.	best	avg.	sec.
$\lfloor n/5 \rfloor$	0.93	1.63	134.37	0.8	1.44	268.73	0.8	1.42	537.46
$\lfloor n/4 \rfloor$	1.12	1.77	134.31	0.7	1.41	268.62	0.7	1.41	537.24
$\lfloor n/2 \rfloor$	2.09	2.76	183.26	0.83	1.42	366.51	0.83	1.41	733.03
n	3.21	4.31	260.38	1.04	1.63	520.75	0.96	1.54	1041.50

as an increased size of the neighborhood increases the computational effort for decision making.

Second, the solution quality generally deteriorates with increased neighborhood size. This effect is partially reversed between $\lfloor n/5 \rfloor$ and $\lfloor n/4 \rfloor$. There we see that the best solutions we found were obtained using $k = \lfloor n/4 \rfloor$. However, we also see that for small computation times $k = \lfloor n/5 \rfloor$ is preferable. This becomes clear from the following observation. In the first iterations pheromone has very little influence on decision making, thus the ants explore the search space. As pheromone is built, the attractiveness list changes adaptively, favoring combinations that were successful in previous iterations, i.e. led to good solutions. In this phase exploration is slowly replaced with exploitation. After a number of iterations the list will be sorted in such a way, that the combinations associated with the $n - m$ largest attractiveness values can all be chosen, i.e. these combinations reflect the best found solution. At that point the algorithm reaches some kind of 'natural' convergence and the ants (almost) only exploit the pheromone information. Thus, the algorithm gradually turns from exploration of the search space in early iterations, to exploitation of the neighborhood of good solutions in later iterations.

The exploration phase in the first iterations of the algorithm depends crucially on the neighborhood size. A small neighborhood, while leading to good solutions quickly, will allow only insufficient exploration of the search space and the algorithm converges too early to a sub-optimal level.

This effect is reduced with increased size k of the neighborhood. However, the other extreme, a neighborhood size of n leads to better exploration but inferior solutions in the first iterations. While it may be able to find very good solutions in the long run its computation times are prohibitive to make it an interesting alternative.

Thus, the conclusion that can be drawn from this analysis is clear cut. The best size of the neighborhood is $k = \lfloor n/4 \rfloor$.

3.4   Comparison between Our New Approach and Existing Meta-heuristics

Now that we know the 'appropriate' size of the neighborhood $k = \lfloor n/4 \rfloor$, we will compare the best results obtained with our new approach (denoted by SavingsAnts) after $2 \cdot n$ iterations using this neighborhood size with the results of the previous Ant System algorithm for the VRP as well as with Tabu Search and Simulated Annealing algorithms. These algorithms are: the ant system algorithm (AS) from [8], the parallel tabu search algorithm (PTS) from [16], the TABUROUTE algorithm (TS) from [17] and the simulated annealing algorithm (SA) from [18]. The results of this comparison are presented in Table 3.[3] First, we present the objective values obtained by the different algorithms for the 14 problems. The last two rows show for all algorithms the relative percentage deviation (RPD) over the best known solution. More specifically, the second to last row gives the average RPD for the random problems (C1-C10), while the last row shows the average RPD for the clustered problems (C11-C14).

Table 3. Comparison of five meta-heuristics

Instance	PTS	TS	SA	AS	SavingsAnts
C1	524.61	524.61	528	524.61	524.61
C2	835.32	835.77	838.62	844.31	838.6
C3	827.53	829.45	829.18	832.32	838.38
C4	1044.35	1036.16	1058	1061.55	1040.86
C5	1334.55	1322.65	1376	1343.46	1307.78
C6	555.43	555.43	555.43	560.24	555.43
C7	909.68	913.23	909.68	916.21	913.01
C8	866.75	865.94	866.75	866.74	870.1
C9	1164.12	1177.76	1164.12	1195.99	1173.42
C10	1420.84	1418.51	1417.85	1451.64	1438.72
C11	1042.11	1073.47	1176	1065.21	1043.89
C12	819.56	819.56	826	819.56	819.56
C13	1550.17	1573.81	1545.98	1559.92	1548.14
C14	866.37	866.37	890	867.07	866.37
RPD (avg.)					
C1-C10	0.71	0.70	1.26	1.76	0.92
C11-C14	0.15	1.28	4.17	0.88	0.16

---

[3] In this table we do not show computation times. A comparison of computation times is not reasonable as all the approaches were tested on different machines. So, while our algorithm consumes by far the smallest computation times, it was also run on the most powerful machine. However, from the execution times reported in Table 2 we are confident, that our algorithm is more than competitive with respect to computational effort.

From Table 3 can be seen that our algorithm shows competitive behavior when compared with the other meta-heuristics. The simulated annealing and previous ant system algorithms are clearly outperformed by our new ant system algorithm. On the other hand, only the parallel tabu search seems to be superior, while the TABUROUTE algorithm shows comparable results. Our SavingsAnts are particularly well suited for the clustered problems as the results for the test instances C11-C14 show. The average percentage deviation of our SavingsAnts' solutions from the best known solutions is only 0.16%.

The reason for the strong performance of the SavingsAnts for clustered problems can be stated as follows. The decision criterion favors the connection of customers that are close to each other and far from the depot. In connection with the simultaneous tour construction this is a good mechanism to identify clusters (c.f. [13]) and to avoid unnecessary connections of customers from two different clusters.

4   Conclusions and Future Research

In this paper we have shown the possible improvements to standard Ant System approaches for VRPs through the use of a problem specific heuristic, namely the Savings algorithm. The computational study performed shows the superior performance of our new approach over the existing Ant System algorithm. In particular, the average results of our approach improves the solution quality of the previous Ant System significantly.

Furthermore, we show that our approach is competitive when compared with other meta-heuristics such as Tabu Search and Simulated Annealing, in particular when applied to clustered problems.

Finally, the average deviation of less than 1.5% over the best known solutions indicates the strength of the proposed algorithm. Moreover, given the performance on the clustered problems this is particularly true for real world problems, which generally exhibit a clustered structure.

Future work will focus on further improvements on the approach. In particular, we will apply our multi-colony approach, as proposed in [19], to the problem. Apart from that we will focus on VRPs with additional features like multiple depots, backhauls and time windows.

Acknowledgments

This work was supported by the Austrian Science Foundation under grant SFB #010 and by the Oesterreichische Nationalbank (OeNB) under grant #8630.

We would also like to thank B.Bullnheimer for drawing our attention to Ant Systems.

Thanks are also due to three anonymous referees for valuable comments on the original version of the papers.

References

1. Garey, M. R. and Johnson, D. S.: Computers and Intractability: A Guide to the Theory of NP Completeness. W. H. Freeman & Co., New York (1979)
2. Colorni, A., Dorigo, M. and Maniezzo, V.: Distributed Optimization by Ant Colonies. In: Varela, F. and Bourgine, P. (Eds.): Proc. Europ. Conf. Artificial Life. Elsevier, Amsterdam (1991)
3. Costa, D. and Hertz, A.: Ants can colour graphs. Journal of the Operational Research Society **48**(3) (1997) 295–305
4. Stützle, T. and Dorigo, M.: ACO Algorithms for the Quadratic Assignment Problem. In: Corne, D., Dorigo, M. and Glover, F. (Eds.): New Ideas in Optimization. Mc Graw-Hill, London (1999)
5. Dorigo, M. and Gambardella, L. M.: Ant Colony System: A cooperative learning approach to the Travelling Salesman Problem. IEEE Transactions on Evolutionary Computation **1**(1) (1997) 53–66
6. Bullnheimer, B., Hartl, R. F. and Strauss, Ch.: A new rank based version of the ant system: a computational study. Central European Journal of Operations Research **7**(1) (1999) 25–38
7. Bullnheimer, B., Hartl, R. F. and Strauss, Ch.: Applying the ant system to the vehicle routing problem. In: Voss, S., Martello, S., Osman, I. H. and Roucairol, C. (Eds.): Meta-Heuristics: Advances and Trends in Local Search Paradigms for Optimization. Kluwer, Boston (1999)
8. Bullnheimer, B., Hartl, R. F. and Strauss, Ch.: An improved ant system algorithm for the vehicle routing problem. Annals of Operations Research **89** (1999) 319–328
9. Gambardella, L. M., Taillard, E. and Agazzi, G.: MACS-VRPTW: A Multiple Ant Colony System for Vehicle Routing Problems with Time Windows. In: Corne, D., Dorigo, M. and Glover, F. (Eds.): New Ideas in Optimization. McGraw-Hill, London (1999)
10. Gutjahr, W. J.: A graph-based Ant System and its convergence. Future Generation Computing Systems. **16** (2000) 873–888
11. Clarke, G. and Wright, J. W.: Scheduling of vehicles from a central depot to a number of delivery points. Operations Research **12** (1964) 568–581
12. Croes, G. A.: A method for solving Traveling Salesman Problems. Operations Research **6** (1958) 791–801
13. Christofides, N., Mingozzi, A. and Toth, P.: The vehicle routing problem. In: Christofides, N., Mingozzi, A., Toth, P. and Sandi, C. (Eds.): Combinatorial Optimization. Wiley, Chicester (1979)
14. Taillard, E. D.: Parallel iterative search methods for vehicle routing problems. Networks **23** (1993) 661–673
15. Rochat, Y. and Taillard, E. D.: Probabilistic Diversification and Intensification in Local Search for Vehicle Routing. Journal of Heuristics **1** (1995) 147–167
16. Rego, C. and Roucairol, C.: A parallel tabu search algorithm using ejection chains for the vehicle routing problem. In: Osman, I. H. and Kelly, J. (Eds.): Meta-Heuristics:Theory and Applications. Kluwer, Boston (1996)
17. Gendreau, M., Hertz, A. and Laporte, G.:A tabu search heuristic for the vehicle routing problem. Management Science **40** (1994) 1276–1290
18. Osman, I. H.: Metastrategy simulated annealing and tabu search algorithms for the vehicle routing problem. Annals of Operations Research **41** (1993) 421–451
19. Doerner, K. F., Hartl, R.F., and Reimann, M.: Are CompetANTS competent for problem solving - the case of a transportation problem. POM Working Paper 01/2001

Updating ACO Pheromones Using Stochastic Gradient Ascent and Cross-Entropy Methods

Marco Dorigo[1], Mark Zlochin[2,*], Nicolas Meuleau[3], and Mauro Birattari[4]

[1] IRIDIA, Université Libre de Bruxelles, Brussels, Belgium
mdorigo@ulb.ac.be
[2] Dept. of Computer Science, Technion – Israel Institute of Technology, Haifa, Israel
zmark@cs.technion.ac.il
[3] IRIDIA, Université Libre de Bruxelles, Brussels, Belgium
nmeuleau@iridia.ulb.ac.be
[4] Intellektik, Darmstadt University of Technology, Darmstadt, Germany
mbiro@intellektik.informatik.tu-darmstadt.de

Abstract. In this paper we introduce two systematic approaches, based on the stochastic gradient ascent algorithm and the cross-entropy method, for deriving the pheromone update rules in the Ant colony optimization metaheuristic. We discuss the relationships between the two methods as well as connections to the update rules previously proposed in the literature.

1 Introduction

The necessity to solve $\mathcal{NP}$-hard optimization problems, for which the existence of efficient exact algorithms is highly unlikely, has led to a wide range of stochastic approximation algorithms. Many of these algorithms, that are often referred to as *metaheuristics*, are not specific to a particular combinatorial problem, but rather present a general approach for organizing the search in the solution space. Examples of well-known metaheuristics are evolutionary algorithms, simulated annealing and tabu search.

At a very abstract level, many metaheuristics can be seen as methods that look for good solutions (possibly optimal ones) by repeating the following two steps:

1. Candidate solutions are constructed using some parameterized probabilistic model, that is, a parameterized probability distributions over the solution space.
2. The candidate solutions are used to update the model's parameters in a way that is deemed to bias future sampling toward low cost solutions.

Recently, a metaheuristic inspired by the foraging behavior of ant has been defined. This metaheuristic, called *ant colony optimization* (ACO), has been successfully applied to the solution of numerous NP-hard problems [1] as well as to dynamic optimization problems [2]. The main innovation of the ACO metaheuristic consists in proposing a novel way to implement the two steps described above:

---

* This work was carried out while the author was at IRIDIA, Université Libre de Bruxelles, Belgium.

S. Cagnoni et al. (Eds.): EvoWorkshops 2002, LNCS 2279, pp. 21–30, 2002.

1. A structure called *construction graph* is coupled with a set of stochastic agents called *artificial ants*, which build new solutions using local information, called *pheromone*,[1] stored in the construction graph.
2. Once ants have built solutions, they use information collected during the construction phase to update the pheromone values.

The exact form of the pheromone update (i.e., the second component of the general approach outlined above) is not strictly defined and a good amount of freedom is left to the algorithm designer concerning how it should be implemented. Various model update rules have been proposed within the ACO framework, but they are all of a somewhat heuristic nature and are lacking a theoretical justification.

On the other hand, the *stochastic gradient ascent* (SGA) [3] and the *cross-entropy* (CE) [4] methods suggest principled ways of updating the parameters associated to the construction graph. In the following we show how the SGA and the CE methods can be cast into the ACO framework. While these two methods have different motivations, we find that in some cases the CE method leads to the same update rule as does SGA. Moreover, quite unexpectedly, some existing ACO updates are re-derived as a particular implementation of the CE method.

The paper is organized as follows. In Section 2 we present the solution construction mechanism used by the ACO metaheuristic. Section 3 presents several typical pheromone update rules among those most used by practitioners and researchers. In Section 4 and 5 we introduce the stochastic gradient ascent and the cross entropy methods respectively, and we show how, once cast into the ACO framework, they can be used to define novel pheromone updating rules. We conclude in Section 6 with a brief summary of the contributions of this short paper.

2   Solution Construction in ACO Metaheuristic

Let us consider a minimization problem[2] $(\mathcal{S}, f)$, where $\mathcal{S}$ is the *set of feasible solutions* and f is the *objective function*, which assigns to each solution $s \in \mathcal{S}$ a cost value $f(s)$. The goal of the minimization problem is to find an optimal solution s^*, that is, a feasible solution of minimum cost. The set of all optimal solutions is denoted by $\mathcal{S}^*$. We assume that the combinatorial optimization problem $(\mathcal{S}, f)$ is mapped on a problem that can be characterized by the following list of items[3]:

- A finite set $\mathcal{C} = \{c_1, c_2, \ldots, c_{N_C}\}$ of *components*.
- A finite set $\mathcal{X}$ of *states* of the problem, defined in terms of all the possible sequences $x = \langle c_i, c_j, \ldots, c_k, \ldots \rangle$ over the elements of $\mathcal{C}$. The length of a sequence x, that is, the number of components in the sequence, is expressed by $|x|$. The maximum length of a sequence is bounded by a positive constant $n < +\infty$.

---

[1] This is the terminology used in the ACO literature for historical reasons. We will keep with this tradition in this paper.

[2] The obvious changes must be done if a maximization problem is considered.

[3] How this mapping can be done in practice has been described in a number of earlier papers on the ACO metaheuristic; see, for example, [1].

- The set of (candidate) solutions S is a subset of $\mathcal{X}$ (i.e., $S \subseteq \mathcal{X}$).
- A set of feasible states $\tilde{\mathcal{X}}$, with $\tilde{\mathcal{X}} \subseteq \mathcal{X}$, defined via a set of *constraints* Ω.
- A non-empty set S^* of optimal solutions, with $S^* \subseteq \tilde{\mathcal{X}}$ and $S^* \subseteq S$.

Given the above formulation, artificial ants build candidate solutions by performing randomized walks on the completely connected, weighted graph $\mathcal{G} = (\mathcal{C}, \mathcal{L}, \mathcal{T})$, where the vertices are the components $\mathcal{C}$, the set $\mathcal{L}$ fully connects the components $\mathcal{C}$, and $\mathcal{T}$ is a vector gathering so-called *pheromone trails* τ.[4] The graph $\mathcal{G}$ is called *construction graph*.

Each artificial ant is put on a randomly chosen vertex of the graph and then it performs a randomized walk by moving at each step from vertex to vertex in the graph in such a way that the next vertex is chosen stochastically according to the strength of the pheromone currently on the arcs.[5] While moving from one node to another of the graph $\mathcal{G}$, constraints Ω may be used to prevent ants from building infeasible solutions. Formally, the solution construction behavior of a generic ant can be described as follows:

ANT_SOLUTION_CONSTRUCTION
for each ant:
- select a start node c_1 according to some problem dependent criterion,
- set $k = 1$ and $x_k = \langle c_1 \rangle$.
- While $(x_k = \langle c_1, c_2, \ldots, c_k \rangle \in \tilde{\mathcal{X}}$ and $x_k \notin S$ and $J_{x_k} \neq \emptyset)$ do:
  at each step k, after building the sequence x_k, select the next node (component) c_{k+1} randomly following

$$
P_{\mathcal{T}}(c_{k+1} = c | x_k) = \begin{cases} \dfrac{F_{(c_k,c)}\big(\tau(c_k,c)\big)}{\displaystyle\sum_{(c_k,y) \in J_{x_k}} F_{(c_k,y)}\big(\tau(c_k,y)\big)} & \text{if } (c_k,c) \in J_{x_k}, \\[4mm] 0 & \text{otherwise;} \end{cases} \tag{1}
$$

where a connection (c_k, y) belongs to J_{x_k} iff the sequence $x_{k+1} = \langle c_1, c_2, \ldots, c_k, y \rangle$ satisfies the constraints Ω (i.e., $x_{k+1} \in \tilde{\mathcal{X}}$) and $F_{(i,j)}(z)$ is some monotonic function (most commonly, $z^\alpha \eta(i,j)^\beta$, where $\alpha, \beta > 0$ and η are heuristic "visibility" values [6]). If at some stage $x_k \notin S$ and $J_{x_k} = \emptyset$, the construction process has reached a dead-end and is therefore abondoned.[6]

After the solution construction has been completed, the artificial ants update the pheromone values. Next we describe several typical updates that were suggested in the past within the ACO framework.

---

[4] Pheromone trails can be associated to components, connections, or both. In the following, unless stated otherwise, we assume that the pheromone trails are associated to connections, so that $\tau(i,j)$ is the pheromone associated to the connection between components i and j. It is straightforward to extend algorithms to the other cases.

[5] It should be noted that the same type of model was later (but independently) used in the CE framework under the name "associated stochastic network" [4,5].

[6] This situation may be prevented by allowing artificial ants to build infeasible solutions as well. In such a case an infeasibility penalty term is usually added to the cost function. It should be noted, however, that in most settings ACO was applied to, the dead-end situation does not occur.

3   Typical ACO Pheromone Updates

Many different schemes for pheromone update have been proposed within the ACO framework. Most of them can be described, however, using the following generic scheme:

GENERIC_ACO_UPDATE
- $\forall s \in \hat{S}_t, \forall (i,j) \in s : \tau(i,j) \leftarrow \tau(i,j) + Q_f(s|S_1, \ldots, S_t)$
- $\forall (i,j) : \tau(i,j) \leftarrow (1 - \rho) \cdot \tau(i,j)$

    where S_i is the sample in the i-th iteration, $\rho, 0 \leq \rho < 1$, is the evaporation rate and $Q_f(s|S_1, \ldots, S_t)$ is some "quality function", which is typically required to be non-increasing with respect to f and is defined over the "reference set" $\hat{S}_t$.

Different ACO algorithms may use different quality functions and reference sets. For example, in the very first ACO algorithm — Ant System [7] — the quality function was simply $1/f(s)$ and the reference set $\hat{S}_t = S_t$. In a more recently proposed scheme, called *iteration best update* [6], the reference set was a singleton containing the best solution within S_t (if there were several iteration-best solutions, one of them was chosen randomly). For the *global-best update* [8,6], the reference set contained the best among all the iteration-best solutions (and if there were more than one global-best solution, the earliest one was chosen). In [7] an *elitist* strategy was introduced, in which the update was a combination of the previous two.

A somewhat different pheromone update was used in *ant colony system* (ACS) [6]. There the pheromones are evaporated by the ants online during the solution construction, hence only the pheromones involved in the construction evaporate.

Another modification of the generic update described above was recently proposed under the name Hyper-Cube (HC) ACO [9]. The HC-ACO, applied to combinatorial problems with binary coded solutions,[7] normalizes the quality function, hence obtaining an automatic scaling of the pheromone values:

$$\tau_i \leftarrow (1 - \rho)\tau_i + \rho \frac{\sum_{\substack{s \in \hat{S}_t \\ s_i = 1}} Q_f(s)}{\sum_{s \in \hat{S}_t} Q_f(s)}. \tag{2}$$

While all the updates described above are of somewhat heuristic nature, the SGA and the CE methods allow to derive pheromone update rules in a more systematic manner, as we show next.

4   Stochastic Gradient Ascent

The construction process described above implicitly defines a probability distribution over the solution space. Let us denote this distribution by $P_\mathcal{T}$, where $\mathcal{T}$ is the vector of

---

[7] Using the notation of Section 2, in such a case the components are bit assignments to the locations, and the pheromone values are associated with the components, rather than with the connections.

pheromone values. Now, the original optimization problem may be replaced with the following equivalent continuous *maximization problem*:

$$\mathcal{T}^* = \underset{\mathcal{T}}{\mathrm{argmax}}\ \mathcal{E}(\mathcal{T}), \tag{3}$$

where $\mathcal{E}(\mathcal{T}) = E_{\mathcal{T}} Q_f(s)$, $E_{\mathcal{T}}$ denotes expectation with respect to $P_{\mathcal{T}}$, and $Q_f(s)$ is a fixed *quality function*, which is strictly decreasing with respect to f. It may be easily verified that $P_{\mathcal{T}^*}$ is greater than zero only over $\mathcal{S}^*$, hence solving problem (3) is equivalent to solving the original combinatorial optimization problem.

One may then search for an optimum (possibly a local one) of problem (3) using a gradient ascent method (in other words, gradient ascent may be used as a heuristic to change $\mathcal{T}$ with the goal of solving (3)):

- Start from some initial guess $\mathcal{T}^0$.
- At stage t, calculate the gradient $\nabla \mathcal{E}(\mathcal{T}^t)$ and update $\mathcal{T}^{t+1}$ to be $\mathcal{T}^t + \alpha_t \nabla \mathcal{E}(\mathcal{T}^t)$.

The gradient can be calculated (theoretically) as follows:

$$\nabla \mathcal{E} = \nabla E_{\mathcal{T}} Q_f(s) = \nabla \sum_s Q_f(s) P_{\mathcal{T}}(s) = \sum_s Q_f(s) \nabla P_{\mathcal{T}}(s)$$

$$= \sum_s P_{\mathcal{T}}(s) Q_f(s) \nabla \ln P_{\mathcal{T}}(s) = E_{\mathcal{T}} Q_f(s) \nabla \ln P_{\mathcal{T}}(s). \tag{4}$$

However, the gradient ascent algorithm cannot be implemented in practice, as for its evaluation a summation over the whole search space is needed. A more practical alternative would be to use *stochastic gradient ascent* [3], which replaces the expectation in Equation 4 by an empirical mean of a sample generated from $P_{\mathcal{T}}$.

The update rule for the stochastic gradient is:

$$\mathcal{T}^{t+1} = \mathcal{T}^t + \alpha_t \sum_{s \in S_t} Q_f(s) \nabla \ln P_{\mathcal{T}^t}(s), \tag{5}$$

where S_t is the sample at iteration t. It remains to be shown how the gradient $\nabla \ln P_{\mathcal{T}^t}(s)$ can be evaluated. The following calculation is a generalization of the one in [10].

From the definition of ANT_SOLUTION_CONSTRUCTION, it follows that, for $s = \langle c_1, c_2, \ldots \rangle$,

$$P_{\mathcal{T}}(s) = \prod_{k=1}^{|s|-1} P_{\mathcal{T}}\big(c_{k+1} \big|\ \mathrm{pref}_k(s)\big), \tag{6}$$

where $\mathrm{pref}_k(s)$ is the k-prefix of s, and consequently

$$\nabla \ln P_{\mathcal{T}}(s) = \sum_{k=1}^{|s|-1} \nabla \ln P_{\mathcal{T}}\big(c_{k+1} \big|\ \mathrm{pref}_k(s)\big). \tag{7}$$

Finally, given a pair of components $(i, j) \in \mathcal{C}^2$, using Equation (1), it is easy to verify that:

– if $i = c_k$ and $j = c_{k+1}$ then

$$\frac{\partial}{\partial \tau(i,j)} \Big(\ln P_\mathcal{T}\big(c_{k+1}|\operatorname{pref}_k(s)\big) \Big) =$$

$$\frac{\partial}{\partial \tau(i,j)} \Big(\ln F\big(\tau(i,j)\big) - \ln \sum_{(i,y)\in J_{x_k}} F\big(\tau(i,y)\big) \Big) =$$

$$\Big(1 - F\big(\tau(i,j)\big) \Big/ \sum_{(i,y)\in J_{x_k}} F\big(\tau(i,y)\big) \Big) \frac{F'\big(\tau(i,j)\big)}{F\big(\tau(i,j)\big)} =$$

$$\Big(1 - P_\mathcal{T}\big(j|\operatorname{pref}_k(s)\big) \Big) G\big(\tau(i,j)\big),$$

where $G(\cdot) = F'(\cdot)/F(\cdot)$ and the subscript of F was omitted for the clarity of presentation.

– if $i = c_k$ and $j \neq c_{k+1}$ then

$$\frac{\partial \ln \Big(P_\mathcal{T}\big(c_{k+1}|\operatorname{pref}_k(s)\big) \Big)}{\partial \tau(i,j)} = -P_\mathcal{T}\big(j|\operatorname{pref}_k(s)\big) G\big(\tau(i,j)\big).$$

By combining these results, the following pheromone update rule is derived:

SGA_UPDATE

$\forall s = \langle c_1, \ldots, c_k, \ldots \rangle \in S_t, 1 \leq k < |s| :$

– $\tau(c_k, c_{k+1}) \leftarrow \tau(c_k, c_{k+1}) + \alpha_t Q_f(s) G\big(\tau(c_k, c_{k+1})\big)$

– $\forall y : \tau(c_k, y) \leftarrow \tau(c_k, y) - \alpha_t Q_f(s) P_\mathcal{T}\big(y|\operatorname{pref}_k(s)\big) G\big(\tau(c_k, y)\big)$

Hence any connection (i, j) used in the construction of a solution is reinforced by an amount $\alpha_t Q_f(s) G\big(\tau(i,j)\big)$, and any connection *considered* during the construction, has its pheromone values evaporated by an amount $\alpha_t Q_f(s) P_\mathcal{T}\big(j|\operatorname{pref}_k(s)\big) G\big(\tau(i,j)\big)$. Note, that if the solutions are allowed to contain loops, a connection may be updated more than once for the same solution.

In order to guarantee the stability of the resulting algorithm, it is desirable for the estimate of the gradient $\nabla \ln P_\mathcal{T}(s)$ to be bounded. This means that a function F, for which $G = F'/F$ is bounded, should be used. Meuleau and Dorigo [10] suggest using $F(\cdot) = \exp(\cdot)$, which leads to $G \equiv 1$. It should be further noted that if, in addition, $Q_f = 1/f$ and $\alpha_t = 1$, the reinforcement part becomes $1/f$, as in Ant System.

5   Cross-Entropy Method

The cross-entropy (CE) method was initially proposed in the stochastic simulation field as a tool for rare events estimation and later adapted as a tool for combinatorial optimization

[4]. In this overview we present a more straightforward derivation of the cross-entropy method (as a combinatorial optimization tool), without reference to the rare events estimation.

Starting from some initial distribution $P_0 \in \mathcal{M}$, the CE method inductively builds a series of distributions $P_t \in \mathcal{M}$, in an attempt to increase the probability of generating low-cost solutions after each iteration. A tentative way to achieve this goal is to set P_{t+1} equal to

$$\hat{P} \propto P_t Q_f, \tag{8}$$

where Q_f is, again, some quality function dependent on the cost value.

If this were possible, after n iteration we would obtain $P_n \propto P_0 Q_f^n$, and as $n \to \infty$, P_n would converge to a probability distribution restricted to $\mathcal{S}^*$. Unfortunately, even if the distribution P_t belongs to the family $\mathcal{M}$, the distribution $\hat{P}$ as defined by (8) does not necessarily remain in $\mathcal{M}$, hence some sort of projection is needed. A natural candidate for P_{t+1}, is the distribution $P \in \mathcal{M}$ that minimizes the *Kullback-Leibler divergence* [11], which is a commonly used measure of misfit between two distributions:

$$D(\hat{P}\|P) = \sum_s \hat{P}(s) \ln \frac{\hat{P}(s)}{P(s)}, \tag{9}$$

or, equivalently, the *cross-entropy*: $-\sum_s \hat{P}(s) \ln P(s)$.

Since $\hat{P} \propto P_t Q_f$, cross-entropy minimization is equivalent to the following maximization problem

$$P_{t+1} = \underset{P \in \mathcal{M}}{\operatorname{argmax}} \sum_s P_t(s) Q_f(s) \ln P(s). \tag{10}$$

It should be noted that, unlike SGA, in the cross-entropy method the quality function is only required to be non-increasing with respect to the cost and may also depend on the iteration index, either deterministically or stochastically, for example, depending on the points sampled so far. One common choice is, for example, $Q_f^t(s) = I(f(s) < f_t)$, where $I(\cdot)$ is an indicator function, and f_t is, for example, some quantile (e.g., lower 10%) of the cost distribution during the last iteration.

Similarly to the gradient ascent algorithm, the maximization problem (10) cannot be solved in practice, as the evaluation of the function $\sum_s P_t(s) Q_f(s) \ln P(s)$ requires summation over the whole solution space, and once again a finite sample approximation is used instead:

$$P_{t+1} = \underset{P \in \mathcal{M}}{\operatorname{argmax}} \sum_{s \in S_t} Q_f(s) \ln P(s), \tag{11}$$

where S_t is a sample from P_t.

Let us now consider problem (11) in more details. At the maximum the gradient must be zero:

$$\sum_{s \in S_t} Q_f(s) \nabla \ln P_T(s) = 0. \tag{12}$$

In some relatively simple cases, for example, when the solution s is represented by an unconstrained string of bits of length n, $(s_1, \ldots, s_n)$, and there is a single parameter τ_i for the i-th position in the string,[8] such that $P_{\mathcal{T}}(s) = \prod_i p_{\tau_i}(s_i)$, the equations system (12) reduces to a set of independent equations:

$$\frac{d \ln p_{\tau_i}}{d\tau_i} \sum_{\substack{s \in S_t \\ s_i=1}} Q_f(s) = -\frac{d \ln(1 - p_{\tau_i})}{d\tau_i} \sum_{\substack{s \in S_t \\ s_i=0}} Q_f(s), \tag{13}$$

which may often be solved analytically. For example, for $p_{\tau_i} = \tau_i$ it can be easily shown that the solution of Equation (13) is simply

$$\tau_i = \frac{\sum_{\substack{s \in S_t \\ s_i=1}} Q_f(s)}{\sum_{s \in S_t} Q_f(s)}. \tag{14}$$

Now, since the pheromone trails τ_i in (14) are random variables, whose values depend on the particular sample, we may wish to make our algorithm more robust by introducing some conservatism into the update. For example, rather than discarding the old pheromone values, the new values may be taken to be a convex combination of the old values and the solution (14):

$$\tau_i \leftarrow (1 - \rho)\tau_i + \rho \frac{\sum_{\substack{s \in S_t \\ s_i=1}} Q_f(s)}{\sum_{s \in S_t} Q_f(s)}. \tag{15}$$

The resulting update is identical to the one used in the Hyper-Cube ACO [9].

In general, however, Equations (12) are coupled and an analytical solution is unavailable. Nevertheless, in the actual implementations of the CE method the update was of the form (14) (with some brief remarks about using (15)) [5], which may be considered as an approximation to the exact solution of the cross-entropy minimization problem (11).

Still, even if the exact solution is not known, some iterative methods for solving this optimization problem may be used. A natural candidate for the iterative solution of the maximization problem (11) is gradient ascent:

- Start with $\mathcal{T}' = \mathcal{T}^t$. (Other starting points are possible, but this is the most natural one, since we may expect $\mathcal{T}^{t+1}$ to be close to $\mathcal{T}^t$.)
- Repeat:
  - $\mathcal{T}' \leftarrow \mathcal{T}' + \alpha \sum_{s \in S_t} Q_f(s) \nabla \ln P_{\mathcal{T}'}(s)$
  - until some stopping criteria is satisfied.
- Set $\mathcal{T}^{t+1} = \mathcal{T}'$.

It should be noted that, since the new vector $\mathcal{T}^{t+1}$ is a random variable, depending on a sample, there is no use in running the gradient ascent process till full convergence. Instead, in order to obtain some robustness against sampling noise, we may use a fixed number of gradient ascent updates. One particular choice, which is of special interest, is the use of a single gradient ascent update, leading to the updating rule:

$$\mathcal{T}^{t+1} = \mathcal{T}^t + \alpha_t \sum_{s \in S_t} Q_f(s) \nabla \ln P_{\mathcal{T}^t}(s) \tag{16}$$

---

[8] This is a particular subtype of models, used in HC-ACO [9], without any non-trivial constraints.

which is identical to the SGA update (5). However, as it was already mentioned earlier, the CE method imposes less restrictions on the quality function (e.g., allowing it to change over time), hence the resulting algorithm may be seen as a generalization of SGA.

To conclude, we have shown that if we use (14) as a (possibly approximate) solution of Equation (11), the Hyper-Cube ACO algorithm is derived. If otherwise we use a single-step gradient ascent for solving (11), we obtain a generalization of the SGA update, in which the quality function is allowed to change over time.

6   Conclusions

The ant colony optimization metaheuristics attempts to solve a combinatorial problem by repeatedly constructing solutions using locally available pheromones, and updating them, so as to increase the probability of generating good solutions in the future. While the construction process, employed in ACO, is well-defined (and, in fact, is one of the distinctive features of ACO metaheuristic), a considerable amount of freedom remains regarding the exact form of the pheromone update rule.

We have described two general approaches, the SGA and the CE methods, for updating pheromone values for the ACO metaheuristic. These two methods provide systematic, theoretically founded ways for deriving the update rules. Moreover, we have also shown that in many cases the updates used by the two methods are quite similar (or even identical in some cases), and sometimes they coincide with existing ACO updates.

While the newly derived pheromone updates do have a solid theoretical motivation, little can be said *a priori* about their performance as compared to the existing update rules. The empirical evaluation of these methods is the subject of ongoing research.

Acknowledgments

Marco Dorigo acknowledges support from the Belgian FNRS, of which he is a Senior Research Associate. Mark Zlochin is supported through a Training Site fellowship funded by the Improving Human Potential (IHP) programme of the Commission of the European Community (CEC), grant HPRN-CT-2000-00032. Nicolas Meuleau is supported by a Marie Curie fellowship funded by the CEC, grant HPMF-CT-2000-00230. Mauro Birattari is supported by a fellowship from the "Metaheuristics Network", a Research Training Network funded by the Improving Human Potential programme of the CEC, grant HPRN-CT-1999-00106. More generally, this work was partially supported by the "Metaheuristics Network", a Research Training Network funded by the Improving Human Potential programme of the CEC, grant HPRN-CT-1999-00106. The information provided in this paper is the sole responsibility of the authors and does not reflect the Community's opinion. The Community is not responsible for any use that might be made of data appearing in this publication.

References

1. M. Dorigo and G. Di Caro. The Ant Colony Optimization meta-heuristic. In D. Corne, M. Dorigo, and F. Glover, editors, *New Ideas in Optimization*, pages 11–32. McGraw Hill, London, UK, 1999.
2. G. Di Caro and M. Dorigo. AntNet: Distributed stigmergetic control for communications networks. *Journal of Artificial Intelligence Research*, 9:317–365, 1998.
3. D. P. Bertsekas. *Nonlinear Programming*. Athena Scientific, Belmont, MA, 1995.
4. R. Y. Rubinstein. The cross-entropy method for combinatorial and continuous optimization. *Methodology and Computing in Applied Probability*, 1(2):127–190, 1999.
5. R. Y. Rubinstein. Combinatorial optimization, cross-entropy, ants and rare events. In S. Uryasev and P. M. Pardalos, editors, *Stochastic Optimization: Algorithms and Applications*. Kluwer Academic Publishers, Dordrecht, The Netherlands, 2001.
6. M. Dorigo and L. M. Gambardella. Ant Colony System: A cooperative learning approach to the traveling salesman problem. *IEEE Transactions on Evolutionary Computation*, 1(1):53–66, 1997.
7. M. Dorigo, V. Maniezzo, and A. Colorni. The Ant System: Optimization by a colony of cooperating agents. *IEEE Transactions on Systems, Man, and Cybernetics – Part B*, 26(1):29–41, 1996.
8. T. Stützle and H. H. Hoos. The MAX-MIN ant system and local search for the traveling salesman problem. In *Proceedings of ICEC'97 - 1997 IEEE 4th International Conference on Evolutionary Computation*, pages 308–313. IEEE Press, Piscataway, NJ, 1997.
9. C. Blum, A. Roli, and M. Dorigo. HC–ACO: The hyper-cube framework for Ant Colony Optimization. In *Proceedings of MIC'2001 – Meta–heuristics International Conference*, volume 2, pages 399–403, Porto, Portugal, 2001. Also available as technical report TR/IRIDIA/2001-16, IRIDIA, Université Libre de Bruxelles, Brussels, Belgium.
10. N. Meuleau and M. Dorigo. Ant colony optimization and stochastic gradient descent. *Artificial Life*, 2002, in press.
11. S. Kullback. *Information Theory and Statistics*. John Wiley & Sons, New York, NY, 1959.

Non-parametric Estimation
of Properties of Combinatorial Landscapes

Anton Eremeev[1] and Colin R. Reeves[2]

[1] Sobolev Institute of Mathematics (Omsk Branch)
Discrete Optimization Laboratory
Omsk, Russia
[2] School of Mathematical and Information Sciences
Coventry University
Coventry, UK

Abstract. Earlier papers [1,2] introduced some statistical estimation methods for measuring certain properties of landscapes induced by heuristic search methods: in particular, the number of optima. In this paper we extend this approach to non-parametric methods which allow us to relax a critical assumption of the earlier approach.
Two techniques are described—the jackknife and the bootstrap—based on statistical ideas of resampling, and the results of some empirical studies are presented and analysed.

1 Introduction

Many heuristic search methods for combinatorial optimization problems (COPs) are based on neighbourhood search or its modern developments (simulated annealing, tabu search etc.). Such methods quickly converge to a local optimum, and the search has to begin afresh—either by starting from a new point, or by using some 'meta'-technique to circumvent the current impasse.

The idea of neighbourhood search (NS) is important, because the idea of a local optimum, and associated properties such as its basin of attraction, can be given a precise meaning. One of the properties that make a problem instance difficult is the number of optima induced by the combination of the fitness function and the neighbourhood operator. However, it is possible to envisage an instance with many local optima whose basins of attraction are minimal in size, while the global optima have very large basins. So the distribution of basin sizes is clearly important too.

In earlier papers [1,2] we showed how, on the assumption of equally-sized basins—an *isotropic* landscape—it was possible to estimate the number of optima using data obtained from random restarts of a NS procedure. This paper will extend the earlier work to deal with the case of non-isotropic landscapes.

1.1 Terminology

In what follows, we are concerned with some finite *search space*, denoted by $\mathcal{X}$, with the objective of optimizing some real-valued function $f : \mathcal{X} \mapsto \mathbb{R}$. Our NS

S. Cagnoni et al. (Eds.): EvoWorkshops 2002, LNCS 2279, pp. 31–40, 2002.

procedure uses a neighbourhood function

$$N : \mathcal{X} \mapsto 2^{\mathcal{X}}$$

to generate a sequence of points $x_0, x_1, \ldots, x_n$, terminating at a local optimum. The strategy followed is assumed to be a 'best improving' one[1], i.e. the search algorithm finds x_{i+1} such that

$$x_{i+1} = \arg \max_{y \in N(x_i)} f(y) \quad \text{and} \quad f(x_{i+1}) > f(x_i),$$

stopping when the second condition cannot be met. We assume that the procedure is used r times, each commencing from a (different) random initial solution. We shall refer to the number of optima in a landscape as ν, and denote by k the number of *distinct* optima seen. The search can be thought of as a function

$$\mu : \mathcal{X} \mapsto \mathcal{X}$$

where if x is the initial point, $\mu(x)$ is the local optimum that it reaches. Each optimum $x_1^*, \ldots, x_\nu^*$ then has a basin of attraction whose normalized size is

$$p_i = \frac{|\{x : \mu(x) = x_i^*\}|}{|\mathcal{X}|}$$

(where the expression $|\cdot|$ means, as usual, the cardinality of a set). As is standard in statistical literature, estimators of a parameter will be denoted by a ^ symbol; maximum likelihood estimators will be denoted by the superscript ML.

2   Parametric Estimation

As explained in [1], the distribution of the random variable K for the isotropic case is

$$P[K = k] = \frac{\nu!}{(\nu - k)!} \frac{S(r, k)}{\nu^r}, \quad 1 \leq k \leq \min(r, \nu),$$

where $S(r, k)$ is the Stirling number of the second kind [3]. This distribution is commonly used in population statistics in ecology, where it is associated with a procedure known as the *Schnabel census*. From this, given observations (k, r), it is possible to use the statistical principle of maximum likelihood [4] to obtain an estimate $\hat{\nu}^{ML}(k, r)$. (Computationally, several standard numerical methods can be used—see [1] for details.) Experiments reported in [1,2] showed that the estimates based on this approach were good for isotropic landscapes, but were quite sharply biased (negatively) when basin sizes are not identical.

Improvements could be made by assuming a basin size distribution for $\{p_i\}$ and using the *frequencies* of observing local optima to fit the parameters of the assumed distribution. However, in [2], tractable distributions (e.g. gamma) appear not to be good models for observed data, while more realistic distributions (e.g. lognormal) are very difficult to use. (On the other hand, Kallel and Garnier [5] have recently reported greater success using the gamma distribution.)

---

[1] Using (say) a 'first improving' strategy does not change the number of optima, although it may affect the effective basin sizes.

3   Non-parametric Estimation

In the last 20 years, statisticians have developed highly sophisticated techniques that bypass the need to estimate the parameters of some unknown distribution. These techniques are generically known as resampling methods. They start from the premise that the best estimator we have for an unknown distribution is the empirical distribution that is embedded in the data.

3.1   Jackknife Estimate

Better estimates appeared to be obtained by using a non-parametric approach—the jackknife [6]. This starts from the obviously biased estimator $\hat{\nu} = k$ (i.e., the actual number of distinct optima found), and assumes that the bias decreases with increasing r. If we now leave out one point of the original sample at a time, we obtain r 're-samples' of the original data, and thus r estimates of ν, which can be combined to give a new estimate

$$\hat{\nu}_{(r-1)} = k - \frac{\beta_1}{r}$$

where β_1 is the number of optima seen only once. Finally, the jackknife estimate is found by using this to correct the estimate $\hat{\nu} = k$: it reduces to

$$\hat{\nu}^{JK} = k + \frac{r-1}{r}\beta_1.$$

Since $k = \sum_{j=1}^{r}\beta_j$, this can be written as

$$\hat{\nu}^{JK} = \sum_{j=1}^{r} a_j\beta_j$$

for some weights $\{a_j\}$. It can be further extended to a generalized estimator of the form

$$\hat{\nu}_s^{JK} = \sum_{j=1}^{r} a_{j,s}\beta_j,$$

corresponding to leaving out $s > 1$ observations at a time. Burnham and Overton [6] provide the appropriate $\{a_{j,s}\}$ values for $s = 1, \ldots, 5$. While the bias is reduced, the variance will increase, and the point where the overall mean squared error ($= \text{bias}^2 + \text{variance}$) reaches a minimum is problem-dependent. A hypothesis test is suggested in [6] whereby the most suitable value for s can be found.

3.2   The Bootstrap

A more powerful version of the resampling technique is provided by the concept of the bootstrap. In general, we wish to estimate a parameter $\theta(X; \mathcal{F})$ using sample data X, representing *iid* random variables drawn from a distribution $\mathcal{F}$.

In order to make statistical statements about θ (for example its bias or variance), the conventional approach would be to assume some particular parametric form of $\mathcal{F}$ from which exact or asymptotic results could be derived.

Efron [7] suggested that, rather than parameterize $\mathcal{F}$, we could use the empirical distribution $\hat{\mathcal{F}}$, and repeatedly sample from it (with replacement), thus generating a sequence of B samples $\hat{X}_b$. Suppose $\hat{\theta}_b = \theta(\hat{X}_b; \hat{\mathcal{F}})$, then

$$\lim_{B \to \infty} \frac{\sum_{b=1}^{B} \hat{\theta}_b}{B} = \theta(X; \hat{\mathcal{F}})$$

where $\theta(X; \hat{\mathcal{F}})$ is the non-parametric maximum likelihood estimator of θ. Provided $\hat{\mathcal{F}}$ is sufficiently 'close' to $\mathcal{F}$, the estimate of θ thus obtained should be a good one. This general approach has become known as a *bootstrap* procedure.

In the context of this problem, we can use the set of r optima found as our data X, and take samples of size r from X (with replacement), estimating the value $\hat{\nu}_b^{ML}$ for each resample. Because the k_b values generated by the resample process cannot exceed the actual k found in X, these estimates will be negatively biased, and because we also know the original $\hat{\nu}$ we can estimate the bias and use this as a bias correction for $\hat{\nu}$.

The drawback to all bootstrap methods is their computational demands, which would clearly be significant in this case. (It is customary to set $B \sim 10^3$— fewer than this can make the procedure unreliable.) Furthermore, this approach takes no explicit account of any information we have as to the values of the $\{p_i\}$.

3.3   The 'Plug-in' Principle

Efron and Tibshirani [8] point out that the bootstrap is actually an example of a general statistical idea that they call the 'plug-in' principle. In order to estimate some parameter θ from sample data, we write θ as a function of $\mathcal{F}$, and 'plug in' the empirical distribution $\hat{\mathcal{F}}$. From this viewpoint we have a sample of r independent outcomes $\{x_{j_1}, \ldots, x_{j_r}\}$ of the search. These correspond to a sample $\{j_1, \ldots, j_r\}$ of realisations of the random variable J, which is the index of the optimum found in a random restart. As the actual probability distribution of J is unknown, we 'plugged in' the empirical distribution of J instead. The application of this principle in estimating ν is described in Figure 1.

However, this principle also leads to another slightly different idea. The data also provide some information on the distribution of $\{p_i\}$. While these data are insufficient to estimate basin sizes directly, we can make some approximations. We assume that each optimum belongs to one of the classes $(C_1, ..., C_r)$, where C_i contains those optima with normalized basin size i/r. (Putting it another way, we assume that all optima seen i times have the same characteristic basin size.) If some classes are not represented in the sample, they are assumed to be empty. In this way we have an empirical distribution, not for the $\{p_i\}$ themselves, but for an aggregated version of this distribution, and this can also be 'plugged in' in order to estimate ν.

The optima in class C_i are assumed to have approximately the same basin size, so that the observations made have arisen from a Schnabel census procedure

Without loss of generality we may assume that $\{j_1, \ldots, j_r\} = \{1, \ldots, k\}$.

1. On the basis of r restarts of the search, construct the empirical distribution $(v_1/r, \ldots, v_k/r)$ for J, where $v_j = |\{i : j_i = j\}|, j = 1, \ldots, k$ is the number of times the optimum j was visited.
2. For $b = 1$ to B do:
   (a) Draw the 'bootstrap sample' $\{j'_1, \ldots, j'_r\}$ where all $j'_i, i = 1, \ldots, r$ are independently sampled from the empirical distribution $(v_1/r, \ldots, v_k/r)$.
   (b) Set $\hat{\nu}_b = \hat{\nu}^{ML}(k_b, r)$, where k_b is the number of distinct entries in the bootstrap sample $\{j'_1, \ldots, j'_r\}$.
3. Calculate the bias estimate
$$\text{bias} = \hat{\nu}^{ML}(k, r) - \frac{\sum_{b=1}^{B} \hat{\nu}_b}{B}$$
4. Output $\hat{\nu}^{BOOT} = \hat{\nu}^{ML}(k, r) + \text{bias}$.

Fig. 1. Algorithm for the bootstrap estimator of ν.

within each class. Thus we can estimate $\hat{\nu}^{ML}(k_i, r_i)$ for each class independently, where
$$k_i = \beta_i \quad \text{and} \quad r_i = i\beta_i.$$

There is one flaw in this idea: when all the optima found are distinct, it is obvious that $\hat{\nu}(r, r) = \infty$. This makes it impossible to give a proper estimate for the number of optima in C_1 if $C_1 \neq \emptyset$. In order to approximate this value, we can merge the observations corresponding to these optima with those of the next non-empty class C_d, where $d = \min_{i>1}(i : C_i \neq \emptyset)$. Thus the 'plug-in' estimate (with $k_d \leftarrow k_1 + k_d$ and $r_s \leftarrow r_1 + r_d$) is
$$\hat{\nu}^{PI} = \sum_{i=d}^{r} \hat{\nu}^{ML}(k_i, r_i).$$

This could also be seen as a non-parametric analogue of the frequency-based parametric approach used (with mixed results) in [2].

4   Experiments

In [1,2], some experiments were carried out: firstly on some N, K landscapes (as introduced by Kauffman [9]) with $N = 15, 16$ and $K = 4, 5$ using a 'bit-flip' neighbourhood, and secondly on some flowshop sequencing problems with 10 jobs and 10 machines, using 5 different neighbourhoods as defined in [10]. Table 1 shows the results of applying the bootstrap based estimators, compared to the results of the parametric maximum likelihood estimate and the jackknife, as reported in [2].

Table 1. Jackknife and bootstrap estimates of ν. In the case of the jackknife, the value in brackets—e.g. (3)—indicates the value of s at which the 'best' estimate was found, according to the Burnham-Overton procedure. The 'double shift' estimates are the actual number found (k), as the estimation procedures tend to break down when the value of k/r is small. The true number of optima was found in each case by complete enumeration.

N, K	True # optima	Estimated # optima			
		MaxL	Jack	Boot	Plug-in
15,4	102	49	88 (3)	63	88
16,4	143	65	153 (4)	84	126
15,5	191	95	180 (4)	133	166
16,5	214	85	189 (5)	119	133
$10/10/P/C_{sum}$					
inversion	810	88	179 (5)	123	161
forward shift	285	71	186 (5)	96	135
exchange	111	31	44 (2)	37	45
backward shift	44	18	22 (1)	20	21
double shift	7	4	4	4	4

It can be seen that the results of the non-parametric estimates are in general much closer to the true value ν than the parametric estimates; the ability to reduce the bias provided by jackknife and bootstrap estimators is clearly seen. The overall impression is that the jackknife estimator gets the closest to ν, but we would hesitate to declare it the winner on the basis of just a few experiments.

4.1 Further N, K Experiments

Some further experiments were run with N, K landscapes of increasing ruggedness. The value of N was kept at 15, but K was increased from 2 to 12 in steps of 1. In each case 1000 NS restarts were made, and the bootstrap estimates also used $B = 1000$. The graph in Figure 2 shows the results.

In this case, it is clear that the non-parametric estimators show similar behaviour, all providing better estimates of ν than $\hat{\nu}^{ML}$. It would appear that any of the three approaches can give quite good estimates of ν for problems where the number of restarts is about 3% of the cardinality of the search space. However, while the bootstrap and plug-in estimators sometimes overestimate the true value, the jackknife seems to have a consistent (but small) negative bias.

4.2 Low Autocorrelation Binary Sequences

N, K landscapes have been extensively investigated, and seem in some sense to be 'well-behaved'. For example, a 'big valley' phenomenon (see [10] or [11] for more on this idea) is often present, which suggests (among other things)

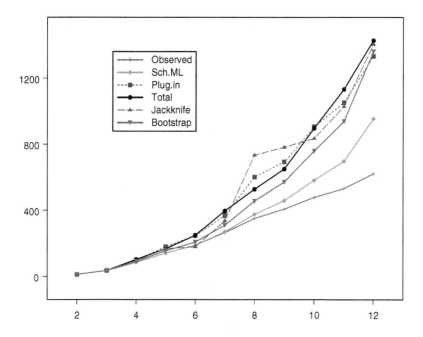

Fig. 2. N, K landscapes for $N = 15$: Estimates of ν based on the Schnabel census/maximum likelihood approach contrasted with the non-parametric estimates. Also shown are the total number of optima obtained by enumeration and the observed number (k) in $r = 1000$ restarts.

that the global optimum has a large basin of attraction, and is located 'near' to other (local) optima. The class of 'low autocorrelation binary sequences' (LABS), however, is notorious for generating problems where the global optima are very hard to find. The task is to find a vector $y \in \{-1, +1\}^N$ that minimizes

$$E = \sum_{p=1}^{N-1} c_p^2$$

where c_p is the autocorrelation of the sequence $(y_1, \ldots, y_N)$ at lag p, i.e.

$$c_p = \sum_{i=1}^{N-p} y_i y_{i+p}$$

An alternative version is often used, where the objective is to *maximize*

$$F = \frac{N^2}{2E}.$$

The obvious neighbourhood to use for this problem is again a 'bit-flip' neighbourhood. For small values of N it is possible to use a complete enumeration to

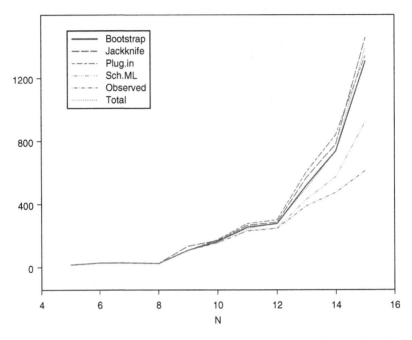

Fig. 3. LABS landscapes for various values of N: Estimates of ν are based on both parametric and non-parametric methods. Also shown are the total number of optima obtained by enumeration (indicated by 'Total') and the observed number (k) in $r = 1000$ restarts.

find the optima. In Figure 3 the results of some initial experiments are compared with a variety of estimators for $N = 5, 6, \ldots, 15$. As in the N, K experiments, the bootstrap used $B = 1000$ resamples.

It can be seen that the non-parametric estimators again stay quite close to the true value up to $N = 15$, but the ML approach underestimates ν throughout. Greater computational resources allowed further experiments with enumeration for $N \leq 24$; only estimates could be made for $N > 24$ (again, using $r = 1000$ restarts). Figure 4 presents the results of these experiments.

Extrapolating from the behaviour for known values of ν, it seems likely that these estimates will be negatively biased. However, there are some interesting differences and similarities in their behaviour. The jackknife method seems to approach an unrealistically low limit as N increases. This is most probably due to the fact that we have not taken the jackknife procedure beyond $s = 5$. Further theoretical work is needed to extend the capability of this method.

It can also be seen that the plug-in estimator is approaching the ML estimator as the problem size increases. In fact this is what one should expect: when k is close to r, most of the optima are seen just once or twice and thus the ML estimate for $C_1 \cup C_d$ becomes the major component in $\hat{\nu}^{PI}$.

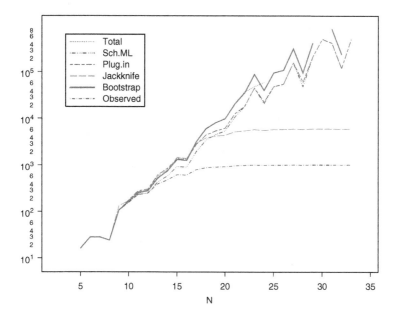

Fig. 4. LABS landscapes for various values of N: Estimates of ν are based on both parametric and non-parametric methods as before, but no true values of ν were available for $N > 24$. Note also that the growth in the number of optima is so fast that this axis is on a log scale.

Generally, the bootstrap estimate has the largest value, and we might conjecture that it has the smallest bias. However, in some of our largest instances of LABS we were unable to compute it owing to overflow when $k = r$. This indicates that the sample size r is insufficient and should be increased to evaluate the landscapes with such a large number of optima. What a sufficient sample size is in general is an open question: does r need to increase exponentially, or sub-exponentially?

An interesting final comment relates to the nature of the LABS landscape. The number of optima does not increase monotonically with N: for $N = 7$ there are 28 optima, yet for $N = 8$ we find $\nu = 24$. A similar dip is observed at $N = 16$. We could conjecture that there is a systematic pattern for $N = 2^q$, for integer $q > 1$. There is conflicting evidence from the estimators, several of which see $\hat{\nu}$ dipping at $N = 24, 28, 32$. The dip at $N = 24$ does not correspond to the real value of ν, so perhaps the same is true for $N = 28$.

5   Conclusion

We have extended earlier work on the estimation of properties of combinatorial landscapes in investigating the question of how many local optima are to be found. Some new methods have been presented, and we have shown experimentally that the non-parametric estimates seem preferable to those based on parametric approaches. The jackknife performs well if r is adequate relative to $|\mathcal{X}|$, but needs further work if it is to be used generally. Both the standard bootstrap and the 'plug-in' method seem fairly robust; of the two, the 'plug-in' method needs less computation. It is clear, however, that more work remains to be done in several areas: for example, in improving the jackknife, in evaluating the number of restarts required to give a reasonable estimate, and in investigating different types of COP, such as (ℓ, θ) landscapes [12], or further examples of permutation problems.

Acknowledgement

Part of this research was carried out when the first author was visiting the UK on a grant from the Royal Society.

References

1. Reeves, C.R.: Estimating the number of optima in a landscape, Part I: Statistical principles. (in review) (2001).
2. Reeves, C.R.: Estimating the number of optima in a landscape, Part II: Experimental investigations. (in review) (2001).
3. Liu, C.L.: *Introduction to Combinatorial Mathematics*. McGraw-Hill, New York (1968).
4. Beaumont, G.P.: *Intermediate Mathematical Statistics*, Chapman & Hall, London (1980).
5. Kallel, L., Garnier, J.: How to detect all maxima of a function. In Kallel, L., Naudts, B., Rogers, A. (eds.): *Theoretical Aspects of Evolutionary Computing*. Springer-Verlag, Berlin (2001) 343-370.
6. Burnham, K.P., Overton, W.S.: Estimation of the size of a closed population when capture probabilities vary between animals. *Biometrika* **65** (1978) 625-633.
7. Efron, B.: Bootstrap methods: another look at the jackknife. *Annals of Statistics* **7** (1979) 1-26.
8. Efron, B., Tibshirani, R.J.: *An Introduction to the Bootstrap*. Chapman & Hall, London (1993).
9. Kauffman, S.: *The Origins of Order: Self-Organization and Selection in Evolution*. Oxford University Press, Oxford (1993).
10. Reeves, C.R.: Landscapes, operators and heuristic search. *Annals of Operational Research* **86** (1999) 473-490.
11. Boese, K.D., Kahng, A.B., Muddu, S.: A new adaptive multi-start technique for combinatorial global optimizations. *Operations Research Letters* **16** (1994) 101-113.
12. Reeves, C.R.: Experiments with tuneable fitness landscapes. In Schoenauer, M., Deb, K., Rudolph, G., Yao, X., Lutton, E., Merelo, J.J., Schwefel, H-P. (eds.) *Parallel Problem-Solving from Nature—PPSN VI*. Springer-Verlag, Berlin Heidelberg New York (2000) 139-148.

Performance of Evolutionary Approaches for Parallel Task Scheduling under Different Representations

Susana Esquivel, Claudia Gatica, and Raúl Gallard

Laboratorio de Investigación y Desarrollo en Inteligencia Computacional, Universidad Nacional de San Luis, Argentina
{esquivel, crgatica, rgallard}@unsl.edu.ar

Abstract. Task scheduling is known to be NP-complete in its general form as well as in many restricted cases. Thus to find a near optimal solution in, at most, polynomial time different heuristics were proposed. The basic Graham´s task graph model [1] was extended to other list-based priority schedulers [2] where increased levels of communication overhead were included [3]. Evolutionary Algorithms (EAs) have been used in the past to implement the allocation of the components (tasks) of a parallel program to processors [4], [5]. In this paper five evolutionary algorithms are compared. All of them use the conventional Single Crossover Per Couple (SCPC) approach but they differ in what is represented by the chromosome: processor dispatching priorities, tasks priority lists, or both priority policies described in a bipartite chromosome. Chromosome structure, genetic operators, experiments and results are discussed.

1 Introduction

A parallel program is a collection of tasks, with precedence constraints to be satisfied among them. It can be seen as a set of tasks whose execution is dictated by given precedence constraints. These precedence constraints between tasks are commonly delineated in a directed acyclic graph known as the *task graph* $G = (T,A)$ where the vertices in $T=\{1,...,n\}$ represent the tasks and the directed arcs indicate the precedence constraints; $(i, j) \in A$ means that task i is an immediate predecessor of task j. The task processing times p_i ($i = 1...n$) are specified by a function $\mu: T \to R^+$. Here, a scheduling policy of the parallel processing system assigns the tasks to the processors and schedules the execution of these tasks so that the precedence relations are satisfied. Consequently for any pair $(i, j) \in A$, the processing starting times S_i and S_j have the following relationship: $S_j \geq S_i + p_i$.

The schedule is such that a task can be assigned to only one processor, and one processor can execute at most one task at a time. Factors, such as number of processors, number of tasks and task precedence make harder to determine a good assignment. Figure 1 shows a task graph with six tasks and a simple single precedence constraint. A schedule can be depicted by a Gantt chart where the initiation and ending times for each task running on the available processors are indicated. By simple observation of the Gantt chart the completion time of the last task abandoning the system, (makespan), the processors utilisation, the evenness of load distribution, the speed up and other performance measures can be easily determined. Diverse objec-

S. Cagnoni et al. (Eds.): EvoWorkshops 2002, LNCS 2279, pp. 41–50, 2002.

tives are of interest for this problem,concerning makespan, an optimal schedule is one such that the total execution time is minimized (see fig.2). Some simple scheduling problems can be solved to optimality in polynomial time while others can be computationally intractable.

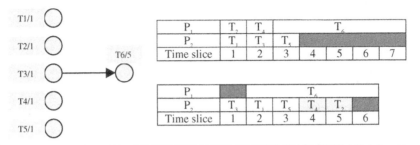

Fig. 1. A task graph

Fig. 2. Two feasible schedules for the task graph of figure 1. Related to the makespan the shortest schedule is optimal

As we are interested in the scheduling of arbitrary task graphs onto a reasonable number of processors we would be content with polynomial time scheduling algorithms that provide good solutions even though optimal ones can not be guaranteed. Most of them are list-scheduling heuristics and were based on the well-known Graham´s List Scheduling Algorithm (LSA). In the following sections we introduce LSA, a set of EAs with different representations, a description of experiments, results and final conclusions.

2 Graham´s List Scheduling Algorithm (LSA)

For a given list of tasks ordered by priority, it is possible to assign tasks to processors by always assigning each available processor to the first unassigned task on the list whose predecessor tasks have already finished execution.

Let us denote,
$T=\{T_1,....,T_n\}$ a set of tasks,
$\mu: T \rightarrow (0, \infty)$ a function which associates an execution time to each task,
$\leq$ a partial order in T and
L a priority list of tasks in T.

Each time a processor is idle, it immediately removes from L the first ready task; that is, an unscheduled task whose ancestors under $\leq$ have all completed execution. In the case that two or more processors attempt to execute the same task, the one with lowest identifier succeeds and the remaining processors look for another adequate task.

Using this heuristic, contrary to the intuition, some anomalies can happen. For example, increasing the number of processors, decreasing the execution times of one or more tasks, or eliminating some of the precedence constraints can actually increase the makespan [1]. We are seeking for heuristics, which are free from these anomalies.

3   Heuristics to Face Task Scheduling Problems

The task allocation problem has been investigated by many researchers [6], [7], [8]. Various heuristic methods have been proposed, such as mincut-based heuristics and orthogonal recursive bisection. List-based priority schedulers have long been the predominant class of static scheduling algorithms [9]. Here, priorities are assigned to tasks following some rule. Most of them are oriented to *critical path, most immediate processor first* priority. List scheduling heuristics were originally devised with zero inter-task communication costs, and even under this simplified assumption the problem remains NP-hard.

Evolutionary Computation techniques also contributed to solve this problem; they can find an acceptable schedule by exploiting relevant areas of the searching space. Regarding solution representation these methods can be roughly categorized as *direct* and *indirect* representations [10]. In direct representation, a complete and feasible schedule is an individual of the evolving population. The only method that performs the search is the evolutionary algorithm because the represented information comprises the whole search space. In the case of indirect representation of solutions the algorithm works on a population of encoded solutions. Because the representation does not directly provide a schedule, a *schedule builder* is necessary to transform a chromosome into a schedule, validate and evaluate it. The schedule builder guarantees the feasibility of a solution and its work depends on the amount of information included in the representation. In this paper indirect representation is implemented by means of *decoders* and permutations. A decoder is a mapping from the representation space into a feasible part of the solution space, which includes mappings between representations that evolve and representations that constitute the input for the evaluation function. This simplifies implementation and produces feasible offspring under different conventional crossover methods, avoiding the use of penalties or repair actions. Permutations are directly evolved but require special genetic operators.

The question now is what should be represented? In the problem we are facing, allocation of parallel related tasks onto processors, we can address the search by looking at the processors or the task priorities, or even by looking at both at the same time.

In this work we propose different EAs using different chromosome representations by means of decoders and permutations. To guide our search towards schedules with optimal makespan, three variants were considered. In the first variant processor dispatching priorities are encoded. In the second one, tasks priority lists are encoded, and in the third scheme a bipartite chromosome represents both processor dispatching priorities and tasks priority lists.

4   Representations

The representation adopted in an EA is crucial to its performance. A particular representation defines the areas of the problem space to be searched making the search more or less effective and determines possible genetic operators to be applied making the process more or less efficient. In the following section we discuss some indirect representations used in this work.

4.1 Indirect-Decode Representations

Under indirect-decode representations conventional genetic operators can be used without applying penalties or repair actions because feasible schedules are generated.

4.1.1 Processor Dispatching Priorities

Here a schedule is encoded in the chromosome in a way such that the task indicated by the gene position is assigned to the processor indicated by the corresponding allele, as shown in fig. 3:

processor → | 1 | 2 | 1 | 2 | 2 | 1 |

task →   1   2   3   4   5   6

Fig. 3. Chromosome structure for processor dispatching priorities

The idea is to use a decoder to reflect which processor has the priority to dispatch a given task. Here the chromosome gives instructions to a decoder on how to build a feasible schedule. In our case the decoder is instructed in the following way: By following the priority list, traverse the chromosome and assign the corresponding task to the indicated processor as soon as the precedence relation is fulfilled. The task priority list is defined at the beginning and remains the same during the search (i.e. canonical order). The processor priorities for dispatching tasks change from individual to individual while searching for an optimal schedule (minimum makespan).

Under this representation simple recombination operators such as one point crossover can be safely used. Also, simplest mutation operators can be implemented by a simple swapping of values at randomly selected chromosome positions or by a random change in an allele. The new allele value identifies any of the involved processors.

4.1.2 Task Priority Lists

In this second proposal each individual in the population represents a list of task priorities. Each list is a permutation of task identifiers. Here a chromosome is an n-vector where the ith component is an integer in the range $1..(n - i + 1)$. The chromosome is interpreted as a strategy to extract items from an ordered list L and builds a permutation. We briefly explain how the decoder works. Given a set of tasks represented by a list L and a chromosome C, the gene values in the chromosome indicate the task positions in the list L. The decoder builds a priority list L' as follows: traversing the chromosome from left to right it takes from L an element whose position is indicated by the gene value, puts this element in L' and deletes it from L (shifting elements to left and reducing L length). It continues in this way until no element remains in L. Fig. 4 shows the list L, the chromosome C and the resulting priority list L'.

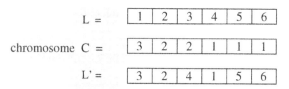

Fig.4. Chromosome structure for task priority lists

Once decoded, LSA as defined in section 2 (giving priority to the processor with lowest identifier when conflict arise), is used to build the schedule. As in the previous case simple genetic operators can be safely applied.

4.1.3 Task Priority Lists and Processor Dispatching Priorities
Under this representation scheme, a task priority list and a processor dispatching priority are encoded in the first and second half of a chromosome (see fig.5).

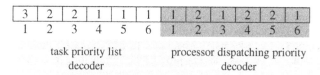

task priority list   processor dispatching priority
decoder   decoder

Fig. 5. Chromosome structure to combine decoders

Here, the left half of the chromosome is decoded first to obtain the corresponding task priority list L'. Then LSA is applied to build the schedule by using L' (3 2 4 1 5 6) and assigning each task to a processor according to the dispatching priority described in the second half of the chromosome. The simple genetic operators are applied now separately to each chromosome half.

4.2 Indirect-Permutation Representations
Under indirect-permutation representations parent's recombination should create offspring permutations. Consequently specialized operators such as partially mapped, ordered or cycle crossover (PMX, OX, CX) should be used.

In our approach we use single permutations as chromosomes to describe processor dispatching priorities and task list priorities, and a bipartite (double permutation) chromosome for the combinations of priorities. In this latter case the genetic operators are, as for decoders, applied independently for each half of the chromosome.

5. Experiments and Results

The experiments implemented five evolutionary algorithms as follows,
EAD1 processor dispatching priorities, indirect-decode representation.
EAD2 task list priorities, indirect-decode representation.

EAD3 processor dispatching priorities and task list priorities, indirect-decode representation.

EAP1 task list priorities, indirect-permutation representation.

EAP2 processor dispatching priorities and task list priorities, indirect-permutation representation.

All of them with randomised initial population of size fixed at 50 individuals for the smaller instances (1 to 8) and 100 individuals for the larger instances (9 to 12). Series of ten runs each were performed on the 12 testing cases under SCPC, using elitism. For indirect-decode representations, one-point crossover and big-creep mutation were applied while for indirect-permutation representations the well-known partially mapped crossover (PMX) and random exchange mutation were used as genetic operators. The maximum number of generations was fixed at 2000, but a stop criterion was used to accept convergence when after 20 consecutive generations, mean population fitness values differing in $\varepsilon \leq 0.001$ were obtained. Probabilities for crossover and mutation were fixed at 0.65 and 0.01, respectively. The testing cases corresponded to:

Instance 1: Task graph G1 (6 tasks and 2 processors)

Instance 2: Task graph G2 (7 tasks and 3 processors)

Instance 3: Task graph G3 (9 tasks and 3 processors)

Instance 4: Task graph G4 (9 tasks and 4 processors)

Instance 5: Task graph G5 (9 tasks and 3 processors, decreasing task's duration)

Instance 6:Task graph G6 (9 tasks and 3 processors, eliminating precedence constraints)

Instance 7: Task graph G7 (10 tasks and 2 processors)

Instance 8: Task graph G8 (13 tasks and 3 processors)

Instance 9: Randomly generated task graph G9 (25 tasks and 5 processors)

Instance 10: Randomly generated task graph G10 (50 tasks and 5 processors)

Instance 11: Randomly generated task graph G11 (25 tasks and 5 processors)

Instance 12: Randomly generated task graph G12 (70 tasks and 10 processors)

In contrast to other scheduling problems such as flow shop or job shop, which have different machine environments and more oriented to production scheduling, after an intensive search in the literature we could find, for this problem, only few benchmarks. The first 8 instances were extracted from the literature [11] and they have known optima. Instances 9 to 12 were generated by random assignment of relations and tasks duration. Their optimum values are unknown. Nevertheless for these instances several trials were run (under this and previous approaches [12]) to determine the best quasi-optimal solution under the whole set of contrasted algorithms. This value will be referred in what follows as an estimated optimal value (the best known value assumed as optimal). Diverse performance variables were considered to contrast the algorithms. To measure the *quality of solutions* provided by EAs we used:

Ebest: (Abs(opt_val – best value)/opt_val)/100. It is the percentile (approximation) error of the best found individual in one run when compared with the known (or a-ssumed) optimum value *opt_val*. It gives us a measure of how far is the quality of the best individual from that opt_val.

Best: is the best (minimum) makespan value found throughout all the runs. This value has been reached by different solutions (schedules).

MEbest: is the mean value of the error, over the total number of runs.

We defined the *versatility* of an evolutionary algorithm as its ability to find diverse schedules of equal minimum makespan. To measure the versatility of the algorithms we used:

TotB defined as the total number of best solutions. It is the mean total number of distinct best solutions found by the algorithm throughout all runs.

In the following tables, quality of solutions and versatility of the algorithms are contrasted.

Quality of Solutions

Table 1 shows results for representations using decoders. Here we can observe that in average all EADs outperform LSA and find the *opt_val* for the smaller instances 1 to 8. For larger instances the *opt-val* is not attained by EAD1 and EAD3 in harder instances 9 and 10, while EAD2 cannot attain the *opt-val* only for instance 10. According to average behaviour the best performer is EAD2 followed by EAD3. Consequently, further experiments with permutations were devoted to task list priorities and to the combined processor-task priorities.

Table 1. Quality of solutions. Minimum makespans (Best) and minimum errors (Ebest) under indirect-decode representations.

Inst.	Opt-value	LSA		EAD1		EAD2		EAD3	
		Best	Ebest	Best	Ebest	Best	Ebest	Best	Ebest
1	6	7	16.60	6	0.00	6	0.00	6	0.00
2	9	9	0.00	9	0.00	9	0.00	9	0.00
3	12	12	0.00	12	0.00	12	0.00	12	0.00
4	12	15	25.00	12	0.00	12	0.00	12	0.00
5	10	16	60.00	10	0.00	10	0.00	10	0.00
6	12	13	8.33	12	0.00	12	0.00	12	0.00
7	31	38	22.58	31	0.00	31	0.00	31	0.00
8	30	33	10.00	30	0.00	30	0.00	30	0.00
9	270	375	38.88	297	10.00	270	0.00	284	5.18
10	549	591	7.65	635	15.66	553	0.72	596	8.56
11	410	412	0.48	410	0.00	410	0.00	410	0.00
12	963	963	0.00	963	0.00	963	0.00	963	0.00
Avg	192.83	207.00	15.79	202.25	2.14	193.17	0.06	197.92	1.15

Table 2 shows results for representations using permutations. Here we can observe that also both EAPs find the optimum for the smaller instances 1 to 8. For larger instances the *opt-val* cannot be attained only for instance 10 by EAP1, but it is very close to it, while EAP2 finds this value.

Table 3 shows mean Ebest values attained by the algorithms for each instance. The combined processor-task priorities with permutations is clearly the winner with an average mean Ebest value of 0.09 %. It also can be observed that indirect-permutation representations are better than indirect-decode representations.

Table 2. Quality of solutions. Minimum makespans (Best) and minimum errors (Ebest) under indirect-permutation representations.

Inst.	Opt-value	EAP1		EAP2	
		Best	Ebest	Best	Ebest
1	6	6	0.00	6	0.00
2	9	9	0.00	9	0.00
3	12	12	0.00	12	0.00
4	12	12	0.00	12	0.00
5	10	10	0.00	10	0.00
6	12	12	0.00	12	0.00
7	31	31	0.00	31	0.00
8	30	30	0.00	30	0.00
9	270	270	0.00	270	0.00
10	549	550	0.18	549	0.00
11	410	410	0.00	410	0.00
12	963	963	0.00	963	0.00
Avg	193.08	192.92	0.02	192.83	0.00

Table 3. Quality of solutions. Mean errors (MEbest) under each representations.

Inst.	EAD1	EAD2	EAD3	EAP1	EAP2
1	0.00	0.00	0.00	0.00	0.00
2	0.00	0.00	0.00	0.00	0.00
3	0.00	0.00	0.00	0.00	0.00
4	0.00	0.00	0.00	0.00	0.00
5	0.00	0.00	0.00	0.00	0.00
6	0.00	0.00	0.00	0.00	0.00
7	0.00	0.00	0.00	0.00	0.00
8	2.60	0.00	0.00	0.00	0.00
9	7.40	2.30	7.50	1.22	0.37
10	8.46	1.85	10.74	0.00	0.00
11	0.00	0.00	0.00	0.85	0.67
12	0.00	0.00	0.00	0.00	0.00
Avg.	1.54	0.35	1.52	0.17	0.09

Versatility

Table 4 shows that except for hard instances 9 and 10, all EAs find more than a single best solution (distinct schedules with the same optimal or quasi optimal makespan). For instance 9, EAD2, EAP1 and EAP2 reach the opt_val (270) while for instance 10 only EAP2 reaches the opt_val (549). It worth saying that these values are new and better than previous values (309 and 591, for instance 9 and 10, respectively) found by means of a multirecombined EA [12]. Average values show higher versatility for decoders (but with inferior quality of solutions in some instances) than for permutations. Also EAs with decoders run reaching the maximum number of generations while with permutation they halt according with the stop criterion. This indicates the need of parameter tuning for decoder-based representations.

Table 4. Versatility of the algorithms, (TotB).

Inst.	LSA	EAD1	EAD2	EAD3	EAP1	EAP2
1	0	2	32	87	12	28
2	1	147	14	358	14	79
3	1	24	82	60	44	12
4	0	74	101	354	80	302
5	0	105	199	287	35	20
6	0	18	129	63	11	13
7	0	4	97	17	10	45
8	0	19	82	3	28	40
9	0	0	1	0	1	1
10	0	0	0	0	0	1
11	0	115	52	41	9	102
12	1	1	6	4	4	102
Avg.	0.25	42.42	66.25	106.17	20.67	62.08

6 Conclusions

The parallel task scheduling problem remains NP-hard even under many simplifying assumptions. Most of the current work is oriented to list-scheduling heuristics where the problem, for a particular instance, is to find the best task priority rule.

In previous works multirecombined EAs, being free of LSA anomalies, showed their effectiveness providing not a single but a set of near optimal solutions to diverse graphs belonging to a selected test suite.

As blind-search algorithms EAs perform the search based only on the quality of solutions found so far, and continuously evolve them with principles of natural selection. The representation issue is of utmost importance to guide this search process through the solution space. This paper explored the effect of different indirect representations based on decoders and permutations of processor dispatching priorities, task (list) priorities and a combination of both priority policies.

Results indicate that permutations work better than decoders, and in this case the combination of priority policies is the best approach providing new minimum makespan values, which could not be reached by more sophisticated (and costly) multirecombined evolutionary algorithms. Permutation representations works better than decoders because they stress the position, the absolute or relative order of genes within the chromosome and suitable crossovers are chosen. These results confirm something that is known in the EC field: by using a better representation less computational effort will be needed to provide high quality solutions.

Further work will be oriented to enhance these algorithms by means of hybridization for fine-tuning and self-adaptation of parameters.

References

1. Graham R. L.: Bounds on Multiprocessing Anomalies and Packing Algorithms. Proceedings of the AFIPS 1972 Spring Joint Computer Conference, pp 205-217, 1972.
2. Adam T.L., Chandy K.M., and Dickson J.R., A comparison of list schedules for parallel processing systems. Communications of the ACM, 17:685-9, 1974.
3. Kruatrachue B., Static task scheduling and grain paccking in parallel processing systems.PhD Thesis, Oregon State University, 1987.
4. Zomaya A., Genetic scheduling for parallel processor systems: Comparative studies and performance issues. IEEE Trans. Parallel and Distributed Systems. Vol. 10, No. 8, 1999.
5. Kidwell M. :Using Genetic Algorithms to Schedule Tasks on a Bus-based System. Proceedings of the 5th International Conference on Genetic Algorithms, pp 368-374, 1993.
6. Cena M.,Crespo M., Gallard R..: Transparent Remote Execution in LAHNOS by Means of a Neural Network Device. ACM Press, Operating Systems Review, Vol. 29, Nr. 1, pp 17-28, 1995.
7. Ercal F.: Heuristic Approaches to Task Allocation for Parallel Computing. Doctoral Dissertation, Ohio State University, 1988.
8. Flower J., Otto S., Salama M.: Optimal mapping of irregular finite element domains to parallel processors. Caltech C3P#292b, 1987.
9. Al_Mouhamed M. and Al_Maasarani A., Performance evaluation of scheduling precedence- constrained on message-passing systems. IEEE Trans. Parallel and Distributed Systems. 5(12):1317-1322, 1994.
10. Bagchi S., Uckum S., Miyabe Y., Kawamura K.: Exploring Problem Specific Recombination Operators for Job Shop Scheduling. Proceedings of the 4th International Conference on Genetic Algorithms, pp 10 – 17, 1991.
11. Pinedo M.,: Scheduling: Theory, Algorithms and Systems. Prentice Hall International Series in Industrial and Systems Engineering, 1995.
12. Esquivel S., Gatica C., Gallard R.: Conventional and Multirecombinative Evolutionary Algorithms for the Parallel Task Scheduling Problem, LNCS 2037: "Applications of Evolutionary Computing", pp. 223 – 232, Springer , April 2001.
13. Bierwirth C., Mattfeld D., Kopfer H: On Permutation representations for Scheduling Problems, PPSN IV, Springer-Verlag. pp310-318, 1996.

A Performance Comparison of Alternative Heuristics for the Flow Shop Scheduling Problem

Susana Esquivel, Guillermo Leguizamón, Federico Zuppa, and Raúl Gallard

Laboratorio de Investigación y Desarrollo en Inteligencia Computacional,
Universidad Nacional de San Luis, Argentina
{esquivel, legui, fede, rgallard}@unsl.edu.ar

Abstract. Determining an optimal schedule to minimise the completion time of the last job abandoning the system (makespan) become a very difficult problem when there are more than two machines in the flow shop. Due, both to its economical impact and complexity, attention to solve this problem has been paid by many researchers. Starting with the Johnson's exact algorithm for the two-machine makespan problem [1], over the past three decades extensive search have been done on pure m-machine flow shop problems. Many researchers faced the Flow Shop Scheduling (FSSP) by means of well-known heuristics which, are successfully used for certain instances of the problem and providing a single acceptable solution. Current trends to solve the FSSP involve Evolutionary Computation and Ant Colony paradigms. This work shows different bio-inspired heuristics for the FSSP, including hybrid versions of enhanced multirecombined evolutionary algorithms and ant colony algorithms [2], on a set of flow shop scheduling instances. A discussion on implementation details, analysis and a comparison of different approaches to the problem is shown.

1 Introduction

The *flow-shop sequencing problem* is generally described as follows: There are m machines and n jobs. Each job consists of m operations and each operation requires a different machine, so n jobs have to be processed in the same sequence on m machines. The processing time of each job in each machine is known. The objective is to find the sequence of jobs minimizing the maximum flow time which is called *makespan*. In our FSSP model we assumed, that each job is processed on all machines in the same order, each machine processes a job at a time, and each job is processed in a single machine at a time. The operations are not preemptable and set-up times are included in the processing times. In a previous work [3] we contrasted the conventional heuristics proposed by Palmer [4], Gupta [5], Campbell et al. [6] and Nawaz et al. (NEH) [7] showing the outstanding behaviour of NEH. In [8,9] we introduced a multirecombined approach in which the idea of a stud (breeding individual) was used. Here, the stud is selected among the multiple intervening parents and mates them, more than once, in multiple crossover operations.

Early experiments with AC algorithms included its application to the Travelling Salesperson Problem, the Quadratic Assignment Problem, as well as the Job Shop Scheduling, Vehicle Routing, Graph Colouring, Telecommunication Network Pro-

S. Cagnoni et al. (Eds.): EvoWorkshops 2002, LNCS 2279, pp. 51–60, 2002.
© Springer-Verlag Berlin Heidelberg 2002

larger diversity is supplied. This can help to avoid premature convergence. Eiben initially used three *scanning crossover* (SX) methods, which essentially take genes from parents to contribute to build the offspring. As defined, SX is not directly applicable to permutations because invalid offspring are created. In our proposal, the SX scanning method is replaced by PMX.

Preliminary steps towards that MCMP approach included MCMP-SOP and MCMP-SRI [8]. In the first one, every member of the mating pool was selected from the existing (old) population. Within this pool, the best individual was selected as the stud and it was coupled more than once with every other member of the pool. In MCMP-SRI only the stud is selected from the population and the rest of the individuals are created randomly (random immigrants). From both MCMP-SOP behaved better.

MCMP-NEH is very similar to MCMP-SOP and MCMP-SRI, but here, only the stud is selected from the population and the rest of the individuals in the mating pool are created in two ways. Part of them are random immigrants and part of them are NEH-based individuals created from randomly chosen population individuals. The NEH heuristic applied is the same as described in section 2, except in its first step, where the jobs are not ranked by the order of the sums of processing times, instead the chromosome provides the rank. The whole process could be seen as an improvement applied to that chromosome. In this implementation, we combined both types of chromosomes, along with the stud selected from the population. In order to decide which method is to be used to generate the individuals for mating, a random number is generated. After generating n_1 mating individuals (including the stud), multiple crossovers are performed. The crossover points are determined in a random way and the stud is combined with the rest of the selected parents. From that multirecombination, $2*(n_1 -1)$ offspring are obtained, but only the best one, which is stored in a temporary structure, survives. After completing all the n_2 crossovers, the best individual in the temporary structure is selected for insertion in the next population (Fig. 1).

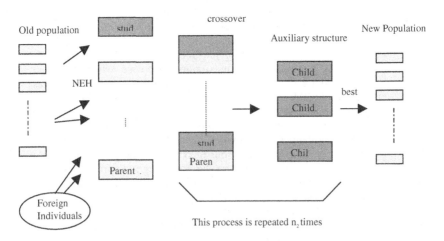

Fig. 1. The multirecombination process in MCMP-NEH

larger diversity is supplied. This can help to avoid premature convergence. Eiben initially used three *scanning crossover* (SX) methods, which essentially take genes from parents to contribute to build the offspring. As defined, SX is not directly applicable to permutations because invalid offspring are created. In our proposal, the SX scanning method is replaced by PMX.

Preliminary steps towards that MCMP approach included MCMP-SOP and MCMP-SRI [8]. In the first one, every member of the mating pool was selected from the existing (old) population. Within this pool, the best individual was selected as the stud and it was coupled more than once with every other member of the pool. In MCMP-SRI only the stud is selected from the population and the rest of the individuals are created randomly (random immigrants). From both MCMP-SOP behaved better.

MCMP-NEH is very similar to MCMP-SOP and MCMP-SRI, but here, only the stud is selected from the population and the rest of the individuals in the mating pool are created in two ways. Part of them are random immigrants and part of them are NEH-based individuals created from randomly chosen population individuals. The NEH heuristic applied is the same as described in section 2, except in its first step, where the jobs are not ranked by the order of the sums of processing times, instead the chromosome provides the rank. The whole process could be seen as an improvement applied to that chromosome. In this implementation, we combined both types of chromosomes, along with the stud selected from the population. In order to decide which method is to be used to generate the individuals for mating, a random number is generated. After generating n_l mating individuals (including the stud), multiple crossovers are performed. The crossover points are determined in a random way and the stud is combined with the rest of the selected parents. From that multirecombination, $2*(n_l-1)$ offspring are obtained, but only the best one, which is stored in a temporary structure, survives. After completing all the n_2 crossovers, the best individual in the temporary structure is selected for insertion in the next population (Fig. 1).

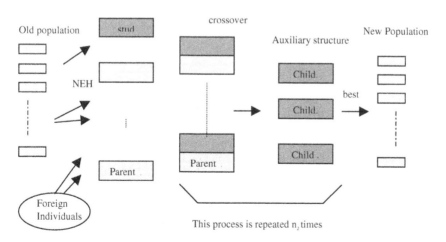

Fig. 1. The multirecombination process in MCMP-NEH

4 The Ant Colony–NEH Approach

The AC approach has shown to be useful for solving combinatorial optimization problems as well as problems for which the objective value changes over time [10]. In AC algorithms a population (colony) of agents (ants) collectively solve the optimization problem under consideration by using a graph representation. Information collected by the ants during the search process is encoded in pheromone trails τ_{ij} associated with edge (i,j). Pheromone trails encode a long-term memory about the whole ant search process. Depending on the problem representation chosen, pheromone trails can be associated with all edges, or only to some of them. Edges can also have an associated heuristic value η_{ij} representing *a priori* information about the problem definition.

FSSP like the TSP is a permutation problem in which a permutation of integers *{1,...,n}* has to be found such that some objective function is minimised. A difference between both problems is that in the TSP, only the relative order of the cities is important. However, for the FSSP, the absolute position of the job is of importance. Thus the position of a job is an attribute of the solution which is represented by τ_{ij}, i.e., the desire of setting job j at position i in the sequence [15]. The heuristic value used in our version is represented by $\eta_{ij}(k) = mk(k) / (new\_mk(k,j) + it(k,j))$, where $mk(k)$ is the objective (makespan) value of the partial solution under construction by the ant k; $new\_mk(k,j)$ and $it(k,j)$, are respectively the new makespan and the accumulated idle time on all machines if job j is selected to be included at position i as the next job in the solution. Thus, the bigger the quantity $(new\_mk(k,j) + it(k,j))$, the smaller is the heuristic value associated to job j at position i. The ants construct the solution by first choosing a job for the first position, then a job for the second position and so on until all jobs are scheduled. The following formula gives the probability selection of the job j at position i.

$$ P_{ij} = \begin{cases} \dfrac{\tau_{ij}^{\alpha}\ \eta_{ij}^{\beta}}{\displaystyle\sum_{h\ \text{not}\ \text{scheduled}} \tau_{ih}^{\alpha}\ \eta_{ih}^{\beta}} & \text{if job } j \text{ is not yet scheduled} \\[4ex] 0 & \text{otherwise} \end{cases} $$

The values of parameters α and β represent respectively the importance of the trail information and the knowledge of the problem in order to drive the search process.

After the whole colony has constructed a complete set of permutations, i.e., they finished a cycle, optionally a process for improving those solutions can be applied and after that, the amount of trails is updated. The solution improving the process applied here is not a local search as in [15], instead we hybridize the AC algorithm as follows: after a cycle is completed, a new set of solutions is obtained. Then, we apply NEH on a randomly chosen subset of these solutions. Only the solutions that get improved after NHE are replaced. According to the behaviour of the AC algorithm, any change on the solutions recently obtained will directly affect the amount of trail to be considered in the next cycle during the item selection process (Pij in Section 4). Thus, improved solutions will guide the search towards more promising regions of the search space. It is important to note that premature convergence is avoided by applying NEH to a small subset of the solutions obtained after each cycle.

Concerning the trail update, in our AC algorithm, only one ant is allowed to update the trail, particularly the best ant in the current cycle. The trail intensities are updated according to $\tau_{ij} = \rho * \tau_{ij} + \Delta\tau_{ij}$, where $1-\rho$ represents the coefficient of evaporation. The amount of trail $\Delta\tau_{ij}$ is equal to $Q/Mbest$ if job j is placed on position i, otherwise zero. $Mbest$ represents the makespan of the ant considered for updating the trails and Q is a parameter of the algorithm. Thus, positions which are often occupied by certain jobs receive a higher amount of pheromone and in the construction phase jobs will be placed preferably on these positions. Similarly to MMAS, our Ant Colony algorithm limits the range of values on the pheromone trails. Thus, the interval for the trail is determined by $[\tau_{min}, \tau_{max}]$ in order to explore alternative positions for the same job.

5 Experiments

In order to contrast the performance of conventional and bio-inspired heuristics, and to determine the effects of multirecombination and/or hibridization on the simpler EAs and ACs versions, we chose to compare the following algorithms:

Single Crossover Per Couple evolutionary algorithm (SCPC), multiple crossovers on multiple parents evolutionary algorithm hibridized with NEH (MCMP-NEH), Ant Colony (AC), Ant Colony hibridized with NEH (AC-NEH), and the conventional NEH heuristic. The algorithms were run on the complete set of the following Taillard's benchmarks [17]: 20 x 5, 20 x 10 and 20x 20 (Jobs x Machines). A series of ten runs were performed for each instance under each algorithm.

To compare the algorithms, the following performance variable was chosen:

$$Ebest = (Abs(opt\_val - best\ value)/opt\_val)100$$

It is the percentile error of the best-found individual when compared with the known, or estimated, optimum value opt_val. It gives us a measure on how far the best individual is from that opt_val.

When the 10-run series were accomplished, mean and minimum $Ebest$ values for each instance were determined and finally average mean and average minimum $Ebest$ values were also determined over all instances.

Parameter setting for evolutionary algorithms: Population size of 100 individuals, a maximum number of generations fixed at 2000. Crossover probability = 0.65 and mutation probability = 0.1. Proportional   Selection, Partially Mapped Crossover (PMX) and Random Exchange Mutation (RXM) were used. In the multirecombined version six crossovers were performed on five parents selected to conform the mating pool. The probability used to decide which method is to be applied to generate the individuals for mating was 0.5.

Parameter setting for the ant colony algorithm: Colony size of 30 ants, a maximum number of cycles fixed at 5000; $\alpha=1$ and $\beta=1$; i.e., a trade-off between the importance of the information trail and knowledge of the problem. The coefficient ρ was set to 0.85 in order to keep past trail information for a longer period of time. The probability used to decide when to apply NEH was 0.01; i.e., about one solution out of 100 solutions generated.

These parameter settings were experimentally determined as the best after a set of many initial trials.

6 Results

In the following tables, the first column identifies the corresponding instance, the following columns show the results under each algorithm and the last row gives the average result values.

20x5 instances

Table 1. Minimun *Ebest* values for 20x5 intances

Instance	SCPC	MCMP-NEH	AC	AC-NEH	NHE
a	0	0	0.01	0	0,63
b	0	0	0.06	0	0,44
c	0	0	0.88	0.09	5,46
d	0.30	0	0.41	0.15	3,63
e	0.56	0	0.57	0	5,58
f	0.50	0	0.21	0	2,76
g	0.96	0	0	0	3,15
h	0	0	0.07	0	2,4
i	0.24	0	0.27	0	4,96
j	0	0	0.29	0	3,88
Avrg.	0.25	0	0.28	0.02	3.29

Table 2. Mean *Ebest* values for 20 x 5 instances

Instance	SCPC	MCMP-NEH	AC	AC-NEH	NEH
a	1.29	0	1.43	0	0,63
b	0.34	0	0.70	0.01	0,44
c	1.23	0,16	3.96	0.10	5,46
d	1.00	0,42	2.25	0.49	3,63
e	0.93	0,25	1.58	0.70	5,58
f	1.10	0,923	1.64	0.16	2,76
g	0.97	0	0.96	0	3,15
h	0.35	0	1.42	0.13	2,4
i	1.48	0	3.39	0.33	4,96
j	1.29	0	1.85	0.16	3,88
Avrg.	1.00	0.18	1.92	0.21	3.29

20 x 10 instances

Table 3. Minimum *Ebest* values for 20x10 instances

Instance	SCPC	MCMP-NEH	AC	AC-NEH	NEH
a	1.70	0.06	1.01	0.25	6,19
b	1.20	0.06	1.20	0.48	4,22
c	0.80	0.33	2.67	0.66	4,08
d	1.16	0.073	1.37	0.21	5,22
e	0.70	**0**	2.83	0.42	5,85
f	0.28	0.22	2.00	0.71	4,01
g	0.60	**0**	1.07	**0**	5,22
h	1.04	0.32	1.82	0.52	4,62
i	1.12	0.06	1.38	0.43	3,39
j	1.38	0.44	2.01	1.06	3,90
Avrg.	1.00	0.16	1.74	0.47	4,67

Table 4. Mean *Ebest* values for 20x10 instances

Instance	SCPC	MCMP-NEH	AC	AC-NEH	NEH
a	2.29	0.28	3.62	0.75	6,19
b	2.56	0.82	3.83	1.29	4,22
c	2.27	0.78	3.99	1.43	4,08
d	2.06	0.67	3.37	1.00	5,22
e	2.42	0.37	4.19	0.95	5,85
f	1.85	1.35	3.27	1.53	4,01
g	1.46	0.16	2.29	0.43	5,22
h	2.65	0.73	4.34	1.51	4,62
i	2.09	0.26	0.07	1.02	3,39
j	2.09	1.00	2.80	1.81	3,9
Avrg.	2.17	0.64	3.18	1.17	4,67

20 x 20 instances

Table 5. Minimum *Ebest* values for 20x20 instances.

Instance	SCPC	MCMP-NEH	AC	AC-NEH	NEH
a	1.21	0.26	2.22	0.43	4,92
b	0.90	0.05	1.09	0.04	2,38
c	1.33	0.27	1.54	0.51	3,65
d	0.31	0.73	0.51	**0**	1,75
e	0.78	0.35	2.13	0.74	4,63
f	0.71	**0**	0.35	0.40	5,53
g	1.14	0.22	1.23	0.30	3,92
h	0.63	0.09	2.00	0.50	2,23
i	0.80	**0**	2.19	0.40	3,08
j	1.56	0.09	1.69	0.36	4,55
Avrg.	0.94	0.21	1.49	0.37	3,66

Table 6. Mean *Ebest* values for 20x20 instance.

Instance	SCPC	MCMP-NEH	AC	AC-NEH	NEH
a	2.12	0.50	3.15	1.00	4,92
b	1.55	0.05	2.60	0.47	2,38
c	1.91	0.80	2.73	1.23	3,65
d	1.69	1.01	3.27	0.43	1,75
e	1.61	0.73	2.62	1.00	4,63
f	1.55	0.32	2.43	1.05	5,53
g	1.83	0.64	2.73	0.98	3,92
h	1.83	0.46	2.79	0.81	2,23
i	1.89	0.17	3.58	0.79	3,08
j	2.94	0.70	3.79	0.85	4,55
Avrg.	1.89	0.54	2.97	0.86	3,66

Tables 1 to 6 show that:

Regarding minimum *Ebest* the simpler evolutionary and ant colony approaches (SCPC and AC) outperform NEH heuristics. Except for some few instances also most mean *Ebest* values are better than those found by the conventional heuristic. On the other hand, it is important to note that the hybridization, of AC and MCMP with NEH, improves considerably the quality of results.

Comparing MCMP-NEH and AC-NEH results for instances of the 20x5 problem size, we see a similar performance to find the best value. Mean *Ebest* values are also alike. For the remaining problem sizes (20x10 and 20X20) they are also similar but MCMP-NEH presents better quality of results than AC-NEH.

7 Conclusions

Following current trends to solve difficult optimisation problems, the FSSP was faced by means of different heuristics. Results obtained from these preliminary approaches deserve the following observations:

MCMP-NEH outperforms significantly the SCPC in every tested problem size. Moreover, according with the results, it can be seen that it has a remarkable trade-off between exploration and exploitation. The same comments can be done about AC-NEH with respect to AC.

Although MCMP-NEH in general outperform AC-NEH it is important to remark some additional aspects. MCMP-NEH requires a greater programming complexity and higher computational effort, slowing too much the search process. After making a few calculations it was determined that the NEH heuristic was performed twice by each individual that was to be inserted in the next population. This adds to 200 each generation and to 400,000 at the end of the run. Further considerations have to be done, in order to apply NEH fewer times on smaller populations to reduce the execution time. Results from AC-NEH are good enough and are available in a small fraction of the time required by MCMP-NEH.

At the light of these preliminary results, research will be oriented to enhance the algorithms performance. In EAs, by means of multirecombination schemes, which preserve jobs adjacency and absolute positions as suggested in [18]. In ACs, these characteristics can be included in their global memory (trails). Also, instead of designing the algorithms with parameter settings resulting as the best after many initial trials we will be focussed to self-adaptation of parameters, when possible.

References

1. Johnson S.: Optimal Two and Three Stage Production Schedule with Setup Times Included. Naval Research Logistic Quarterly, Vol. 1, pp 61-68, 1954.
2. Dorigo M., V. Maniezzo & A. Colorni: The Ant System: Optimization by a Colony of Cooperating Agents. IEEE Transactions on Systems, Man, and Cybernetics-Part B, 26(1), pp 29-41, 1996.
3. Esquivel S., Zuppa F., Gallard R.: Contrasting Conventional and Evolutionary Approaches for the Flow Shop Scheduling Problem, Second International ICSC Symposium of Intelligent Systems , University of Paisley, Scotland U.K, pp 340-345, 2000.
4. Palmer D.: Sequencing Jobs through a Multistage Process in the Minimun Total Time. A Quick Method of Obtaining a near Optimun., Operational Research Quarterly 16, pp 101-107, 1965.
5. Gupta J.: A Functional Heuristic Algorithm for the Flow Shop Scheduling Problem, Operational Research Quarterly 22, pp 39-48, 1971.
6. Campbell H., Dudek R., Smith M.: A Heuristic Algorithm for the n Job m Machines Sequencing Problem. Management Science 16, pp 630-637, 1970.
7. Nawaz M., Enscore E., Ham I.: A Heuristic Algorithm for the m-machine n-job Flow Shop Sequencing Problem. Omeea, Vol. II, pp 11-95, 1983.
8. Esquivel S., Zuppa F., Gallard R.: Using the Stud in Multiple Parents for the Flow Shop Scheduling Problem, invited session on Current Trends in Evolutionary Computation to Face Scheduling Problems, 4th. International ICSC Symposium on Soft Computing and Intelligent Systems for Industry, presentado y publicado, Paisley, Scotland, United Kingdom, pp133, 2001.
9. Esquivel S., Zuppa F., Gallard R.: Multiple Crossover, Multiple Parents and the Stud for Optimization in the Flow Shop Scheduling Problem, World Multiconference on Systemics, Cybernetics and Informatics, Vol III: Emergent Computing and Virtual Engineering, pp 388 – 392, Orlando, Florida, USA. 2001.
10. Corne D, Dorigo M,and Glover F.,, editors: New Ideas in Optimization. Advanced topics in computer science series. McGraw-Hill, 1999.
11. Colorni A., Dorigo M., Maniezzo V. and Trubian M.: Ant system for Job-shop Scheduling. JORBEL - Belgian Journal of Operations Research, Statistics and Computer Science, 34(1), pp 39-53, 1994.
12. Dorigo M., Maniezzo V. and Colorni A.: The Ant System: Optimization by a Colony of Cooperating Agents. IEEE Transactions on Systems, Man, and Cybernetics-Part B, 26(1), pp 29, 1996.
13. Marques C., Zwaan S. Ant Colony Optimization for Job Shop Scheduling, Procedings 3rd Workshop of Genetic Algorithms & Artificial Life, GAAL 99, Lisboa, 1999.
14. Stützle T., den Besten M. and Dorigo M.: Ant Colony Optimization for the Total Weighted Tardiness Problem. In Deb et al, editors, Proceedings of PPSN-VI, Sixth International Conference on Parallel Problem Solving from Nature, volume 1917 of LNCS, pages 611-620, 2000.
15. Stützle T: An Ant Approach to the Flow Shop Problem, Proceedings of EUFIT'98, Aachen, pp 1560-1564, 1998.

16. Eiben A.E., Raué P-E., and Ruttkay Zs., *Genetic algorithms with multi-parent recombination*. In Davidor, H.-P. Schwefel, and R. Männer, editors, Proceedings of the 3rd Conference on Parallel Problem Solving from Nature, number 866 in LNCS, pages 78-87. Springer-Verlag, 1994
17. Taillard E.: Benchmarks for Basic Scheduling Problems, European Journal of Operational Research, Vol. 64, pp 278-285, 1993.
18. Chen S. and Smith S.: Improving Genetic Algorithms by Search Space Reductions (with Applications to Flow Shop Scheduling), Proceedings of the Genetic and Evolutionary Computation Conference, GECCO-99, USA, Morgan Kauffman 1999.

Exploiting Fitness Distance Correlation
of Set Covering Problems

Markus Finger[1], Thomas Stützle[1], and Helena Lourenço[2]

[1] Darmstadt University of Technology, Intellectics Group,
Alexanderstr. 10, 64283 Darmstadt, Germany
[2] Universitat Pompeu Fabra, Department of Economics and Business,
R. Trias Fargas 25-27, 08005 Barcelona, Spain

Abstract. The set covering problem is an $\mathcal{NP}$-hard combinatorial optimization problem that arises in applications ranging from crew scheduling in airlines to driver scheduling in public mass transport. In this paper we analyze search space characteristics of a widely used set of benchmark instances through an analysis of the fitness-distance correlation. This analysis shows that there exist several classes of set covering instances that show a largely different behavior. For instances with high fitness distance correlation, we propose new ways of generating core problems and analyze the performance of algorithms exploiting these core problems.

1 Introduction

The set covering problem (SCP) is a well-known $\mathcal{NP}$-hard combinatorial optimization problem. Its importance lies in the large number of real applications ranging from crew scheduling in airlines [1], driver scheduling in public transportation [2], and scheduling and production planning in several industries [3] that can be modeled as SCPs.

The SCP consists in finding a subset of columns of a zero-one $m \times n$ matrix such that it covers all the rows of the matrix at minimum cost. Let $M = \{1, 2, .., m\}$ and $N = \{1, 2, .., n\}$ be, respectively, the set of rows and the set of columns. Let $A = (a_{ji})$ be a zero-one matrix and $c = (c_i)$ be a n-dimensional integer vector that represents the cost of column i. We say that a column i covers a row j if $a_{ji} = 1$. The problem can be formally stated as

Minimize $\quad v(SCP) = \sum_{i=1}^{n} c_i x_i$
s.t. $\quad \sum_{i=1}^{n} a_{ji} x_i \geq 1, \ j = 1, ..., m,$
$\quad x_i \in \{0, 1\}, \ i = 1, ..., n,$

where $x_i = 1$ if the column is in the solution, and $x_i = 0$, otherwise.

Because of ist enormous practical relevance, a large number of solution approaches, including exact and approximate algorithms, were proposed. Exact algorithms can solve instances with up to a few hundred rows and few thousand columns; a comparison of exact algorithms can be found in [4]. Because of the large size of SCP instances, for which exact algorithms are not anymore feasible, a large number of approximate algorithms was proposed. While pure greedy

S. Cagnoni et al. (Eds.): EvoWorkshops 2002, LNCS 2279, pp. 61–71, 2002.
© Springer-Verlag Berlin Heidelberg 2002

construction heuristics perform rather poorly, Lagrangian-based heuristics are very popular with the most efficient ones being the approach by Ceria, Nobili and Sassano [5] and the CFT heuristic by Caprara, Fischetti and Toth [6]. Since a few years, also applications of metaheuristics to the SCP have increased. So far, the best results are mainly due to Genetic Algorithms [7,8], Simulated Annealing algorithms, the most recent being by Brusco, Jacobs and Thompson [9], and the very high performing iterated local search [10] algorithms by Marchiori and Steenbeck [11] and by Yagiura, Kishida and Ibaraki [12]. These latter two algorithms iteratively move through construction/destruction phases and local search phases.

Despite the many algorithmic approaches to the SCP, no insights into the search space characteristics of the SCP for approximate algorithms have been obtained. Yet, such insights are valuable to understand algorithm behavior as well as they may propose new ways of attacking a problem. In this article, we analyze search space characteristics of the SCP by analyzing the fitness distance correlation [13] and show that very strong differences among the search space characteristics exist between different types of instances. Based on our insights from the analysis, we propose new ways of generating core problems, that is a much smaller SCP containing a subset of the columns that are most likely to appear in optimal solutions. Experimental results with a known SA algorithm [9] that was modified to work on our core problems proves the viability of our approach leading to results competitive to other codes.

The paper is organized as follows. The next section presents the details of the fitness-distance analysis, Section 3 gives computational results for an algorithm using a new definition of core problems and we conclude in Section 4.

2   Fitness Distance Analysis

Central to the search space analysis of combinatorial optimization problems is the notion of *fitness landscape* [14,15]. Intuitively, the fitness landscape can be imagined as a mountainous region with hills, craters, and valleys. The performance of metaheuristics strongly depends on the shape of this search space and, in particular, on the ruggedness of the landscape, the distribution of the valleys, craters and the local minima in the search space, and the overall number of the local minima.

Formally, the fitness landscape is defined by (i) the set of all possible solutions $\mathcal{S}$, (ii) an objective function that assigns to every $s \in \mathcal{S}$ a fitness value $f(s)$, and (iii) a distance measure $d(s, s')$ which gives the distance between solutions s and s'. The fitness landscape determines the shape of the search space as encountered by a local search algorithm.

For the investigation of the suitability of a fitness landscape for adaptive multi-start algorithms like the best performing metaheuristics for the SCP [9], [11], [12], the analysis of the correlation between solution costs and the distance between solutions or to globally optimal solutions has proved to be a useful tool [16,13]. The fitness distance correlation (FDC) [13] measures the correlation of the solution cost and the distance to the closest global optimum. Given a set of cost values $C = \{c_1, \ldots, c_m\}$ and the corresponding distances $D = \{d_1, \ldots, d_m\}$ to the closest global optimum the correlation coefficient is defined as:

$$r(C, D) = \frac{c_{CD}}{s_C \cdot s_D}, \text{where } c_{CD} = \frac{1}{m} \sum_{i=1}^{m} (c_i - \bar{c})(d_i - \bar{d})$$

where $\bar{c}$, $\bar{d}$ are the average cost and the average distance, s_C and s_D are the standard deviations of the costs and distances, respectively.

One difficulty for applying the FDC analysis to the SCP is that no straightforward distance measure exists, because of the SCP being a sub-set problem. Therefore, we rather use the closeness between solutions (based on the closeness, also a measure for the distance between solutions may be defined):

Definition 1. *Let $s, s' \in S$ be two feasible solutions for an SCP instance, then we define the closeness $n(\cdot, \cdot)$ and the distance $d(\cdot, \cdot)$ between s and s' as*

$$n(s, s') = \text{number of same columns in } s \text{ and } s'$$
$$d(s, s') = \max(|s|, |s'|) - n(s, s')$$

A maximal distance $d(s, s') = \max(|s|, |s'|)$ means that both solutions do not have any column in common, if $d(s, s') = 0$ we have $s = s'$.

As a first step, we generated a number of best-known solutions for all the instances under concern using a variant of the SA of Brusco, Jacobs, and Thompson [9]. For instances where optimal solutions are not available, it is conjectured that the best-known solutions are actually optimal. In a second step, we generated 1000 locally optimal solutions, starting from random initial solutions, using the local search algorithm given in Figure 1. The function $SN_1(s, j)$ performs the following steps: (i) it removes the column j from the current solution s, (ii) it completes s by iteratively choosing randomly a still uncovered row and adding a column with best value of $c_i/cv(i)$, where $cv(i)$ is the cover value of column i, that is, the number of still uncovered rows that is covered by column i, and (iii) it removes redundant columns.

For the fitness distance analysis we focused on two sets of benchmark instances. The first set, available from ORLIB at http://mscmga.ms.ic.ac.uk/info.html, is composed of randomly generated instances with varying density and size. Here, we only present results for the instances classes C–H, where classes C, D have $m = 400, n = 4000$ and a matrix density of 2% and 5%, respectively; classes E, F have $m = 500, n = 5000$ and a matrix density of 10% and 20%, respectively; and classes G, H have $m = 1000, n = 10000$ and a matrix density of 2% and 5%, respectively. Each class contains five instances. In these instances the column costs are uniformly distributed in the interval $[1, 100]$, each column covers at least one row, and each row is covered by at least two columns.

The second set with instances aa03-aa06 and aa11-aa20 stems from the paper of Balas and Carrera [17]. There are 14 instances with m varying from 105 to 272 and n varying from 3095 to 8661; the density of these instances are around 4%

Procedure $IV_{\mathcal{N}_1}^f$;
    $s :=$RandomSolution;
    while (improvement) **do**
        $r := s$;
        while ($r \neq \emptyset \wedge$ no improvement found so far) **do**
            choose randomly a column $j \in r$ and $r := r \setminus \{j\}$;
            $s' := \mathrm{SN}_1(s,j)$;
            if $(f(s') < f(s))$ **then** $s := s'$;
        end while;
    end while;
    return s;
end $IV_{\mathcal{N}_1}^f$;

Fig. 1. First-improvement local search procedure $IV_{\mathcal{N}_1}^f$.

for instances aa03-aa06 and around 2.6% for instances aa11-aa20; additionally, the instances differ in the range of the column weights, where weigths are either in a range from 91–3619 (instances aa03-aa06) or in the range of 35–2966.

2.1   Results on ORLIB Instances

Results on the FDC analysis are plotted in Figure 2 and detailed results are available in Table 1. In the plots the points are stratified according to the solution quality. The reason is that only few different values are possible for the solution quality. For example, the instance E4 has an optimal solution of 28 and therefore, a solution of 29 has a deviation of around 3.5% from the optimal solution.

The high values for the correlation coefficient (see Table 1) confirm the observation of a high, positive correlation between the solution quality and the closeness to optimal solutions in the plots in Figure 2. Only for one instance (C.3) the correlation coefficient is below 0.25. The solution quality of the local optima is relatively high and for some of the instances, the best local optima among the 1000 generated even matched the best known solutions. Particularly interesting are two observations. First, the average percentage closeness between the local optima and the total number of all different columns in the 1000 local optima depends strongly on the size of the instance as well as their densities: the larger the density the smaller is the average closeness among local optima, the less columns are in local optima and the less is the number of distinct columns in the 1000 local optima. Second, there is a direct relationship between problem size and local optima statistics: The larger the instance, the more distinct columns are encountered in the local optima (compare instance classes C and G and instance classes D and H–they have the same densities). Interestingly, for some instance classes (with same density) the average closeness of the local optima actually increases, like it is the case for C and G. This may indicate that these problem do not really become intrinsically harder with instance size.

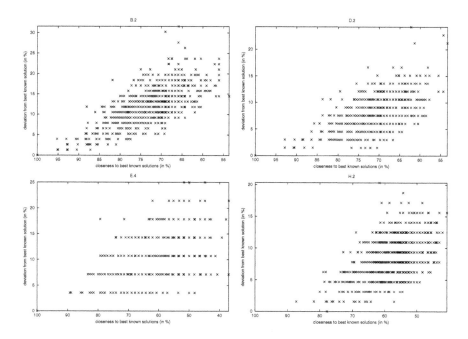

Fig. 2. Fitness distance graphs for some selected ORLIB instances.

2.2 Results on Instances from Balas and Carrera

The FDC plots of some instances from Balas and Carrera (BC) are given in Figure 3, detailed results are given in Table 2. The search space analysis of these instances shows a very different behavior. First, the closeness of the instances to the best known solutions is very low; this can be seen when inspecting the plots: all points are in the range of a closeness between 0-35%; whereas for the ORLIB instances the closeness was mostly larger than 0.5. Second, there is only a very small correlation between fitness and closeness to the best known solutions, showing that the objective function value provides much less guidance towards the global optima than on the ORLIB instances. Third, the average closeness of the local optima is very low, leading to the fact that the overall number of distinct columns in the local optima is extremely high.

These results suggest that the BC instances are much harder to solve for local search than the ORLIB instances because less guidance is given by the objective function and a much larger number of columns actually appear in good solutions, leading to a larger *effective* search space.

3 Core Problems

The results of the search space analysis can be exploited to find systematic and well justified ways for reducing instance size by defining small core problems:

generate local optima according to the algorithm of Figure 1 and define the core problem as the union of the columns contained in the local optima. Such an approach is interesting in a situation as observed for ORLIB problems, where the structure of local optima correlates strongly with that of global optima. In such a case, the columns of global optima are very likely to occur in many of the local optima.

Table 1. Results of the FDC analysis of ORLIB instances. Given are the instance identifier (PI), the best known solutions, the correlation coefficient r, the number of distinct local optima, the best, average, and worst solution found, the average closeness, the number of different columns in the 1000 local optima, and the number of best known solutions (GO) used in the FDC analysis.

| PI | best known sol. | r | no. LO | sol. quality local minima | | | $\overline{C(LO)}$ (in %) | $\sum$ col (LO) | GO |
				avg.	best	worst			
C.1	227	0.5955	762	236.67	229	272	74.58	252	1
C.2	219	0.5741	964	234.76	221	262	65.66	292	11
C.3	243	0.1355	989	266.55	252	296	64.07	310	1
C.4	219	0.6400	945	237.70	225	268	64.02	290	113
C.5	215	0.7409	870	224.48	215	270	71.02	257	7
D.1	60	0.7238	776	64.31	60	79	29.75	152	9
D.2	66	0.6306	864	70.93	66	82	31.68	166	42
D.3	72	0.4653	929	78.18	73	89	31.81	191	25
D.4	62	0.5522	664	66.73	62	79	35.61	160	3
D.5	61	0.7868	297	65.74	61	76	35.22	148	2
E.1	29	0.7202	836	30.90	29	36	17.65	119	163
E.2	30	0.2944	992	33.33	30	40	13.54	151	1
E.3	27	0.4699	962	29.92	27	35	14.39	132	1
E.4	28	0.5516	961	30.68	28	35	16.44	123	6
E.5	28	0.7602	844	30.08	28	38	17.61	126	38
F.1	14	0.6475	979	15.51	14	18	5.78	91	55
F.2	15	0.7462	868	16.26	15	19	7.22	90	98
F.3	14	0.4428	925	16.24	14	19	6.53	84	1
F.4	14	0.5802	991	15.68	14	18	5.42	90	28
F.5	13	0.2661	997	15.17	14	17	4.44	88	1
G.1	176	0.7667	995	189.31	178	204	85.93	421	66
G.2	154	0.5016	1000	165.96	158	183	82.42	404	2
G.3	166	0.4355	999	177.11	170	192	90.60	396	5
G.4	168	0.3989	1000	181.18	173	201	82.26	422	1
G.5	168	0.4897	999	181.68	173	198	81.39	439	33
H.1	63	0.3646	1000	69.57	64	77	33.11	297	1
H.2	63	0.5301	1000	69.34	64	77	32.79	298	2
H.3	60	0.4107	1000	65.90	61	77	31.57	275	1
H.4	58	0.5582	1000	63.66	59	69	31.02	277	45
H.5	55	0.6344	1000	60.25	56	66	33.58	258	45

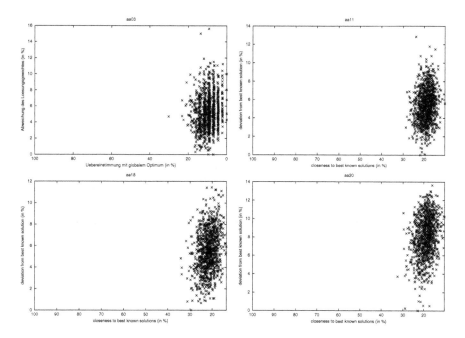

Fig. 3. Fitness distance graphs for some selected instances of Balas and Carrera.

Table 2. Results of the FDC analysis of ORLIB instances. For a description of the entries see Table 1.

PI	best known sol.	r	no. LO	sol. quality local minima avg.	best	worst	$\overline{C(LO)}$ (in %)	$\sum$ col (LO)	GO
aa03	33155	0.1486	1000	36319.21	34485	39873	2.89	3565	257
aa04	34573	0.2072	1000	37650.96	34731	40803	3.89	3057	704
aa05	31623	0.0071	1000	33027.26	32235	36237	4.43	2915	155
aa06	37464	0.1262	1000	40701.95	38762	44345	4.47	2741	458
aa11	35384	0.1702	1000	39843.31	37804	42658	15.00	3733	64
aa12	30809	0.1936	1000	35385.19	32962	38114	13.82	3576	1202
aa13	33211	0.2092	1000	38143.74	36260	40836	12.04	3424	339
aa14	33219	0.2114	1000	37209.92	35441	40318	13.67	3374	48
aa15	34409	0.1943	1000	38950.44	36903	41390	14.65	3043	128
aa16	32752	0.1865	1000	37983.32	35302	40494	13.86	3137	271
aa17	31612	0.2480	1000	36608.98	34487	39334	14.26	2866	400
aa18	36782	0.2407	1000	41182.28	39013	43482	16.96	2806	1024
aa19	32317	0.2464	1000	36893.85	34395	39345	10.60	2774	4356
aa20	34912	0.2583	1000	40114.32	37116	42179	15.79	2663	55

For the generation of core problems two issues are important: First, a sufficient number of different local optima has to be generated. Second, the overall time of generating core problems and subsequently solving an instance should be smaller than solving an instance without using core problems.

We tested this idea using a variant of the Simulated Annealing algorithm by Brusco et al. [9] on the ORLIB instances which is run on the initial problem (SA) and on the core problems SA$_{core}$. For the generation of the core problems, we construct 25 starting solutions for the local search in Figure 1 but starting from solutions generated by a randomized greedy construction heuristic instead of random initital solutions and the core problem then contains all columns member of any of the 25 local optima. The greedy construction heuristic iteratively selects first an still uncovered row and in a second step it chooses randomly among the five highest ranked columns, where the rank of a column is determined by its cost divided by the cover value. (The use of a greedy algorithm has the advantage that the subsequent local search is much faster than from random starting solutions.)

The results of SA$_{core}$ are in Table 4, the results of SA together with a tabu search variant (IVT$_{\mathcal{N}_1}^b$) and a zero temperature SA algorithm (SA$_{T_0}$) are in Table 3. All results in Tables 3 and 4 are based on 25 independent trials; in the case of the results on the core problems, for each trial a new core problem

Table 3. Computational results with Tabu Search, Simulated Annealing, and zero temperature annealing.

PI	best known sol.	IVT$_{\mathcal{N}_1}^b$ average LG	no. opt	time	SA average LG	no opt	time	SA$_{T_0}$ average LG	no opt	time
E.1	29	29	25	11.69	29	25	1.60	29	25	0.53
E.2	30	30.84	4	147.42	30	25	0.26	30	25	0.24
E.3	27	27.08	23	325.86	27	25	0.22	27	25	0.21
E.4	28	28	25	249.21	28	25	7.47	28	25	8.06
E.5	28	28	25	50.94	28	25	2.07	28	25	1.35
F.1	14	14	25	120.61	14	25	4.85	14	25	6.36
F.2	15	15	25	19.03	15	25	0.50	15	25	1.78
F.3	14	14	25	421.88	14	25	39.17	14	25	64.40
F.4	14	14	25	212.54	14	25	12.49	14	25	17.95
F.5	13	13.88	3	65.65	13.24	19	77.80	13.92	2	14.35
G.1	176	177.16	8	402.16	176	25	16.49	177.32	11	23.78
G.2	154	154.56	13	425.52	155	0	28.25	156.72	0	10.48
G.3	166	167.52	0	290.98	167.32	2	38.90	167.6	0	28.92
G.4	168	169.8	5	325.94	168.68	15	43.81	170.88	3	29.35
G.5	168	169	11	372.57	168	25	18.64	168.8	14	25.62
H.1	63	64.12	0	756.83	63.92	2	126.11	64.2	0	116.98
H.2	63	64.24	19	631.37	63.16	21	179.74	63.96	2	73.76
H.3	59	60.64	9	580.45	59.28	19	255.89	60.28	18	177.98
H.4	58	59.04	6	498.56	58	25	198.00	58.6	11	164.52
H.5	55	55.36	16	425.97	55	25	35.92	55.28	20	67.30

Table 4. Computational results with Simulated Annealing using core problems.

Problem	best known sol.	SA_{core} average sol.qual	solution quality best	solution quality no. opt.	solution quality worst	time (sec)	no. LO	$time_{LO}$	Core-problem m	Core-problem No. columns
E.1	29	29	29	25	29	0.008	25	2.315	500	83.2
E.2	30	31	31	25	31	0.073	25	3.155	500	91.72
E.3	27	27.04	27	24	28	0.000	25	2.616	500	86.16
E.4	28	28	28	25	28	0.101	25	2.035	500	85.4
E.5	28	28	28	25	28	0.055	25	2.508	500	83.96
F.1	14	14	14	25	14	0.020	25	1.961	500	53.08
F.2	15	15	15	25	15	0.006	25	2.184	500	49.28
F.3	14	14	14	25	14	0.110	25	2.622	500	51.4
F.4	14	14	14	25	14	0.081	25	1.645	500	49.12
F.5	13	14	14	25	14	0.000	25	1.450	500	47.56
G.1	176	176	176	25	176	3.152	25	20.246	1000	301.48
G.2	154	155.04	154	1	156	5.776	25	17.730	1000	279.24
G.3	166	167.52	166	3	169	16.414	25	17.202	1000	283.08
G.4	168	168.48	168	19	170	11.646	25	19.640	1000	286.16
G.5	168	168.12	168	23	170	2.997	25	22.417	1000	296.32
H.1	63	64	64	25	64	2.663	25	31.347	1000	183
H.2	63	63.12	63	22	64	7.883	25	31.473	1000	179.36
H.3	59	59.56	59	15	61	6.015	25	27.839	1000	185.76
H.4	58	58.28	58	18	59	7.227	25	28.519	1000	179
H.5	55	55	55	25	55	0.537	25	28.436	1000	175.6

was generated; all algorithms were limited to 2000 iterations. The programs were coded with C++ and run on a Pentium III 700MHz CPU.[1]

The computational results show that the SA working on core problems is significantly faster, which is most visible for the largest instances of the class H. The speed-up only comes with a very minor loss in solution quality on some few instances when compared to the SA working on all columns; for some few instances even better performance could be obtained. We expect that on the few instances, where SA_{core} did not match the best-known solutions, this could be achieved by increasing the number of local optima for generating core problems. Interestingly, the largest part of the computation time is spent on the generation of the core problems and not by the SA. The main reason is certainly that the number of columns in the core problems is very small (see last column of Table 4).

---

[1] Let us remark here that our results (by SA or SA_{core}) to a large extent match the results obtained with the best performing algorithms for the SCP; only on some few problems slightly better average solution qualities have been reported in [9,6,11,12]. Note that the best performing approaches are using ad-hoc defined core problems that are typically modified at solution time. We refer to [6,11,12] for details.

4   Conclusions

We have analyzed some search space characteristics of the SCP and, based in this analysis we proposed new, systematic ways of generating core problems for the SCP. Computational results on the SCP instances from ORLIB showed that the resulting core problems are very small, but for the ORLIB instances they often contain all columns necessary to find the best-known solutions. In fact, some of the core problems were solved using an exact algorithm and we could verify, that the best solutions found by an Simulated Annealing algorithm working on core problems, identified consistently the optimal solutions for the core problems.

There are several ways how this work can be extended. First, we could extend our analysis to other instances from real applications. Second, faster ways of identifying core problems would be interesting, because this task consumes the largest part of the computation time of SA_{core}. Third, our approach could be enhanced by dynamically changing the core problems during the run of the algorithm as done in [6,11]. Fourth, the same ideas of generating core problems could also be applied to other problems that show significant fitness distance correlation. The results on the SCP suggest that this last idea is very promising for a number of combinatorial problems where similar high fitness distance correlations are observed.

References

1. E. Housos and T. Elmoth. Automatic optimization of subproblems in scheduling airlines crews. *Interfaces*, 27(5):68–77, 1997.
2. H. Lourenço, R. Portugal, and J.P. Paix ao. Multiobjective metaheuristics for the bus-driver scheduling problem. *Transportation Science*, 35(3):331–343, 2001.
3. F.J. Vasko and F.E. Wolf. Optimal selection of ingot sizes via set covering. *Operations Research*, 15:115–121, 1988.
4. A. Caprara, M. Fischetti, and P. Toth. Algorithms for the set covering problem. Technical Report OR-98-3, DEIS, University of Bologna, Italy, 1998.
5. S. Ceria, P. Nobili, and A. Sassano. A lagrangian-based heuristic for large-scale set covering problems. *Mathematical Programming*, 81:215–228, 1998.
6. A. Caprara, M. Fischetti, and P. Toth. A heuristic method for the set covering problem. *Operations Research*, 47:730–743, 1999.
7. J.E. Beasley and P.C. Chu. A genetic algorithm for the set covering problem. *European Journal of Operational Research*, 94:392–404, 1996.
8. A.V. Eremeev. A genetic algorithm with a non-binary representation for the set covering problem. In *Proceedings of OR'98*, pages 175–181. Springer Verlag, 1999.
9. M.J. Brusco, L.W. Jacobs, and G.M. Thompson. A morphing procedure to supplement a simulated annealing heuristic for cost- and coverage-corrleated set covering problems. *Annals of Operations Research*, 86:611–627, 1999.
10. H.R. Lourenço, O. Martin, and T. Stützle. Iterated local search. In F. Glover and G. Kochenberger, editors, *Handbook of Metaheuristics*. Kluwer Academic Publishers, Boston, MA, USA, 2002. to appear.
11. E. Marchiori and A. Steenbeek. An evolutionary algorithm for large scale set covering problems with application to airline crew scheduling. In *Real World Applications of Evolutionary Computing*, volume 1083 of *Lecture Notes in Computer Science*, pages 367–381. Springer Verlag, Berlin, Germany, 2000.

12. M. Kishida M. Yagiura and T. Ibaraki. A 3-flip neighborhood local search for the set covering problem. submitted for publication, 2001.
13. T. Jones and S. Forrest. Fitness distance correlation as a measure of problem difficulty for genetic algorithms. In L.J. Eshelman, editor, *Proc. of the 6th Int. Conf. on Genetic Algorithms*, pages 184–192. Morgan Kaufman, 1995.
14. P.F. Stadler. Towards a theory of landscapes. Technical Report Technical Report SFI–95–03–030, Santa Fe Institute, 1995.
15. E.D. Weinberger. Correlated and uncorrelated fitness landscapes and how to tell the difference. *Biological Cybernetics*, 63:325–336, 1990.
16. K.D. Boese, A.B. Kahng, and S. Muddu. A New Adaptive Multi-Start Technique for Combinatorial Global Optimization. *Operations Research Letters*, 16:101–113, 1994.
17. E. Balas and M.C. Carrera. A dynamic subgradient-based branch and bound procedure for set covering. *Operations Research*, 44:875–890, 1996.

A Population Based Approach for ACO

Michael Guntsch[1] and Martin Middendorf[2]

[1] Institute for Applied Computer Science and Formal Description Methods
University of Karlsruhe, Germany
`guntsch@aifb.uni-karlsruhe.de`
[2] Computer Science Group
Catholic University of Eichstätt-Ingolstadt, Germany
`martin.middendorf@ku-eichstaett.de`

Abstract. A population based ACO (Ant Colony Optimization) algorithm is proposed where (nearly) all pheromone information corresponds to solutions that are members of the actual population. Advantages of the population based approach are that it seems promising for solving dynamic optimization problems, its finite state space and the chances it offers for designing new metaheuristics. We compare the behavior of the new approach to the standard ACO approach for several instances of the TSP and the QAP problem. The results show that the new approach is competitive.

1 Introduction

The ability of real ants to find short paths between their nest and food sources is based on indirect communication by the use of pheromones. This behavior of ant colonies has motivated the design of the Ant Colony Optimization (ACO) metaheuristic (Dorigo et al. [1,2]). The ACO metaheuristic has been applied to several combinatorial optimization problems for which the construction of a solution can be described as a path in a decision graph (see [3] for an introduction and overview).

An ACO algorithm is an iterative search process where in every iteration (generation) each of m (artificial) ants constructs one solution by following a path through the decision graph. For each decision on which edge in the decision graph the ant should follow it uses pheromone information which stems from former ants that have found good solutions. As an example consider a simple ACO algorithm for the Traveling Salesperson problem (TSP). The pheromone information is stored in a so called pheromone matrix where element τ_{ij}, indicates how good it seems to move from city i to city j, for $i, j \in [1 : n] = \{1, \ldots, n\}$. The probabilistic decisions of the ants are made according to a local decision rule where the probabilities of the possible outcomes are correlated to the relative amounts of pheromone on the corresponding edges in the decision graph. In addition an ant may also use heuristic information for its decisions. Thus, the pheromone information and the heuristic information are indicators of how good it seems to follow the corresponding edge.

S. Cagnoni et al. (Eds.): EvoWorkshops 2002, LNCS 2279, pp. 72–81, 2002.

In this paper a modification of the way pheromone information is updated is proposed and tested on the Traveling Salesperson Problem (TSP) and the Quadratic Assignment Problem (QAP). To limit the amount of pheromone in the matrix and gradually make old updates weaker, the standard approach is to multiply all values τ_{ij} with a factor $(1-\rho)$, $0 < \rho < 1$ after every iteration before any positive reinforcement is done on good solutions. The method we propose is to keep track of all good solutions up to a certain age and thereafter remove their influence explicitly from the pheromone matrix through a negative update on the τ_{ij} values they had modified. Hence, the value with which an element of the pheromone matrix was initialized is at the same time it's minimum value, since negative updates occur only to cancel out previous positive updates of the same magnitude. The effect of having a minimum pheromone value has been applied successfully for ACO algorithms by Stützle and Hoos [4,5].

One motivation for introducing a population based scheme is the planned application to dynamic optimization problems. An ACO approach for a dynamic TSP problem where instances change at certain intervals through the deletion and insertion of cities has been studied in [6,7]. Strategies for pheromone modification in reaction to changes of the problem instance together with an elitist strategy for use in dynamic environments have been proposed. A standard elitist strategy for ACO algorithms is that an elitist ant updates the pheromone values in every generation according to the best solution found so far. But after a change of the problem instance the best solution found so far will usually no longer represent a valid solution to the new instance. Instead of simply forgetting the old best solution, it was adapted so that it becomes a reasonably good solution for the new instance. Clearly, this is only reasonable when the changes of the instance are not too strong.

Since there exist often good strategies to adapt solutions for an old instance to a new instance we propose here to maintain in addition to a possible elitist solution a population of other solutions. As members of this population we chose the last k solutions that have been used to update the pheromone matrix. Since in dynamic environments pheromone that stems from very old instances is usually less valuable than pheromone of ants that worked on recent instances we propose to skip all the very old pheromone. That means we skip all pheromone corresponding to solutions that are not in the actual population. In other words, every pheromone value depends only on solutions that are in the actual population and for every solution in the population there is some amount of pheromone on the corresponding elements in the pheromone matrix.

Often in dynamic real time environments it is important to have an algorithm that can react to changes quickly. Therefore we propose a fast pheromone update process. Instead of applying evaporation to every pheromone value the amount of pheromone that corresponds to a solution in the population is not changed. Only when a solution leaves the population its amount of pheromone is subtracted from the corresponding pheromone values.

Since after each generation of ants one solution enters the population and one solution leaves the population the pheromone update is done only for pheromone values corresponding to this two solutions. In case of an n city TSP this means that the number of pheromone values that have to be changed is at most $2n$ for the new method (using one addition or subtraction operation for each value) compared to n^2 for most versions of the standard method (using one multiplication operation for each value).

It has to be mentioned that Maniezzo [8] proposed a method for pheromone update that is also fast. In his method an ant that found a solution that is better than the average solution quality of the k former ants adds a certain amount of pheromone to the corresponding pheromone values. Otherwise, if the solution is worse than this average, pheromone is subtracted from the corresponding pheromone values. In case of the TSP this means that for every ant n additions respectively subtractions are done (plus a constant number of arithmetic operation to determine the actual amount of pheromone that is added/subtracted).

To ensure that every pheromone value has a minimum value we do not remove the initial amount of pheromone (cmp. [4,5]). Otherwise it could happen that pheromone values become zero because there is no solution in the population that puts pheromone on them. But then the ants cannot make the corresponding decision.

Another motivation for introducing a population based approach is that it might allow a different approach to handle it theoretically due to the finite state space. Moreover, a population based approach offers interesting possibilities for designing new metaheuristics.

In this study we show that the population based approach performs at least not worse than the standard method on static problems. Hence, there are good chances that it is useful in particular for dynamic problems and for using it in new metaheuristics. The population based approach is tested experimentally on several instances of the TSP and the Quadratic Assignment problem (QAP). Moreover it is investigated how the size of the population influences the optimization behavior.

The paper is organized as follows. In Section 2 we describe a standard ACO algorithm and the new population based approach. The test setup is given in Section 3. The experimental results are discussed in Section 4, and conclusions are drawn in Section 5.

2   ACO Approaches

In this section we describe ACO algorithms that work according to the standard approach and according to the new population based approach for the TSP and the QAP problem. It should be mentioned that the aim of this study is not to find the best ACO algorithm for our test problems but to compare the basic approaches. It is clear that the algorithms for both approaches can be improved. For an example the ants can change the pheromone values during their search for a solution ([9]) or local search can be applied to the solutions found by the ants (see [10,11,5]).

2.1   Standard Approach

The ACO algorithm that uses the standard approach for the TSP uses ideas from Dorigo et al. [2,9]. In every generation each of m ants constructs one tour through all the given n cities. Starting at a random city an ant selects the next city using heuristic information as well as pheromone information. For each pair (i, j), $i, j \in [1 : n]$, $i \neq j$ of cities there is heuristic information, denoted by η_{ij}, and pheromone information, denoted by τ_{ij}. The heuristic value is $\eta_{ij} = 1/d_{ij}$ where d_{ij} is the distance between city i and city j.

   With probability q_0, where $0 \leq q_0 < 1$ is a parameter of the algorithm, an ant at city i chooses next city j from the set S of cities that have not been visited so far which maximizes $\tau_{ij}^{\alpha} \cdot \eta_{ij}^{\beta}$, where α and β are constants that determine the relative influence of the heuristic values and the pheromone values on the decision of the ant. With probability $1 - q_0$ the next city is chosen according to the probability distribution over S determined by

$$p_{ij} = \frac{\tau_{ij}^{\alpha} \cdot \eta_{ij}^{\beta}}{\sum_{h \in S} \tau_{ih}^{\alpha} \cdot \eta_{ih}^{\beta}} \tag{1}$$

   Before doing the pheromone update some of the old pheromone is evaporated for all i, j according to

$$\tau_{ij} \mapsto (1 - \rho) \cdot \tau_{ij}$$

where parameter $\rho \in (0, 1)$ determines the evaporation rate. For pheromone update an elitist strategy is used where one elitist ant updates pheromone along the best solution found so far, i.e. for every city i some amount of pheromone is added to element (i, j) of the pheromone matrix when j is the successor of i in the so far best found tour. A weight $0 \leq w_e \leq 1$ determines the relative amount of pheromone placed by the elitist ant compared to the amount placed by the best ant in the current generation (the extreme cases are $w_e = 0$ and $w_e = 1$ where no respectively all of the new pheromone is from the elitist ant). For initialization we set $\tau_{ij} = \tau_{init}$ where $\tau_{init} > 0$ for every edge (i, j).

   For the QAP problem there are n facilities, n locations, and $n \times n$ matrices $A = [a_{ij}]$ and $B = [b_{ij}]$ where a_{ij} is the distance between locations i and j and b_{hk} is the flow between facilities h and k. The problem is to find an assignment of facilities to locations, i.e., a permutation f of $[1 : n]$ such that the following measure is minimized $\sum_{i=1}^{n} \sum_{j=1}^{n} a_{f(i)f(j)} b_{ij}$. The ACO algorithm for QAP uses pheromone values τ_{ij}, $i, j \in [1 : n]$ which refer to the assignment of facility i to location j. No heuristic is used for the QAP.

Max-Min version. Since the pheromone values in the population based approach are never zero we use also a version of the standard ACO algorithm were the pheromone values that are used by the ants for their decisions have a minimal value. This version of the algorithm is called the Max-Min ACO algorithm (Note, that the idea to use minimum values stems from the MAX-MIN ACO algorithm proposed in [4,5]). The Max-Min ACO algorithm used the same pheromone matrix as the standard ACO algorithm. Instead of using the pheromone values τ_{ij}

directly to determine the probability of a decision as in (1) every ant of the Max-Min ACO algorithm determines the probability using the values $\tau_{ij} + \tau_+ \cdot \tau_{init}$ were $\tau_+ > 0$ is a parameter. Hence, the minimal value is $\tau_+ \cdot \tau_{init}$ and (1) is replaced by

$$p_{ij} = \frac{(\tau_{ij} + \tau_+ \cdot \tau_{init})^\alpha \cdot \eta_{ij}^\beta}{\sum_{h \in S}(\tau_{ih} + \tau_+ \cdot \tau_{init})^\alpha \cdot \eta_{ih}^\beta} \tag{2}$$

2.2   Population Based Approach

In the population based approach the first generation of ants works in the same way as in the standard approach, i.e., the ants search solutions using the initial pheromone matrix and the best ant in the generation adds pheromone to the pheromone matrix. But no pheromone evaporation is done. The best solution is then put in the (initially empty) solution population. After k generations there are exactly k solutions in the population. From generation $k + 1$ on one solution in the population is removed from the population and the corresponding amount of pheromone is subtracted from the elements of the pheromone matrix. That means for all pheromone values corresponding to the removed solution the same amount of pheromone is subtracted that was added when this solution had entered the population. In this paper we chose to remove the oldest solution from the population. We call this algorithm FIFO-Queue ACO algorithm. Observe that pheromone values are never zero since the initial amount of pheromone remains. A parameter τ_{max} determines the maximum amount of pheromone. The amount of pheromone that an ant updates is then $(1 - w_e)(\tau_{max} - \tau_{init})/k$.

3   Test Setup

We evaluated the standard, the Max-Min and the FIFO-Queue ACO algorithms on the Traveling Salesperson Problem (TSP) and the Quadratic Assignment Problem. The following test-instances (from TSPLIB [12] respectively QAPLIB [13]) were used for a comparison of the standard and the FIFO-Queue ACO algorithm:

eil101, 101 cities, symmetric (euclidean)   wil50, size 50, symmetric
kroA100, 100 cities, symmetric (euclidean) wil100, size 100, symmetric
d198, 198 cities, symmetric (euclidean)   tai100a, size 100, asymmetric
kro124p, 100 cities, asymmetric   tai100b, size 100, asymmetric
ftv170, 170 cities, asymmetric

We used parameters values $\rho = 0.1, 0.05, 0.02, 0.01, 0.005$ with a corresponding number of iterations $t = 2000, 5000, 10000, 20000, 50000$ for the standard ACO algorithm. The number of iterations was chosen to ensure that it had converged in (nearly) each of it's runs. The FIFO-Queue ACO algorithm was evaluated with maximum pheromone values $\tau_{max} = 1.0, 3.0, 10.0$ [1] and population

---

[1] For the symmetric TSPs, the actual maximum pheromone-value is $\tau_{max}/2$ because of symmetric updating.

sizes $k = 1, 5, 25$ and ran for $t = 50000$ iterations for all parameter combinations, since convergence is not possible. Furthermore, these parameter-settings were combined with $q_0 = 0.0, 0.5, 0.9$ and $w_e = 0.0, 0.25, 0.5, 0.75, 1.0$.

A (so far) limited comparison was done with the Max-Min ACO algorithm. All combinations of $\tau_+ \in \{0.1, 0.3, 1.0\}$, $\rho \in \{0.01, 0.05, 0.25\}$, and $w_e \in \{0.0, 0.25, 0.5, 0.75, 1.0\}$ were tested on the TSP instances eil101, kroA100, and kro124p, and on the QAP instances tai100a and wil50. Since neither type of algorithms converges to a point where no further improvement to the solution can be expected, only t=50000 was evaluated.

The other parameters for all three ACO algorithms were $m = 10$ ants per iteration, $\alpha = 1$, and $\beta = 5$, $\tau_{init} = 1/(n-1)$ ($\beta = 0$, $\tau_{init} = 1/n$) for TSP (respectively QAP). Each parameter combination was tested with 10 different random seeds for the standard and FIFO-Queue ACO algorithm and 5 different random seeds for the Max-Min ACO algorithm and the average was taken.

4   Results

4.1   Comparison with Standard Approach

In order to determine whether the FIFO-Queue ACO algorithm is a competitive alternative to the standard one, we ranked the different parameter combinations (altogether 210, 75 for the standard and 135 for the FIFO-Queue ACO algorithm) according to their average performance on the individual problem instances. From these instance specific ranks, we took the averages for the two problem classes to generate one class-specific rank for each parameter combination. Of course, this method cannot be used for a detailed comparison of solution quality, but the statement "a better average rank implies at least as good as a performance on average" holds, enabling us to ascertain whether the FIFO-Queue ACO algorithm is competitive. Tables 1 and 2 show the average ranks of the best respective parameter combination for the standard and FIFO-Queue ACO algorithm after different number of iterations and for different values of ρ. The number of iterations used in the tables are $t_0 = t/100$, $t_1 = t/50$, $t_2 = t/25$, $t_3 = t/10$, $t_4 = t/5$, $t_5 = t/2.5$, and $t_6 = t$ for a total number of iterations t that the standard ACO algorithm ran.

For both tables, with the exception of using $\rho = 0.005$ at time t_5 for QAP, the best combination of parameters of the FIFO-Queue ACO algorithm always achieves a better average rank than the standard ACO algorithm for any t_i, indicating that it's performance is indeed competitive.

The average rank (for all t_i) over all parameter combinations was 66.5 for TSP and 53.2 for QAP. The tables show that the average rank of the best parameter combination was significantly lower for both the standard and the FIFO-Queue ACO algorithm on QAP than on TSP. This seems to indicate that more of a distinction between good and bad parameter sets can be made for QAP, whereas for the TSP the different parameter combinations perform closer to one another in terms of quality. Also the use of a heuristic for TSP might result in

Table 1. Average ranks for standard and FIFO-Queue ACO algorithm on TSP

		t_0	t_1	t_2	t_3	t_4	t_5	t_6
$\rho = 0.1$	Standard	27.0	34.8	50.6	57.0	77.0	89.4	101.6
	Queue	9.0	17.8	16.8	19.4	20.2	16.2	14.0
$\rho = 0.05$	Standard	47.4	50.6	49.4	54.2	71.8	85.2	92.6
	Queue	18.8	18.0	19.6	19.2	16.6	14.4	11.6
$\rho = 0.02$	Standard	65.2	64.4	47.8	43.2	53.0	62.2	77.4
	Queue	17.8	19.8	19.0	17.0	15.2	11.8	13.6
$\rho = 0.01$	Standard	69.2	68.2	48.0	43.4	55.8	61.0	69.8
	Queue	18.6	18.4	16.0	15.4	11.6	14.0	14.6
$\rho = 0.005$	Standard	75.2	66.2	48.6	51.2	58.2	64.2	71.6
	Queue	17.2	15.8	14.8	12.2	13.6	15.0	18.8

Table 2. Average ranks for standard and FIFO-Queue ACO algorithm on QAP

		t_0	t_1	t_2	t_3	t_4	t_5	t_6
$\rho = 0.1$	Standard	27.4	25.8	14.4	4.6	2.4	6.6	25.8
	Queue	4.0	4.2	3.0	3.6	3.6	4.2	5.2
$\rho = 0.05$	Standard	38.2	26.2	17.0	7.0	5.0	8.6	27.4
	Queue	2.8	3.4	2.6	3.2	5.2	5.8	5.2
$\rho = 0.02$	Standard	40.8	41.6	28.2	15.8	8.0	7.0	17.0
	Queue	3.4	2.6	2.6	4.8	5.8	5.8	6.2
$\rho = 0.01$	Standard	46.8	44.6	31.8	19.6	8.2	7.0	18.0
	Queue	2.6	2.6	3.6	5.2	5.6	6.2	7.4
$\rho = 0.005$	Standard	44.8	41.8	31.2	14.0	9.0	3.0	14.8
	Queue	2.2	4.4	5.0	5.6	6.4	7.8	7.2

a more similar behavior of different parameter combinations. For TSP, the best parameter combination seems to be more instance-specific than for QAP.

Note that for any row in tables 1 and 2 the combination of parameters that is best does not necessarily stay the same. Instead, it undergoes a characteristic development depending on at what time t_i best performance was sought. Tables 3 and 4 give examples for the parameter-development of the best combinations that takes place in such a row. For the standard ACO algorithm, the typical development towards parameters that focus more on exploration than on exploitation when given ample time can be seen with declining q_0 and w_e, for TSP as well as QAP.

For the FIFO-Queue ACO algorithm, the situation is less clear. For TSP, all best combinations from Table 1 used population size $k = 1$, which means that the only guidance for the ants of a given generation are the solutions of the elite ant and of the best working ant from the previous iteration, and the heuristic. It seems that the ability to quickly explore the neighborhood of a promising solution (last iteration's best), combined with a drive toward the best solution found so

Table 3. Parameter-development over time for the best respective combinations on TSP (top 3 rows for standard ACO algorithm, bottom four for FIFO-Queue), $\rho = 0.01$.

$\rho = 0.01$	200	400	800	2000	4000	8000	20000
q_0	0.5	0.5	0.0	0.0	0.0	0.0	0.0
w_e	0.75	0.75	0.75	0.5	0.5	0.25	0.25
τ_{max}	1.0	1.0	1.0	1.0	1.0	1.0	1.0
q_0	0.5	0.5	0.5	0.5	0.5	0.5	0.5
w_e	0.25	0.25	0.25	0.25	0.25	0.25	0.25
τ_{max}	3.0	1.0	1.0	1.0	1.0	1.0	1.0
k	1	1	1	1	1	1	1

Table 4. Parameter-development over time for the best respective combinations on QAP (top 3 rows for standard ACO algorithm, bottom four for FIFO-Queue), $\rho = 0.01$.

$\rho = 0.01$	200	400	800	2000	4000	8000	20000
q_0	0.9	0.5	0.5	0.0	0.0	0.0	0.0
w_e	1.0	0.75	1.0	0.75	0.5	0.25	0.25
τ_{max}	1.0	1.0	1.0	1.0	1.0	1.0	1.0
q_0	0.5	0.9	0.9	0.9	0.9	0.9	0.9
w_e	0.75	0.5	0.5	0.5	0.5	0.75	0.75
τ_{max}	10.0	3.0	3.0	3.0	3.0	3.0	10.0
k	1	5	5	5	5	25	1

far and a strong heuristic influence to prevent choices that seem drastically bad, works best. Of the best five FIFO-Queue combinations for the scenario of Table 3, four used population size $k = 1$ and one $k = 5$. A positive elite weight $w_e \in \{0.25, 0.5, 0.75\}$ was also used by all these combinations, suggesting that some degree of enforcement of the best solution found so far is beneficial. These traits are combined with $q_0 = 0.5$ and at first $\tau_{max} = 3$, after t_1 $\tau_{max} = 1.0$. $q_0 = 0.9$ and $\tau_{max} = 1.0$ achieve a similar effect in the FIFO-Queue ACO algorithm, both determining how likely it is that an ant will choose the best (q_0) or one of the best (τ_{max}) choices according to pheromone and heuristic information. The medium q_0 value and low τ_{max} indicate that deviating often from the best path found so far produces good results.

For QAP, the parameter combinations which performed well are somewhat different from those for TSP. The only similarity is the strong guidance provided by an elite ant with $w_e \in \{0.5, 0.75\}$. However, both a higher q_0 and a higher τ_{max} value suggest that it is beneficial to stay closer to the best placement(s) encoded in the pheromone matrix. Also, the larger population size of $k = 5$ and once even $k = 25$ point out that a greater diversity in solutions encoded into the matrix is helpful. The probable reason for this is the absence of a heuristic to generate an uneven probability distribution.

4.2   Comparison with Max-Min ACO Algorithm

To compare the FIFO-Queue ACO algorithm with the Max-Min ACO algorithm use a rank-based comparison as in Subsection 4 (altogether 84 different parameter combinations, 45 for the Max-Min and 39 for the FIFO-Queue ACO algorithm).

Table 5. Average ranks for Max-Min and FIFO-Queue ACO algorithm on TSP

	500	1000	2000	5000	10000	20000	50000
Max-Min	20.6	14.6	21.6	14.0	22.3	17.3	17.0
Queue	2.6	4.6	11.6	11.3	10.6	11.3	12.6

Table 6. Average ranks for Max-Min and FIFO-Queue ACO algorithm on QAP

	500	1000	2000	5000	10000	20000	50000
Max-Min	18.5	17.5	14	13	11	8.5	8
Queue	12	13.5	13	9	4.5	7.5	10

Tables 5 and 6 show the average rank of the best parameter combination for FIFO-Queue and Max-Min ACO algorithm. The main observation than can be made is that for all cases but one (QAP after 50000 itertaions), the average rank of the /nfigFIFO-Queue ACO algorithm is lower than that of the Max-Min ACO algorithm, indicating that the former's performance is at least as good as the latter's. This suggests that the solution quality which is achieved by the FIFO-Queue ACO algorithm is not only due to the implicit minimum and maximum pherome values, but also to the quick adaptation a small explicit population of solutions is able to perform.

5   Conclusion

The aim of this paper was to establish a population based ACO algorithm as an alternative to standard ACO algorithms. The performance measured on the two problem classes (TSP and QAP) show that the new approach is competitive.

The new approach offers several possibilities for future work. First, it seems well suited for coping with dynamic optimization problems which require a fast re-orientation of the pheromone information. Second, due to the discrete nature of the population and hence the pheromone matrix of the FIFO-Queue ACO algorithm, it offers starting points for a theoretical analysis. Third the approach offers possibilities for new metaheuristics that use the population of solutions.

References

1. M. Dorigo. *Optimization, Learning and Natural Algorithms* (in Italian). PhD thesis, Dipartimento di Elettronica , Politecnico di Milano, Italy, 1992. pp. 140.
2. M. Dorigo, V. Maniezzo, and A. Colorni. The ant system: Optimization by a colony of cooperating agents. *IEEE Trans. Systems, Man, and Cybernetics – Part B*, 26:29–41, 1996.
3. M. Dorigo and G. Di Caro. The ant colony optimization meta-heuristic. In D. Corne, M. Dorigo, and F. Glover, editors, *New Ideas in Optimization*, pages 11–32. McGraw-Hill, 1999.
4. T. Stützle and H. Hoos. Improvements on the ant system: Introducing MAX(MIN) ant system. In *Proc. of the International Conf. on Artificial Neutral Networks and Genetic Algorithms*, pages 245–249. Springer Verlag, 1997.
5. T. Stützle and H.H. Hoos. MAX-MIN ant system. *Future Generation Computer Systems Journal*, 16(8):889–914, 2000.
6. M. Guntsch and M. Middendorf. Pheromone modification strategies for ant algorithms applied to dynamic TSP. In E.J.W. Boers et al., editor, *Applications of Evolutionary Computing: Proceedings of EvoWorkshops 2001*, number 2037 in Lecture Notes in Computer Science, pages 213–222. Springer Verlag, 2000.
7. M. Guntsch, M. Middendorf, and H. Schmeck. An ant colony optimization approach to dynamic TSP. In L. Spector et al., editor, *Proceedings of the Genetic and Evolutionary Computation Conference (GECCO-2001)*, pages 860–867. Morgan Kaufmann Publishers, 2001.
8. V. Maniezzo. Exact and approximate nondeterministic tree-search procedures for the quadratic assignment problem. *INFORMS Journal on Computing*, 11(4), 1999.
9. M. Dorigo and L.M. Gambardella. Ant colony system: A cooperative learning approach to the traveling salesman problem. *IEEE Transactions on Evolutionary Computation*, 1:53–66, 1997.
10. L.-M. Gambardella, E.D. Taillard, and M. Dorigo. Ant colonies for the quadratic assignment problem. *Journal of the Operational Research Society*, 50:167–76, 1999.
11. T. Stützle and M. Dorigo. ACO algorithms for the quadratic assignment problem. In D. Corne, M. Dorigo, and F. Glover, editors, *New Ideas in Optimization*, pages 33–50. McGraw-Hill, 1999.
12. http://www.iwr.uni-heidelberg.de/groups/comopt/software/tsplib95.
13. http://www.opt.math.tu-graz.ac.at/qaplib.

Comparing Classical Methods for Solving Binary Constraint Satisfaction Problems with State of the Art Evolutionary Computation

Jano I. van Hemert

Leiden Institute of Advanced Computer Science, Leiden University
jvhemert@cs.leidenuniv.nl

Abstract. Constraint Satisfaction Problems form a class of problems that are generally computationally difficult and have been addressed with many complete and heuristic algorithms. We present two complete algorithms, as well as two evolutionary algorithms, and compare them on randomly generated instances of binary constraint satisfaction problems. We find that the evolutionary algorithms are less effective than the classical techniques.

1 Introduction

A *constraint satisfaction problem* (CSP) is a set of variables all of which have a corresponding domain. As well as variables, it contains a set of constraints that restricts certain simultaneous value assignments to occur. The objective is to assign each variable one of the values from its domain without violating any of the restricted simultaneous assignments as set by the constraints.

Constraint satisfaction has been a topic of research in many different forms. Lately, research has been concentrating on binary constraint satisfaction problems, which is a more abstract form of constraint satisfaction. This general model has resulted in a quest for faster algorithms that solve CSPs. Just like particular constraint satisfaction problems, such as k-graph colouring and 3-SAT, binary CSPs have a characteristic phase-transition, which is the transition from a region in which almost all problems have many solutions to a region in which almost all problems have no solutions. The problem instances found in this phase-transition are among the hardest to solve, making them interesting candidates for experimental research [1].

Backtracking kind of algorithms were the first to be used for tackling CSPs. One of the main advantages of these *classical algorithms* is that they are sound (i.e., every result is indeed a solution) and complete (i.e., they make it possible to find all solutions) [2]. Unfortunately, such algorithms need enormous amounts of constraint checks to solve the problem instances inside the phase-transition. To counter for this, many techniques have been developed and most of these keep the properties of soundness and completeness intact.

Compared with the classical algorithms, the evolutionary computation technique is a newcomer in constraint satisfaction. During the past ten years much study has gone into improving the speed and accuracy of evolutionary algorithms

S. Cagnoni et al. (Eds.): EvoWorkshops 2002, LNCS 2279, pp. 82–91, 2002.

on solving CSPs. Since evolutionary algorithms are stochastic, they are neither sound nor complete.

In this paper we first give more details on binary constraint satisfaction problems. Then, in Section 3 and Section 4 we present the algorithms. Section 5 provides the experimental setup and the results, followed by conclusions in Section 6.

2   Binary Constraint Satisfaction Problems

In general, a CSP has a set of constraints where each constraint can be over any subset of the variables. Here, we focus on binary CSPs: a model that only allows constraints over a maximum of two variables. At first, this seems a restriction, but Tsang has shown [2] this is not the case by proving that any CSP can be rewritten into a binary CSP. Solving the general CSP corresponds then to finding a solution in the binary form. Although Bacchus and van Beek [3] argue that the binary form may, depending on the problem, not be the most efficient way of handling a CSP, it still remains a popular object of study.

The binary model has led to a study where the object is to find problem instances that are difficult to solve. Collections of these problem instances serve as a good test bed for experimental research on solving methods. The idea behind finding difficult problem instances is that these are most likely to be the problem instances having only one solution, yet without being over-constrained and without any other structure that would make the solution easy to find. The first models were made using four parameters in the model, the number of variables n, the overall domain size m, the density of the constraints p_1 and the average tightness of the constraints p_2. Smith [4] estimated the number of solutions by using these parameters in a predictor shown in Equation 1.

$$E(solutions) = m^n (1 - p_2)^{\frac{n(n-1)p_1}{2}} \qquad (1)$$

The predictor helps us in visualising the total space of problem instances. If we fix two of the four parameters in the model, in our case the number of variables and the domain size, we are left with the parameters density and tightness. Using Equation 1 we can plot these latter parameters against the predicted number of solvable problem instances $E(solutions)$. Figure 1 shows this plot for the case where we set the number of variables and domain size to fifteen. To make the plot easier to read we cut off $E(solutions)$ above one.

Figure 1 clearly shows three regions. First, it shows a region where the number of solvable problem instances is equal to or higher than one ($E(solutions) \geq 1$). Second, it shows a region where the number of solvable problem instances is equal to zero ($E(solutions) = 0$) and third, it shows a region in between the other regions, which is called the *mushy region* ($0 < E(solutions) < 1$). Some problem instances that have the density and tightness settings within the mushy region have no solutions, while others have. To make the region more visible a contour plot is drawn in the x,y-plane. This is the area of parameter settings where we expect to find the most difficult to solve problem instances.

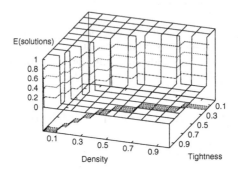

Fig. 1. The expected number of solutions with $E(solutions) \leq 1$ for fixed $n = 15$ and $m = 15$ plotted against density and tightness

3   Classical Algorithms

The term *classical algorithm* is used in this paper for algorithms that are not based on evolutionary computation. Algorithms in this section have been tested with an adapted version of *csplib*[1] where we have added a model for randomly generating binary CSPs.

3.1   Chronological Backtracking

Chronological backtracking (BT) is a simple algorithm dating back to 1965 [5] which is easy to construct using recursion. This algorithm is selected because it is simple and most algorithms studied in the field of constraint programming are based on some form of backtracking. These algorithms rely on recursion to move them through the search space, guided by all sorts of heuristics to find solutions as quickly as possible.

```
bool Backtrack(solution, current)              bool Consistent(solution, current)
   if current > number_of_variables then          for i = 1 ... current − 1 do
      return true;                                    constraint_checks++;
   else                                               if Conflict(current, i, solution[current],
      foreach value ∈ domain_current do                  solution[i]) then
         solution[current] = value;                        return false;
         if Consistent(solution, current) then      return true;
            if Backtrack(solution, current+1) then
               return true;
   return false;
```

Fig. 2. Pseudo code for chronological backtracking

The chronological backtracking algorithm is shown in pseudo code in Figure 2. This code uses the function `Conflict` to check whether the value assign-

[1] available at http://www.lpaig.uwaterloo.ca/~vanbeek/software/csplib.tar.gz

ment of two variables (*current* and *i*) raises a constraint conflict. A global variable *constraint_checks* is used to keep track of the number of constraint checks performed during a run of the algorithm. Chronological backtracking consists of two functions. First, the recursive function `Backtrack` that tries, for every variable, all values of the corresponding variable's domain. Second, a function called `Consistent` is used to check if a partial solution causes a constraint violation involving any of the previous assigned variables.

3.2 Forward Checking with Conflict-Directed Backjumping

The method *forward checking with conflict-directed backjumping* (FC-CBJ) is constructed using two techniques. Forward checking, which originates from 1980 [6] and constraint-directed backjumping, which was added in 1993 [7].

Forward checking instantiates a current variable and then checks forwards all the uninstantiated (future) variables. During this process all values incompatible with the current variable are removed from the domains of the corresponding uninstantiated variables. The process continues until it has instantiated all variables, i.e., found a solution, or until it checks a variable that has an empty domain left. In the last case the effects of the forward checking, i.e., the shrinking of domains, is undone and a new value for the current variable is tried.

Conflict-directed backjumping tries to improve the speed of forward checking by jumping over previous conflict checks that are unnecessary to repeat. To make this possible more bookkeeping is needed. Every variable is assigned its own conflict set which contains future variables that have failed consistency checks with the value assigned to the current variable. More precisely, every time a consistency check fails between an instantiation of the current variable and an instantiation of a future variable, the future variable is added to the conflict set of the current variable. When a domain annihilation of a future variable occurs the variables in the conflict set are added to the current variable's conflict set. When we run out of values to try for our current variable we turn to its conflict set joined with the conflict set of the annihilated variable and pick the variable that has been assigned earliest in the search. Joining the conflict sets of the new current and old current variable, and the past set of the old current (minus the new current) ensures that we keep our information up-to-date.

4 Evolutionary Algorithms

Two evolutionary algorithms will represent the current state of evolutionary computation on solving binary CSPs. These two algorithms are at this moment reported as performing the best on solving binary CSPS. The algorithms presented here all try to minimise the fitness function, where a fitness of zero equals to finding a solution.

Before we present the two algorithms in detail we want to point out a potential problem. Before an evolutionary algorithm can solve a constraint satisfaction problem another problem has to be overcome. Basically, every evolutionary algorithm is a stochastic optimisation algorithm that can be applied to minimise a (fitness) function. However, a constraint satisfaction problem is a decision problem which offers no help to our evolutionary algorithms. To bypass this problem we often measure another property, such as the number of violated constraints. By minimising this property we arrive at a solution for our decision problem when we arrive at zero constraint violations.

The algorithms here share a common feature. They both adapt the fitness function during the run. This means that over time the same candidate solutions can have different fitness values. The rationale behind this adaptive scheme is to let the algorithms gain information on the problem instance they are currently solving, in order to better guide their search for a solution.

4.1 Microgenetic Iterative Descent Method

The *microgenetic iterative descent* (MID) method is a rather technical evolutionary algorithm. It uses a small population, hence the term microgenetic. The original idea is from Dozier et al. [8,9]. This study involves gradually improving the general idea of using heuristics and breakout mechanisms to solve CSPs. The implementation of Eiben et al. is a re-implementation of the versions described in [8,9].

The core of MID is the breakout mechanism used to reduce the chance of getting stuck in local optima. Basically the breakout mechanism is a bookkeeper for pairs of values that have been involved in a constraint violation when the algorithm previously got stuck in a local optimum. The collection of breakouts with corresponding weights that is constructed during the search is used in the fitness function of the evolutionary algorithm. Equation 2 shows that the fitness function comprises of two parts. First, the sum of all violated constraints is counted per variable. Second, the sum of all weights of breakouts that are used by the solution. This discourages the appearance of solutions that make use of pairs of values known to appear in local optima:

$$\text{fitness}(solution) = \sum_{v \in \text{variables}} V(v) + \sum_{b \in \text{breakouts}} B(b), \qquad (2)$$

where $V(v)$ is the number of violated constraints corresponding with variable v and where $B(b)$ is the weight corresponding with breakout b or 0 if b does not exist or when b is not used in the *solution*.

4.2 Stepwise Adaptation of Weights

The *stepwise adaptation of weights* (SAW) method [10,11] is an extension for evolutionary algorithms that is intended to increase the efficiency of the search process. The earlier evolutionary algorithms used a fitness function that only counted the number of constraint violations. Obviously, this is a blind way of searching that ignores possible knowledge of, for instance, the internal structure of a problem. The SAW method is a technique for adding this kind of knowledge to the fitness function.

The SAW method works by assigning a weight w_c to each constraint c of the CSP instance to solve. These weights are all initialised to one. When the evolutionary algorithm is running, its main loop will be paused every 250 evaluations to update the weights. Within an update each weight w_c is incremented with one if the corresponding constraint c is violated by the best solution in the current population.

By having this weight vector included into the fitness function of the evolutionary algorithm we hope to force the evolutionary algorithm into focusing more on constraints that seem hard to satisfy during a search. Equation 3 shows how the weights are included in the fitness function. Note that keeping the weights w_c set to one during the run of an evolutionary algorithm would result in a standard evolutionary algorithm, that is, without SAW. Hence,

$$\text{fitness}(solution) = \sum_{c \in constraints} w_c \cdot C(c) \tag{3}$$

$$\text{where } C(c) = \begin{cases} 1 & \text{if } c \text{ is violated by } solution, \\ 0 & \text{otherwise.} \end{cases}$$

The evolutionary algorithm in [10,11] uses an order-based representation, i.e., a permutation of the variables of the binary CSP. A greedy decoder is used to decode this permutation into a value assignment.

5   Experiments and Results

We randomly generate problem instances of binary CSPs, using RandomCsp [12]. The process whereby we create these instances is called Model B [13], which first determines how many constraints and conflicts an instance will contain and then distributes both of these uniform randomly over the instance. Every instance is generated with the same domain size for each of its variables, this domain size is fixed to fifteen. In the first experiment we will vary the constraint density and the constraint tightness, while in the second experiment we will vary the number of variables.

The results for the evolutionary algorithms used in this paper are taken from a study by Eiben et al. [11], where the same set of randomly generated binary CSPs is used.

We measure the *success rate* (SR) and *the average number of constraint checks* (ACS) for each pair of density and tightness. The classical algorithms are sound and therefore the percentage of solutions presented is the actual percentage of solutions in the test set. When an evolutionary algorithm reaches the same SR it has actually found a solution to every CSP that had one. Whenever an evolutionary algorithm is not able to find a solution within a fixed number of generated proposed solutions it is terminated. This number is fixed to 100,000 for both evolutionary algorithms.

5.1   Fixed Variables and Domain Sizes

The test set consists of problem instances created with two parameters fixed and two parameters that vary: the number of variables and the overall domain size are set to fifteen and the constraint density and average tightness are varied from 0.1 to 0.9 with a step size of 0.2. Thereby, creating 25 different pairs of density and tightness. For each of these pairs we randomly generate 25 different problem instances. Because of the stochastic nature of evolutionary algorithms we let each of them do ten independent runs on each instance, which follows the procedure in [11].

Table 1. Success ratio and average number of constraint checks, with standard deviations within brackets for 25 pairs of density and tightness settings. Every result is averaged over 25 instances

density,		tightness				
algorithm		0.1	0.3	0.5	0.7	0.9
0.1	BT	1.00 1.08e2 (3e0)	1.00 1.18e2 (9e0)	1.00 1.37e2 (1e1)	1.00 3.52e2 (9e2)	**0.84 7.99e3 (2e4)**
	FC-CBJ	1.00 1.53e3 (2e1)	1.00 1.44e3 (4e1)	1.00 1.35e3 (6e1)	1.00 1.25e3 (1e2)	**0.84 8.64e3 (2e4)**
	MID	1.00 1.01e1 (3e-1)	1.00 5.07e1 (2e1)	1.00 2.25e2 (6e1)	1.00 0.48e2 (3e2)	**0.84 2.09e5 (4e5)**
	SAW	1.00 1.00e1 (0)	1.00 1.00e1 (0)	1.00 1.75e1 (2e1)	1.00 8.91e1 (8e1)	**0.36 6.74e5 (5e5)**
0.3	BT	1.00 1.21e2 (1e1)	1.00 1.03e3 (4e3)	1.00 6.86e3 (1e4)	**0.56 1.60e5 (4e5)**	0.00 3.41e4 (7e3)
	FC-CBJ	1.00 1.4e3 (3e1)	1.00 1.08e3 (7e1)	1.00 9.42e2 (3e2)	**0.56 1.48e4 (2e4)**	0.00 3.41e4 (7e3)
	MID	1.00 1.01e1 (3e1)	1.00 1.74e3 (3e2)	1.00 1.2e4 (4e3)	**0.41 2.35e6 (8e5)**	0.00 3.15e6 (0)
	SAW	1.00 3.10e1 (0)	1.00 4.48e1 (4e1)	1.00 1.26e3 (7e2)	**0.17 2.78e6 (8e5)**	0.00 3.15e6 (0)
0.5	BT	1.00 1.31e2 (1e1)	1.00 1.07e3 (2e3)	**1.00 2.11e5 (5e5)**	0.00 1.59e4 (1e4)	0.00 2.64e4 (4e3)
	FC-CBJ	1.00 1.27e3 (4e1)	1.00 8.66e2 (6e1)	**1.00 2.29e4 (4e4)**	0.00 7.3e3 (2e3)	0.00 2.64e4 (4e3)
	MID	1.00 5.43e2 (1e2)	1.00 9.33e3 (2e3)	**0.93 1.65e6 (7e5)**	0.00 5.25e6 (0)	0.00 5.25e6 (0)
	SAW	1.00 5.22e1 (1e0)	1.00 3.82e2 (3e2)	**0.69 2.01e6 (2e6)**	0.00 5.25e6 (0)	0.00 5.25e6 (0)
0.7	BT	1.00 1.49e2 (2e1)	**1.00 2.43e4 (9e4)**	0.00 1.64e5 (9e4)	0.00 6.38e3 (3e3)	0.00 2.49e4 (3e3)
	FC-CBJ	1.00 1.16e3 (5e1)	**1.00 9.92e2 (5e2)**	0.00 3.97e4 (1e4)	0.00 4.42e3 (8e2)	0.00 2.49e4 (3e3)
	MID	1.00 1.55e3 (2e2)	**1.00 5.49e4 (2e4)**	0.00 7.35e6 (0)	0.00 7.35e6 (0)	0.00 7.35e6 (0)
	SAW	1.00 7.48e1 (9e0)	**1.00 5.28e3 (3e3)**	0.00 7.35e6 (0)	0.00 7.35e6 (0)	0.00 7.35e6 (0)
0.9	BT	1.00 3.29e2 (2e4)	**1.00 1.94e5 (2e5)**	0.00 7.40e4 (9e3)	0.00 3.38e5 (5e2)	0.00 2.28e4 (3e3)
	FC-CBJ	1.00 1.09e3 (5e1)	**1.00 1.47e4 (2e4)**	0.00 2.42e4 (2e3)	0.00 3.67e4 (4e2)	0.00 2.28e4 (3e3)
	MID	1.00 3.29e3 (5e2)	**1.00 7.4e5 (3e5)**	0.00 9.45e6 (0)	0.00 9.45e6 (0)	0.00 9.45e6 (0)
	SAW	1.00 1.02e2 (4e1)	**1.00 2.54e5 (1e5)**	0.00 9.45e6 (0)	0.00 9.45e6 (0)	0.00 9.45e6 (0)

A quick look at the results in Table 1 shows three distinct "regions" of parameter settings, which correspond to the theoretical expectation presented in the x,y-plane in Figure 1. First, the upper left where the density and tightness are low shows that all problem instances have solutions that may easily be found by every algorithm. Second, opposite of the left corner where density and tightness are high we see that none of the problem instances can be solved. Third, between these two regions a mushy region exists where not for every pair of density and tightness we may solve all of the problem instances. Furthermore, here the algorithms need significantly more constraint checks to determine the solution or to determine that no solution exists. Also we see that for those settings not every instance is solvable as is the case for density/tightness pairs 0.1/0.9 and 0.3/0.7.

Looking at the lower right corner we see the large difference in the number of constraint checks for the classical algorithms and for the evolutionary algorithms. The explanation for this difference is twofold. First, the large numbers for evolutionary algorithms occur because, in general, evolutionary algorithms do not know when to quit. Second, the small numbers for the classical algorithms are caused by the problem instances becoming over-constrained, making it easy to reduce the search space.

The results of SAW in the upper left region seem a little strange. As evolutionary algorithms are generally started with a random population we would not expect them to find solutions quickly. Nevertheless, SAW is able to find a solution (on average) with fewer than 400 constraint checks for all of these pairs. Again the explanation is twofold. Firstly, this version of SAW uses a very small population size, thus losing little at the first evaluation of the whole population. Secondly, its greedy decoder is successful for easy problems that have many solutions, thus able to solve them in a few attempts.

When observing the results we see that except for easy problems classical algorithms are better. When we focus on the classical algorithms we see that the winner is FC-CBJ. Although it does not beat the other algorithms in the upper left and lower right regions it makes up for this small loss in the mushy region.

To verify the significance of our results we performed a number of statistical tests [14]. When we look at the results and take these as data for statistical analysis, we are confronted with a problem. The data is not normally distributed according to KS-Liliefors tests. Even if they were normally distributed the standard deviations are too far apart. This implies that the spread in the results is too large to perform statistical tests like student or ANNOVA. We turn towards a Wilcoxon Rank test to do our analysis as this test is able to handle data that is not normally distributed.

Table 2. Results of the Wilcoxon Rank test for paired up algorithms, where H_0^1 is the one-sided hypothesis, H_0^2 the two-sided hypothesis and z the normally distributed variable

	FC-CBJ	MID	SAW		MID	SAW		SAW
H_0^1	0.0005	0.0000	0.0000		0.0000	0.0000		0.0000
H_0^2 BT	0.0011	0.0001	0.0000	FC-CBJ	0.0000	0.0000	MID	0.0000
z	-3.2642	-4.0314	-13.0.99		-9.7333	-13.153		-19.634

We list the results of each instance, averaging the 10 runs for the evolutionary algorithms. This gives us 625 lines. Then we pair up two algorithms and perform the Wilcoxon's Rank Pairs Signed Rank Test [15] for pairwise ascertained data sets. Here the null-hypothesis (H_0) is that the both results are to be treated equally. We test the null-hypothesis for one-side (H_0^1) and for two-sides (H_0^2). In Table 2 we see that this test shows the results of our comparison is significant for every pair as the largest chance that the null-hypothesis is true is only 0.0011.

5.2   Scale-up tests

Besides a large comparison for a fixed number of variables and domain size, we also perform a scale-up test whereby we increase the number of variables. We want to see whether evolutionary algorithms are able to outperform classical methods when the problem size increases. Again we randomly generate 25 different problem instances per settings. We start with 10 variables and increase with steps of 5 up until 45. Every instance used in this experiment has a solution.

Because for $n \leq 45$ every run of a classical algorithm produces a solution, we only present results on the average number of constraint checks. The graph on the right in Figure 3 shows, on a logarithmic scale, the results from the table on the left. Chronological backtracking scales up very poorly. Although, for a small number of variables chronological backtracking is able to compete with the other three algorithms, we see that from $n = 25$ onwards the number of constraint checks increases exponentially. The two evolutionary algorithms are able to keep up a good performance for small number of n, but eventually they have to give in. Very different is FC-CBJ that shows an almost constant performance until the number of variables has reached as much as 35.

Fig. 3. Results for the scale-up test where the number of variables is varied, while the domain size is fixed to 15 and the constraintness and tightness are both set to 0.3. Note that for $n = 45$ the maximum number of evaluations is set to 200,000

n	BT	FC-CBJ	MID		SAW	
	ACS	ACS	SR	ACS	SR	ACS
10	5.90e1	5.49e2	1.00	1.30e2	1.00	1.30e1
15	1.71e2	1.10e3	1.00	1.61e3	1.00	6.20e1
20	4.14e2	1.75e3	1.00	9.29e3	1.00	2.85e2
25	5.31e4	2.63e3	1.00	3.69e4	1.00	2.70e3
30	3.10e6	3.75e3	1.00	1.35e5	1.00	2.47e4
35	5.43e7	9.32e3	1.00	6.16e5	1.00	2.61e5
40	9.57e7	3.88e5	1.00	4.04e6	1.00	4.37e6
45	4.10e9	3.70e7	0.44	4.70e7	0.24	5.30e7

The last value ($n = 45$) is the first time that the evolutionary algorithms have trouble finding a solution within the maximum of 100,000 evaluations. MID has a success rate of 0.21 and SAW has 0.12. We increase this maximum to 200,000 because FC-CBJ is using more constraint checks than the previous maximum allows. MID's success rate rises to 0.44 and SAW's success rate rises to 0.24. The number of evaluations of both is larger than that of FC-CBJ.

6　Conclusions

In this study we have made a comparison based on a test set of randomly generated binary constraint satisfaction problems between two classical algorithms and two evolutionary algorithms. This comparison shows that evolutionary algorithms have to improve their speed by a considerable factor if they want to compete with an algorithm as simple as chronological backtracking.

In our comparison, the idea that evolutionary algorithms might outperform classical methods when the problem at hand is scaled-up holds only partially. Although both evolutionary algorithms easily outperform the simple backtracking algorithm, they are not fast enough to take on the more modern FC-CBJ.

One reason evolutionary algorithms are not promising is that they fail to realize when a problem has no solution — unlike the classical algorithms that, aided by preprocessing methods such as arc-consistency, are very efficient in pointing out the impossible.

References

1. C.P. Williams and T. Hogg. Exploiting the deep structure of constraint problems. *Artificial Intelligence*, 70:73–117, 1994.
2. E. Tsang. *Foundations of Constraint Satisfaction*. Academic Press, 1993.
3. F. Bacchus and P. van Beek. On the conversion between non-binary and binary constraint satisfaction problems. In *Proceedings of the 15th International Conference on Artificial Intelligence*. Morgan Kaufmann, 1998.
4. B.M. Smith. Phase transition and the mushy region in constraint satisfaction problems. In A. G. Cohn, editor, *Proceedings of the 11th European Conference on Artificial Intelligence*, pages 100–104. Wiley, 1994.
5. S.W. Golomb and L.D. Baumert. Backtrack programming. *A.C.M.*, 12(4):516–524, October 1965.
6. R. Haralick and G. Elliot. Increasing tree search efficiency for constraint-satisfaction problems. *Artificial Intelligence*, 14(3rd):263–313, 1980.
7. P. Prosser. Hybrid algorithms for the constraint satisfaction problem. *Computational Intelligence*, 9(3):268–299, August 1993.
8. J. Bowen and G. Dozier. Solving constraint satisfaction problems using a genetic/systematic search hybride that realizes when to quit. In L.J. Eshelman, editor, *Proceedings of the 6th International Conference on Genetic Algorithms*, pages 122–129. Morgan Kaufmann, 1995.
9. G. Dozier, J. Bowen, and D. Bahler. Solving randomly generated constraint satisfaction problems using a micro-evolutionary hybrid that evolves a population of hill-climbers. In *Proceedings of the 2nd IEEE Conference on Evolutionary Computation*, pages 614–619. IEEE Press, 1995.
10. A.E. Eiben, J.K. van der Hauw, and J.I. van Hemert. Graph coloring with adaptive evolutionary algorithms. *Journal of Heuristics*, 4(1):25–46, 1998.
11. A.E. Eiben, J.I. van Hemert, E. Marchiori, and A.G. Steenbeek. Solving binary constraint satisfaction problems using evolutionary algorithms with an adaptive fitness function. In *Proceedings of the 5th Conference on Parallel Problem Solving from Nature*, number 1498 in LNCS, pages 196–205, Berlin, 1998. Springer.
12. J.I. van Hemert. Randomcsp, a library for generating and handling binary constraint satisfaction problems, Version 1.5, 2001. http://www.liacs.nl/~jvhemert/randomcsp.
13. E. M. Palmer. *Graphical Evolution*. John-Wiley & Sons, New York, 1985.
14. D.S. Moore and G.P. McCabe. *Introduction to the Practice of Statistics*. W.H. Freeman and Company, New York, 3rd edition, 1998.
15. R.J. Barlow. *Statistics: a guide to the use of statistical methods in the physical sciences*. John Wiley & Sons, 1995.

Application of Genetic Algorithms in Nanoscience: Cluster Geometry Optimization

Roy L. Johnston[1], Thomas V. Mortimer-Jones[1], Christopher Roberts[1], Sarah Darby[1], and Frederick R. Manby[2]

[1] School of Chemical Sciences, University of Birmingham,
Edgbaston, Birmingham B15 2TT, UK
roy@tc.bham.ac.uk
http://www.bham.ac.uk/~roy/
[2] School of Chemistry, University of Bristol,
Cantocks Close, Bristol BS8 1TS, UK

Abstract. An account is presented of the design and application of Genetic Algorithms for the geometry optimization (energy minimization) of clusters and nanoparticles, where the interactions between atoms, ions or molecules are described by a variety of potential energy functions (force fields). A detailed description is presented of the Birmingham Cluster Genetic Algorithm Program, developed in our group, and two specific applications are highlighted: the use of a GA to optimize the geometry and atom distribution in mixed Cu-Au clusters; and the use of an energy predator in an attempt to identify the lowest six isomers of C_{40}.

1 Introduction: Clusters and Cluster Modelling

Clusters are aggregates of anything from a few to many millions of atoms or molecules [1]. One of the main areas of interest in clusters is the study of the evolution of cluster structures and properties, as a function of their size. Clusters are formed by most chemical elements and can be found in a number of environments – for example: copper, silver and gold clusters in stained glass windows; silver clusters on photographic films; molecular clusters in the atmosphere and carbon clusters in soot.

Since, for large clusters (of hundreds or thousands of atoms) *ab initio* calculations are unfeasible, there has been much interest in developing empirical atomistic potentials for the simulation of such species [2]. For a given potential energy function and cluster size, we generally wish to find the arrangement of atoms (or ions or molecules) corresponding to the lowest potential energy – *i.e.* the global minimum (GM) on the potential energy hypersurface. As the number of minima rises exponentially with increasing cluster size, finding the global minimum is a particularly difficult problem and traditional Monte Carlo and Molecular Dynamics Simulated Annealing approaches often encounter difficulties finding global minima for clusters. It is for this reason that Genetic Algorithms have found increasing use in the area of cluster geometry optimization.

S. Cagnoni et al. (Eds.): EvoWorkshops 2002, LNCS 2279, pp. 92–101, 2002.

2   Overview of Previous GAs for Cluster Optimization

The Genetic Algorithm (GA) [3, 4] is a search technique which is based on the principles of natural evolution. It uses operators that are analogues of the evolutionary processes of genetic crossover, mutation and natural selection to explore multi-dimensional parameter spaces.

The use of GAs for optimizing cluster geometries was pioneered in the early 1990s by Hartke (for small silicon clusters) [5] and Xiao and Williams (for molecular clusters) [6]. In both cases the cluster geometries were binary encoded, with the genetic operators acting in a bitwise fashion on the binary strings. An important stage in the evolution of GAs for cluster optimization occurred when Zeiri [7] introduced a GA that operated on the real-valued Cartesian coordinates of the clusters. This approach allowed for a representation of the cluster in terms of continuous variables and removed the requirement for encoding and decoding binary genes.

The next significant step in the development of GAs for cluster optimization was due to Deaven and Ho [8], who performed a gradient driven local minimization of the cluster energy after each new cluster was generated. The introduction of local minimization effectively transforms the cluster potential energy hypersurface into a stepped surface, where each step corresponds to a basin of attraction of a local minimum on the potential energy surface. This simplification of the surface, which greatly facilitates the search for the global minimum by reducing the space that the GA has to search, also underpins the Basin Hopping Monte Carlo method developed by Wales and the "Monte Carlo plus energy minimization" approach of Scheraga [9]. In the GA context, such local minimization corresponds to Lamarckian, rather than Darwinian evolution.

Deaven and Ho also introduced the 3-dimensional "cut and splice" crossover operator [8]. This operator, which has been employed in most subsequent cluster GA work, gives a more physical meaning to the crossover process. In this crossover mechanism, good schemata correspond to regions of the parent clusters which have low energy local structure. Because the genetic operators in the Deaven and Ho GA act on the clusters themselves, in configuration space, rather than on a string representation of the problem, the crossover operation is "phenotypic", rather than "genotypic" [10].

A number of GAs have now been developed for cluster geometry optimization and a comprehensive review of this work has been presented elsewhere [11].

3   The Birmingham Cluster Genetic Algorithm Program

The basic features of our cluster geometry optimization GA program [12] are summarised below.

3.1   Initialization

For a given cluster size (nuclearity N), a number of clusters, N_{clus} (typically ranging from 10 to 30) are generated at random to form the initial population (the "zeroth generation") – using real-valued Cartesian coordinates picked to lie within a sphere of radius proportional to $N^{\frac{1}{3}}$ (with constraints placed on minimum and maximum nearest-neighbour distances). All of the clusters in the initial population are then relaxed into the nearest local minima, by minimizing the cluster potential energy as a function of the cluster coordinates, using the quasi-Newton L-BFGS routine [13].

The GA operations of mating, mutation and selection (on the basis of fitness) are performed to evolve one generation into the next. In this discussion, we will use the term "mating" to refer to the process by which two parent clusters are combined to generate offspring. The mechanism, at the chromosome level, by which genetic material is combined, will be termed "crossover".

3.2   Fitness

The fitness of a string gives a measure of its quality. In this case, the lowest energy (most negative V_{clus}) clusters have the highest fitness and the highest energy (least negative V_{clus}) clusters have the lowest fitness. The cluster GA uses dynamic fitness scaling, based on a normalised value of the energy, ρ:

$$\rho_i = \frac{V_i - V_{min}}{V_{max} - V_{min}} \tag{1}$$

where V_{min} and V_{max} are the lowest and highest energy clusters in the current population, respectively. We have employed exponential, linear, power and hyperbolic tangent fitness functions, which differ in the way in which they discriminate between high and low energy structures.

3.3   Mating/Crossover

Parents are selected by either roulette wheel or tournament selection [4]. In the roulette wheel method, a cluster is chosen at random and selected for mating if its fitness value (F_i) is greater than a randomly generated number between 0 and 1 (*i.e.* if $F_i > R[0,1]$). Crossover takes place when two parents have been accepted for mating. The tournament selection method picks a number of clusters at random from the population to form a "tournament" pool. The two lowest energy clusters are then selected as parents from this tournament pool. In both of these selection schemes, low energy clusters (with high fitness values) are more likely to be selected for mating and therefore to pass their structural characteristics on to the next generation. Once a pair of parents have been selected, they are subjected to the crossover operation.

Crossover is carried out using a variant of the cut and splice crossover operator of Deaven and Ho [8]. Random rotations (about two perpendicular axes) are performed on both parent clusters and then both clusters are cut horizontally about one or two positions, parallel to the xy plane, and complementary fragments are spliced together, as shown in Fig. 1. Several different crossover routines have been developed that create an offspring by making one or two

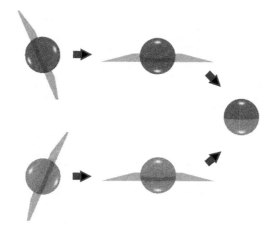

Fig. 1. The Deaven-Ho cut and splice crossover operation

cuts and pasting together complementary slices. For the single cut method, the cutting plane can be chosen at random, it can be defined to pass through the middle of the cluster, or weighted according to the relative fitnesses of the two parents.

Mating continues until a predetermined number of offspring (N_{off}) have been generated. The number of offspring is generally set to approximately 80% of the population size. Each offspring cluster is then relaxed into the nearest local minimum.

3.4   Mutation

While the crossover operation leads to a mixing of genetic material in the off-spring, no new genetic material is introduced. In an attempt to maintain population diversity, a mutation operator is introduced. Each string has a probability (P_{mut}) of undergoing mutation, which perturbs some or all of the atomic positions within a cluster. After mutation, each "mutant" cluster is locally minimized.

A number of mutation schemes have been adopted:

Atom Replacement Mutation. involves replacing the atomic coordinates of a certain number of the atoms with randomly generated values. The number of atomic coordinates replaced, is set to be approximately one third of the total number of atoms, N.

Twist Mutation. rotates the upper half of the cluster about the z axis by a random angle, relative to the bottom half.

Cluster Replacement Mutation. involves the replacement of an entire cluster with a new, randomly generated cluster. The cluster is generated in an identical way to that used for the generation of the initial population.

Atom Permutation Mutation. swaps the atom types of a pair of atoms without perturbing the structure of the cluster. Approximately $\frac{N}{3}$ atom label swaps are performed per cluster mutation.

These mutations generally correspond to relatively large perturbations of the genetic code and we are currently investigating reducing the magnitude of mutations, especially in later generations, so as to increase the occurrence of high fitness mutations.

3.5   Subsequent Generations

The new population is generated from the N_{clus} lowest energy (highest fitness) clusters, selected from the old population, the new offspring clusters and the mutated clusters. The inclusion of clusters from the previous generation makes the GA elitist. The whole process of mating, mutation and selection is repeated for a specified number (N_{gen}) of generations or until the population is deemed to have converged – if the range of cluster energies in the population does not change for a prescribed number of generations.

4   Applications

We have applied our cluster optimization GA to a number of different types of clusters, ranging from simple model clusters held together by the pair-wise additive Morse potential [14], to ionic MgO clusters [15], carbon clusters [16] and metallic clusters, including mixed Cu-Au clusters [17]). We have also developed a parallel implementation of the program and have introduced a predator operator, to enable low-lying metastable cluster isomers to be identified [16]. Here, we present a brief account of two of these studies: namely the use of a GA to find the lowest energy structures in mixed Cu-Au clusters; and the application of a predator to find the six lowest energy isomers of C_{40}. Further details can be found in the above mentioned papers and in a recent review [11].

Since, except for very small clusters, it is unfeasible to make a systematic grid search of all available coordinate space, one can never be absolutely certain that the GM has actually been found. However, by repeated runs (from different starting populations) one can increase the likelihood of finding the GM. In the following discussion, for conciseness, I will use the phrase "global minimum" to denote the lowest energy cluster found after repeated runs of the GA.

4.1   Cu-Au Nanoalloy Clusters

Bimetallic "nano-alloy" clusters are of interest for their catalytic and materials applications and because their properties may be tuned by varying the composition and atomic ordering, as well as the size of the clusters. There have been

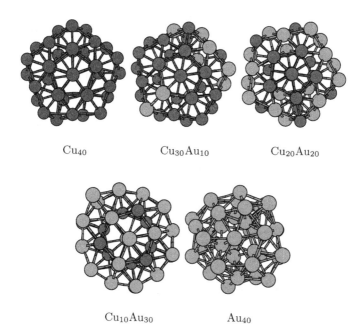

Fig. 2. Structures of 40-atom Cu, Au and Cu-Au clusters (dark atoms = Cu, light atoms = Au)

a number of studies of alloy clusters using many-body potentials [18]. Here, we will discuss the application of the cluster GA to study Cu-Au clusters modelled by the Gupta many-body potential [17].

The Gupta potential [19] is the sum of repulsive (V^r) pair and attractive many-body (V^m) terms:

$$V_{clus} = \sum_i^N \{V^r(i) - V^m(i)\} \tag{2}$$

where

$$V^r(i) = \sum_{j \neq i}^N A \cdot \exp\left(-p\left(\frac{r_{ij}}{r_0} - 1\right)\right) \tag{3}$$

and

$$V^m(i) = \left[\sum_{j \neq i}^N \zeta^2 \cdot \exp\left(-2q\left(\frac{r_{ij}}{r_0} - 1\right)\right)\right]^{\frac{1}{2}}. \tag{4}$$

r_{ij} is the distance between atoms i and j and the parameters A, r_0, ζ, p and q are fitted to experimental values of the cohesive energy, lattice parameters and independent elastic constants for the crystal structure at 0 K. For Au_xCu_y alloy clusters, the parameters take different values for each of the different types (Cu-Cu, Au-Cu and Au-Au) of interaction.

We have studied clusters with the compositions of the common bulk Au-Cu alloy phases: $(AuCu_3)_N$; $(AuCu)_N$; and $(Au_3Cu)_N$ and have compared them with pure copper and gold clusters, also modelled by Gupta potentials. Standard GA parameters and operators were used, except for the introduction of atom permutation mutation for the alloy clusters – which was found to greatly improve the reproducibility of the results and the likelihood of finding the GM. Jellinek has introduced the term "homotops" to describe A_aB_b alloy cluster isomers, for fixed number of atoms (N) and composition (a/b ratio), which have the same geometrical arrangement of atoms, but differ in the way in which the A and B-type atoms are arranged [18]. As the number of homotops rises combinatorially with cluster size, global optimization (in terms of both geometrical isomers and homotops) is an extremely difficult task.

Using our GA, pure copper clusters were found to adopt regular, symmetric structures based on icosahedral packing, while gold clusters have a greater tendency towards amorphous structures. In many cases (*e.g.* for 14, 16 and 55 atoms), the replacement of a single Au atom by Cu was found to change the structure of the most stable cluster to that of the pure Cu cluster.

The lowest energy structures (which I shall term GM for convenience) found by the GA for 40-atom clusters of varying composition – Cu_{40}; $(AuCu_3)_{10}$ (= $Cu_{30}Au_{10}$); $(AuCu)_{20}$ (= $Cu_{20}Au_{20}$); $(Au_3Cu)_{10}$ (= $Cu_{10}Au_{30}$); and Au_{40} – are shown in Fig. 3, from which it is apparent that the clusters Cu_{40}, $Cu_{30}Au_{10}$ and $Cu_{20}Au_{20}$ have the same decahedral geometry. In the decahedral alloy clusters, the Au atoms generally lie on the surface, while the Cu atoms are encapsulated. Although the Au_{40} cluster has a low symmetry, amorphous structure, the gold-rich $Cu_{10}Au_{30}$ cluster has a structure which is more symmetrical than the Au_{40} GM, but it is not decahedral. The $Cu_{10}Au_{30}$ GM has a flattened (oblate) topology, in which all of the Au atoms, except one, lie on the surface of the cluster and 7 of the Cu atoms occupy interior sites. The observed atomic segregation in the alloy clusters may be driven by lowering the surface energy and/or relieving the internal strain of the cluster.

4.2   Use of a Predator Operator to Find the Lowest Energy Isomers of C_{40}

GAs have proven to be very successful in determining global minima, but in a number of physical applications, structures corresponding to higher local minima may be at least as important as the GM. The biologically active forms of proteins, for example, do not always correspond to global minima.

Recently, we have taken the analogy with natural evolution one step further by considering the use of a "predator" to remove unwanted (although otherwise potentially optimal) individuals or traits from the population [16]. Sometimes unwanted members of a population can be removed by imposing a constraint on the fitness function, however, in seeking minima other than the GM, a suitable modification of the fitness function is not always possible. In principle, predation can be carried out using any property of the cluster, for example a shape selective predator could be used to remove individuals from the population (with a certain probability) if they show (or fail to show) certain topological features, such as sphericity, ring size etc..

The simplest predator is the energy predator, in which clusters with energies at or below a certain value are removed from the population. The energy predator can thus be used to search for low energy minima other than the global minimum, or to enhance the efficiency of the GA by removing specific low energy non-global minima that the GA may be incorrectly converging towards. In the energy predator, a cluster is "killed" by the predator if its potential energy is less than $V_{targ} + \delta V$, where V_{targ} is the target energy and δV is a small energy interval (typicaly 1×10^{-6} eV). The GA is first run without the predator to find the GM cluster and its energy. Then the predator is invoked to remove the GM, so that the GA finds the next lowest energy isomer (the first metastable structure). The energy of the first excited state is then used as the target energy which ensures that both the GM and first metastable isomer are predated. This cycle is continued until the required number of low-lying isomers have been found.

The energy predator operator has been used to find the six lowest energy isomers of C_{40}, with the bonding in the clusters described by the Murrell-Mottram (MM) many-body potential [2]. The geometries of the six most stable isomers of C_{40} are shown in Fig. 4, in order of increasing energy, where C_{40}^1 is the lowest energy cluster (the proposed GM for this model potential) found for C_{40}.

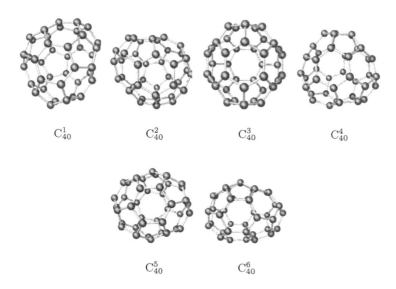

C_{40}^1 C_{40}^2 C_{40}^3 C_{40}^4

C_{40}^5 C_{40}^6

Fig. 3. Geometries of the six lowest energy isomers of C_{40} obtained using the predator

Albertazzi *et al.* [20] have carried out detailed studies of the isomers of C_{40} using molecular mechanics, tight binding and *ab initio* methods. The aim of their work was to confirm that the minimization of pentagon adjacency is a major factor in determining the relative stability of carbon clusters of this type (the so-called "fullerenes"). The cluster isomer of C_{40} with the smallest pentagon adjacency count is the D_2 cage, corresponding to isomer C_{40}^5 found by the GA (for the MM potential). Albertazzi *et al.* found that the D_2 cage was predicted by 11 out of the 12 methods used to be the lowest energy configuration. Although the potential function that we have used does not give the true (experimental) global minimum, this reflects an inadequacy of the model potential, rather than the GA, and the true global minimum is found among the top 6 isomers. This emphasizes the utility of the predator in searching for low energy local minima for a given empirical potential, which may then be checked (recalculated) at a higher level of theory.

5   Conclusions

The studies reported here and in the literature [11] show that the GA is a powerful technique for finding (with a high probability) global minima and other low energy structures for clusters of different types – within the limitations of the model potential energy function adopted.

A number of further studies are currently being planned – for example: comparing the efficiency of GA with Simulated Annealing; extending the approach to the study of molecular clusters (*e.g.* water nano-droplets); and introducing the adaptive control of mating and mutation within the GA.

Acknowledgements

The authors wish to thank Dr Lesley Lloyd and Dr Nicholas Wilson for helpful discussions. Our research has been supported by funding from: EPSRC (Ph.D. studentship for CR); HEFCE (JREI Grant for computing equipment); The Royal Society (University Research Fellowship for FRM), the University of Birmingham (Ph.D. studentship for TVM-J) and COMPAQ Computer Corp. (Galway, Ireland – sponsorship of TVM-J).

References

1. Haberland, H. (ed.): Clusters of Atoms and Molecules. Springer-Verlag, Berlin (1994).
2. Cox, H., Johnston, R.L., Murrell, J.N.: Empirical Potentials for Modelling Solids, Surfaces and Clusters. Journal of Solid State Chemistry **145** (1999) 517–540.
3. Holland, J.: Adaptation in Natural and Artificial Systems. University of Michigan Press, Ann Arbor, MI (1975).
4. Goldberg, D.E.: Genetic Algorithms in Search, Optimization and Machine Learning. Addison-Wesley, Reading, MA (1989).

5. Hartke, B.: Global Geometry Optimization of Clusters Using Genetic Algorithms. Journal of Physical Chemistry **97** (1993) 9973–9976.
6. Xiao, Y., Williams, D.E.: Genetic Algorithm – a New Approach to the Prediction of the Structure of Molecular Clusters. Chemical Physics Letters **215** (1993) 17–24.
7. Zeiri, Y.: Prediction of the Lowest Energy Structure of Clusters Using a Genetic Algorithm. Physical Review E **51** (1995) 2769–2772.
8. Deaven, D.M., Ho, K.M.: Molecular-Geometry Optimization with a Genetic Algorithm. Physical Review Letters **75** (1995) 288–291.
9. Wales, D.J., Scheraga, H.A.: Global optimization of clusters, crystals, and biomolecules. Science **285** (1999) 1368–1372.
10. Hartke, B.: Global Cluster Geometry Optimization by a Phenotype Algorithm with Niches: Location of Elusive Minima, and Low-order Scaling with Cluster Size. Journal of Computational Chemistry **20** (1999) 1752–1759.
11. Johnston, R.L., Roberts, C.: Genetic Algorithms for the Geometry Optimization of Clusters and Nanoparticles. In: Cartwright, H., Sztandera, L. (eds.): Soft Computing Approaches in Chemistry. Physica-Verlag, Heidelberg (2002) in press.
12. Johnston, R.L., Roberts, C.: Cluster Geometry Optimization Genetic Algorithm Program. University of Birmingham (1999).
13. Byrd, R.H., Lu, P.H., Nocedal, J., Zhu, C.Y.: A Limited Memory Algorithm for Bound Constrained Optimization. SIAM Journal on Scientific Computing **16** (1995) 1190–1208.
14. Roberts, C., Johnston, R.L., Wilson, N.T.: A Genetic Algorithm for the Structural Optimisation of Morse Clusters. Theoretical Chemistry Accounts **104** (2000) 123–130.
15. Roberts, C., Johnston, R.L.: Investigation of the Structures of MgO Clusters Using a Genetic Algorithm. Phys.Chem.Chem.Phys. **3** (2001) 5024–5034.
16. Manby, F.R., Johnston, R.L., Roberts, C.: Predatory Genetic Algorithms. MATCH (Communications in Mathematical and Computational Chemistry) **38** (1998) 111–122.
17. Darby, S., Mortimer-Jones, T.V., Johnston, R.L., Roberts, C.: Theoretical Sudy of Cu-Au Nanoalloy Clusters Using a Genetic Algorithm. Journal of Chemical Physics (in press).
18. Jellinek, J., Krissinel, E.B.: Alloy Clusters: Structural Classes, Mixing, and Phase Changes. In: Jellinek, J. (ed.): Theory of Atomic and Molecular Clusters. Springer, Berlin (1999) 277–308.
19. Cleri, F., Rosato, V.: Tight-Binding Potentials for Transition-Metals and Alloys. Physical Review B **48** (1993) 22–33.
20. Albertazzi, E., Domene, C., Fowler, P.W., Heine, T., Seifert, G., van Alsenoy, C., Zerbetto, F.: Pentagon Adjacency as a Determinant of Fullerene Stability. Phys.Chem.Chem.Phys. **1** (1999) 2913–2918.

A Memetic Algorithm
for Vertex-Biconnectivity Augmentation

Sandor Kersting, Günther R. Raidl, and Ivana Ljubić

Institute of Computer Graphics and Algorithms, Vienna University of Technology,
Favoritenstraße 9–11/186, 1040 Vienna, Austria
{kersting|raidl|ljubic}@ads.tuwien.ac.at

Abstract. This paper considers the problem of augmenting a given graph by a cheapest possible set of additional edges in order to make the graph vertex-biconnected. A real-world instance of this problem is the enhancement of an already established computer network to become robust against single node failures. The presented memetic algorithm includes an effective preprocessing of problem data and a fast local improvement strategy which is applied during initialization, mutation, and recombination. Only feasible, locally optimal solutions are created as candidates. Empirical results indicate the superiority of the new approach over two previous heuristics and an earlier evolutionary method.

1 Introduction

Robustness against failure is an important issue when designing a commercial computer network. It is often not acceptable that the failure of a single service node – be it a computer, router, or other device – leads to a disconnection of others. Redundant connections need to be established to provide alternative routes in case of a (temporary) break of any one node. We represent such a network by an undirected graph. It is said to be *vertex-biconnected* if at least two nodes need to be removed together with their incident edges in order to separate the graph into disconnected components. In a connected graph that is not vertex-biconnected a critical node whose removal would disconnect the graph is called *cut-point*. We say that we *cover a cut-point* when we add a set of edges to the graph which ensures that the removal of this vertex no longer disconnects the graph. Our global aim is to identify a set of edges with minimum total costs that covers all existing cut-points.

A more formal definition of the vertex-biconnectivity augmentation problem for graphs (V2AUG): Let $G = (V, E)$ be a vertex-biconnected, undirected graph with node set V and edge set E representing all possible connections. Each edge $e \in E$ has associated $cost(e) > 0$. A connected, spanning, but not vertex-biconnected subgraph $G_0 = (V, E_0)$ with $E_0 \subset E$ represents a fixed, existing network, and $E_a = E \setminus E_0$ is the set of edges that may be used for augmentation. The objective is to determine a subset of these candidate edges $E_s \subseteq E_a$ so that

S. Cagnoni et al. (Eds.): EvoWorkshops 2002, LNCS 2279, pp. 102–111, 2002.

the augmented graph $G_s = (V, E_0 \cup E_s)$ is vertex-biconnected and

$$cost(E_s) = \sum_{e \in E_s} cost(e) \tag{1}$$

is minimal.

The next section summarizes former approaches to this problem. Then a new memetic algorithm is presented. Section 3 explains its preprocessing, which creates needed data structures and reduces the size of the problem in general considerably by fixing or eliminating certain edges in safe ways. Section 4 describes the main algorithm, including a new local improvement algorithm for creating locally optimal offspring only. In Sect. 5 empirical results are presented and compared to two previous heuristics and a hybrid genetic algorithm. Conclusions are drawn in Sect. 6.

2 Previous Work

Eswaran and Tarjan [1] originally investigated the V2AUG problem. They showed it to be NP-hard. An exact polynomial-time algorithm could only be found for the special case when G is complete and each edge has unit costs [4].

Frederickson and Jàjà [2] provided an approximation algorithm for the general case which finds a solution within a factor 2 of the optimum. The algorithm includes a preprocessing step that transforms the fixed graph G_0 into a *block-cut tree*, see Sect. 3. Each potential augmentation edge from E_a is superimposed on the block-cut tree, and certain redundant edges are identified and eliminated. In the main part of the algorithm, the block-cut tree is directed toward an arbitrarily chosen leaf as root node, and each tree-edge is assigned zero costs. Each cut-point is substituted by star-shaped structures including new dummy-nodes in order to guarantee that strongly connecting the block-cut tree implies vertex-biconnectivity of the underlying fixed graph G_0. All superimposed augmentation edges are also directed. A minimum out-branching algorithm as described by Gabow et al. [3] is then applied to the block-cut tree including the superimposed augmentation edges to identify the solution's ede-set E_s. The computational effort of the algorithm is $O(|V|^2)$.

An improved variant of this approximation algorithm has been developed by Khuller and Thurimella [5]. It exhibits a time complexity of only $O(|E| + |V| \log |V|)$, but has the same approximation factor of 2.

An iterative approach based on Khuller and Thurimella's algorithm has been proposed by Zhu et al. [10]. In each step, a *drop*-heuristic measures the gain of each augmentation edge if it would be included in a final solution. This is achieved by calling the branching algorithm for each edge once with its cost set to zero and once with its original cost. The edge with the highest gain is then fixed, and its cost are permanently set to zero. The process is repeated until the obtained branching has zero total costs. Furthermore, the whole algorithm is applied with each leaf of the block-cut tree becoming once the root, and the overall cheapest solution is the final one. Although the theoretical approximation factor remains 2, practical results are usually much better than when applying Khuller and Thurimella's algorithm. However, time requirements are raised significantly.

A straight-forward hybrid genetic algorithm for the V2AUG problem has been proposed by Ljubić and Kratica [6]. This algorithm is based on a binary encoding in which each bit corresponds to an edge in E_a. Standard uniform crossover and bit-flip mutation are applied. Infeasible solutions are repaired by a greedy algorithm which temporarily removes cut-points one by one and searches for suitable augmentation edges that reconnect separated components.

Another, "lighter" kind of connectivity property is edge-biconnectivity, which means that a graph remains connected after the removal of any single edge. While vertex-biconnectivity implies edge-connectivity, the reverse is not true. Similar algorithms as for V2AUG have been applied to the edge-biconnectivity augmentation problem (E2AUG). From the algorithmic point-of-view, E2AUG is easier to deal with, since it does not require the special block-cut tree.

Recently, Raidl and Ljubić described in [7,9] an effective evolutionary algorithm for E2AUG. A compact edge set encoding and special initialization and variation operators that include a local improvement heuristic are applied. In this way, the space of locally optimal solutions is searched only. The approach belongs to the broader class of so-called *local-search-based memetic algorithms* [8].

Based on this algorithm for E2AUG, the memetic algorithm for V2AUG presented in this article has been developed. Major differences lie in the underlying data structures (e.g. the now necessary block-cut tree), the preprocessing, and the local improvement algorithm. While it is relatively easy to check and eventually establish the coverage of a single critical edge in case of E2AUG, this is significantly harder to achieve for a node in the V2AUG-case, especially in an efficient way: A fixed edge can always be covered by a single augmentation edge, and it is obvious which augmentation edges are able to cover the fixed edge. On the other side, a combination of multiple augmentation edges is in general necessary to cover a cut-point completely.

3   Preprocessing

During preprocessing, a block-cut tree is derived from the fixed graph G_0 according to [1], and other supporting data structures are created. They are all needed for an efficient implementation of the main algorithm. Furthermore, several deterministic rules are applied in order to reduce E_a in a safe way. The following paragraphs describe these mechanisms in detail.

3.1   The Block-Cut Tree

A block-cut tree $T = (V_T, E_T)$ with node set V_T and edge set E_T is an undirected tree that represents the connections between already vertex-biconnected components (called *blocks*) and cut-points of the underlying fixed graph G_0.

Two types of nodes form V_T: cut-nodes and block-nodes. Each cut-point in G_0 is represented by a corresponding cut-node in V_T, each maximal vertex-biconnected block in G_0 by a unique block-node in V_T. A block-node is associated

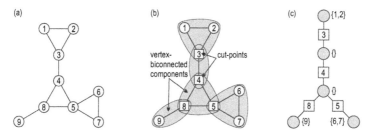

Fig. 1. The derivation of a block-cut tree: (a) given graph G_0, (b) identified blocks (shaded areas) and cut-points (square nodes), and (c) the block-cut tree.

with all nodes of the represented block in G_0 that are no cut-points. If the represented block consists of cut-points only, the block-node is not associated with any node from V_0.

A cut-node and a block-node are connected by an edge in E_T iff the corresponding cut-point is part of the block in G_0. Thus, cut-nodes and block-nodes always alternate on any path in T. The resulting structure is always a tree, since otherwise, the nodes forming a cycle can be shrinked into a single, larger block. Figure 1 illustrates the derivation of the block-cut tree.

After T has been derived from G_0, all potential augmentation edges in E_a are superimposed on T forming a new edge-set E_A: For each edge $(u, v) \in E_a$, a corresponding edge (u', v') is created with $u', v' \in V_T$ being the nodes that are associated with u, respectively v. The mapping from E_A to E_a is stored in order to be finally able to derive the original edges of an identified solution. Note that $G_A = (V_T, E_T \cup E_A)$ may be a multi-graph containing self-loops and multiple edges between two nodes; however, the reductions described in Sect. 3.3 will make this graph simple.

3.2 When Is a Cut-Point Covered?

A block-cut tree's edge $e \in E_T$ is said to be covered by an augmentation edge $e_A = (u, v) \in E_A$ iff e is part of the unique path in T connecting u with v. In order to cover a cut-node $v_c \in V_T$ completely, all its incident edges need to be covered, but this is in general not a sufficient condition.

If v_c would be removed from T, the tree will fall into k disconnected components $C_1^{v_c}, \ldots, C_k^{v_c}$, where k is the degree of v_c; we call them cut-components of v_c. We say an augmentation edge $e_A = (u, v) \in E_A$ contributes in covering the cut-node v_c, iff two edges incident to v_c are covered by e_A. Such an augmentation edge is obviously not incident to v_c and unites two cut-components $C_i^{v_c}$ and $C_j^{v_c}$. To cover v_c completely, exactly $k - 1$ augmentation edges are needed, and they must unite all components $C_1^{v_c}, \ldots, C_k^{v_c}$ into one.

For any cut-node v_c, let $A(v_c) \subseteq E_A$ be the set of all augmentation edges that contribute in covering v_c by uniting two of its cut-components. Furthermore, for each $e_A \in E_A$, let $R(e_A)$ be the set of all cut-nodes to whose covering e_A

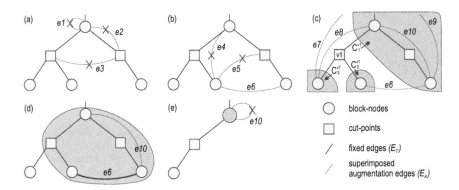

Fig. 2. Examples for preprocessing: (a) Edges that do not contribute in covering a cut-point are removed. (b) When $cost(e6) \leq cost(e4)$ and $cost(e6) \leq cost(e5)$, $e4$ and $e5$ are discarded. (c) As $e6$ is the only edge that connects C_2^{v1} to any other cut-component of $v1$, it is fixed; (d) the cycle caused by fixing $e6$ is shrinked into a new block; (e) $e10$ becomes a self-loop and is finally also discarded.

contributes. The proposed memetic algorithm explicitly computes and stores all sets $A(v_c)$ for all cut-nodes and the sets $R(e_A)$ for all augmentation edges as supporting data structures for its main part.

Later, we need to check efficiently if a certain cut-node is covered by a subset of augmentation edges $S \subset E_A$. This check is in general performed in $O(|S|)$ time with the aid of a union-find data structure. However, in most cases the degree of the cut-node is less than four, and then it is sufficient just to check whether each cut-component of the considered cut-node is connected to any other cut-component.

3.3 Reducing the Search Space

From E_A we discard all edges that do not contribute in covering any cut-node ($R(e_A) = \{\}$). In particular, edges forming self-loops or edges connecting a cut-node with an adjacent block-node or with another cut-node adjacent to the same block-node are removed in this way; see Fig. 2(a). Furthermore, from multiple edges connecting the same nodes in T, only one with minimum weight is retained. In this way, $G_A = (V_T, E_T \cup E_A)$ becomes a simple graph.

In addition to these simple reductions, we apply the following more sophisticated steps which are partly adopted from [9].

Edge Elimination: If there are two edges $e_A, e'_A \in E_A$, $cost(e_A) \leq cost(e'_A)$, and e_A covers all those edges which are covered by e'_A (in addition to others), e'_A is obsolete and can be discarded. All such edges can be identified in $O(|V|^2)$ time as a byproduct from a dynamic programming algorithm that computes distance values needed for the algorithm from Frederickson and Jájá [2]; see Fig. 2(b).

Fixing of Edges: An edge $e_A \in E_A$ must be included in any feasible solution to the V2AUG problem, when it represents the only possibility to connect a cut-node's cut-component $C_i^{v_c}$ to any other cut-component of v_c. In more detail, we process for each cut-node v_c its set $A(v_c)$ and look for such edges, which are then fixed by moving them from E_A to E_T; see Fig. 2(c). The corresponding original augmentation edges from E_a are permanently marked to be included in any future solution.

Shrinking: By fixing an edge, a cycle is introduced in T. This cycle forms a new vertex-biconnected component that can be shrinked into a single block-node, see Fig. 2(d) and 2(e). After shrinking all cycles, all changes in T are reflected to the supporting data structures. Owing to these changes, more edges may become available for elimination. Therefore, all reduction steps are repeated until no further shrinking is possible.

4   The Memetic Algorithm

The main part of the new approach is a steady-state evolutionary algorithm, in which in each iteration one new candidate solution is always created by selecting two parents in k-ary tournaments with replacement, recombining them and applying mutation. Such a solution replaces the worst solution in the population with one exception: To maintain a minimum diversity, a new candidate that resembles a solution already contained in the population is discarded.

A solution is represented by directly storing the set of its augmentation edges $S \subseteq E_A$ in the form of a hash-table. In this way, only $O(|S|) = O(|V|)$ space is needed, since $|S| < |V|$ in any locally optimal solution, in general becoming $|S| \ll |V|$ with increasing number of nodes, and an edge can be added, deleted or checked for existence in constant time.

Local Improvement: For the creation of initial solutions and the variation operators that derive new solutions, the following local improvement method plays a central role. From a feasible solution S, it removes redundant edges until the solution becomes locally optimal in the sense that no further edge can be removed without including others or making the solution infeasible.

For the cut-components of each cut-node, it is first determined how often each of them is connected to any other by the edges in S. Edges that provide the only connection for any cut-component must always be included in a feasible solution and are therefore not redundant.

The remaining edges in S are then processed one-by-one in decreasing cost-order. All cut-nodes in whose coverage a certain augmentation edge $e \in S$ participates ($R(e)$) are checked if they remain covered when e is removed, see Sect. 3.2. If this is the case, this edge is actually redundant and removed from S.

In the worst case, the computational effort of this local improvement may be $O(|V|^3)$, however, it is much lower on average.

Initialization: A member of the initial population is created by randomly selecting edges from E_A without replacement and including each in the initially empty edge-set S if it is not redundant. This process stops when all cut-nodes are covered.

The selection of edges for inclusion is biased toward cheaper edges by sorting E_A according to costs, and choosing an edge via the following random-rank:

$$rank = \lfloor |\mathcal{N}(0, s)| \cdot |V| \rfloor \bmod |E_A|, \tag{2}$$

$\mathcal{N}(0, s)$ is a normally distributed random variable with zero mean and standard deviation s, a strategy parameter that determines the strength of biasing.

Recombination: This operator was designed with the aim to provide highest possible heritability. First, edges common in both parents S_1 and S_2 are always adopted: $S = S_1 \cap S_2$. Then, while not all cut-nodes are covered, an edge is randomly selected from the set of remaining parental edges $((S_1 \cup S_2) \setminus (S_2 \cap S_1))$ and included in the offspring S if it provides a new cover. To emphasize the inclusion of low-cost edges, they are selected via binary tournaments with replacement. As final step, local improvement is called.

Mutation: Having created a new offspring via recombination, mutation is applied with a certain probability in order to introduce new edges that were not available in the parents. From S, one edge is selected randomly and removed. This makes one or more cut-nodes uncovered. These cut-nodes are identified and newly covered in random order: For each cut-node v_c, the edges from $A(v_c)$ that would actually help in covering v_c anew are determined. From this set, edges are repeatedly chosen at random and included in S until v_c is completely covered. Finally, local improvement is applied again.

The selection of the edge to be removed is biased toward more expensive edges: A pair of edges is drawn at random and a binary tournament selection with replacement decides among them.

5   Empirical Results

To compare the presented approach with other algorithms we have used test instances of different size and structure. Since shrinking can always reduce the problem of augmenting a general connected graph G_0 to the problem of augmenting a tree, G_0 is always a spanning tree in these test instances. Table 1 shows the number of nodes, the number of augmentation edges, and the number of cut-points (CP) before and after applying the memetic algorithm's preprocessing.

The first eight instances have been created with Zhu's generator [10] and were already used in [6,7,9]. The remaining ones are derived from Euclidean instances of Reinelt's TSP-library[1] in the following way: G is the graph containing all nodes of the TSP-instance and edges for each node to its nearest k neighbors, where k is the number shown in parentheses in Table 1; $k = \infty$ represents the complete graph. Edge costs are always the Euclidean distances rounded to nearest integer values. From G, a minimum spanning tree is derived and fixed as G_0.

---

[1] www.iwr.uni-heidelberg.de/groups/comopt/software/TSPLIB95

Table 1. Problem instances and results of the memetic algorithm's preprocessing.

instance	$|V|$	$|E_a|$	$cost(e) \in$	$CP(G_0)$	$|V_T|$	$|E_A|$	$CP(T)$
B1	60	55	$\{1, 2, ..., 1770\}$	34	1	0	0
C3	100	149	$\{1, 2, ..., 4950\}$	53	59	66	33
M1	70	290	$\{0, 10, ..., 1000\}$	37	70	227	37
M2	80	327	$\{0, 10, ..., 1000\}$	41	80	242	41
M3	90	349	$\{0, 10, ..., 1000\}$	42	90	262	42
N1	100	1104	$\{10, 11, ..., 50\}$	50	100	687	50
N2	110	1161	$\{10, 11, ..., 50\}$	56	110	734	56
R1	200	9715	$\{1, 2, ..., 100\}$	115	200	3995	115
a280 (50)	280	7654	TSP-lib Eucl.	218	280	1561	218
a280 (100)	280	15769	TSP-lib Eucl.	220	280	3322	220
a280 (∞)	280	38781	TSP-lib Eucl.	217	280	10182	217
rd400 (100)	400	22610	TSP-lib Eucl.	306	400	4959	306
rd400 (200)	400	46444	TSP-lib Eucl.	306	400	9368	306
pr439 (100)	439	26865	TSP-lib Eucl.	364	439	6095	364
pr439 (200)	439	55111	TSP-lib Eucl.	362	439	11890	362
pr439 (∞)	439	95703	TSP-lib Eucl.	362	439	19574	362
rat575 (∞)	575	164451	TSP-lib Eucl.	436	575	34514	436
pcb1173 (30)	1173	18321	TSP-lib Eucl.	953	1173	4492	953
d2103 (50)	2103	55630	TSP-lib Eucl.	1963	2103	6657	1963

Table 2. Results of the heuristics from Khuller and Thurimella (KT), Zhu et al. (ZKR), the genetic algorithm from Ljubić and Kratica (LK), and the memetic algorithm (MA).

		KT	ZKR	LK		MA				
instance	C^*	gap	gap	gap	evals	gap	σ	t [s]	evals	SR [%]
B1	15512.0*	8.6	0.0	0.0	900	0.0	0.0	< 1	1	100.0
C3	59129.0*	10.4	1.1	0.0	7300	0.0	0.0	< 1	906	100.0
M1	2940.0*	8.2	0.0	0.0	2700	0.0	0.0	2	2509	100.0
M2	4600.0*	5.9	0.0	0.0	8700	0.0	0.0	2	1667	100.0
M3	4980.0*	6.2	0.4	0.0	8100	0.0	0.0	4	2387	100.0
N1	390.0*	31.5	4.1	4.9	27800	0.0	0.0	10	4671	100.0
N2	429.0*	39.2	4.0	2.3	90000	0.0	0.0	12	7777	100.0
R1	121.4*	16.8	1.9	–	–	0.1	0.2	169	27738	36.7
a280 (50)	474.0*	25.1	–	–	–	0.0	0.1	15	8831	80.0
a280 (100)	473.0*	29.4	–	–	–	0.1	0.4	27	13568	93.3
a280 (∞)	489.0*	21.9	–	–	–	0.9	0.6	77	16445	3.3
rd400 (100)	4001.0	28.9	–	–	–	0.2	0.2	138	36694	20.0
rd400 (200)	3990.0	28.6	–	–	–	0.6	0.5	263	43521	3.3
pr439 (100)	27907.0*	20.5	–	–	–	0.7	0.5	181	29286	3.3
pr439 (200)	28289.0	19.1	–	–	–	1.1	0.5	296	43157	6.7
pr439 (∞)	27810.0	20.5	–	–	–	2.8	1.1	470	42512	3.3
rat575 (∞)	1546.0	33.4	–	–	–	1.3	0.6	1259	28668	3.3
pcb1173 (30)	11464.0	28.1	–	–	–	0.2	0.1	3202	49737	13.3
d2103 (50)	7333.0	9.6	–	–	–	0.0	0.0	14388	17974	63.3

Results of preprocessing document that a fixing of edges is only possible in shallower graphs like B1 and C3, where the number of cut-points could be dramatically reduced. In case of dense graphs, edge-elimination was highly effective. On average, the number of augmentation edges could be reduced to about a quarter.

The following setup was used for the memetic algorithm as it proved to be robust in preliminary tests. Population size: 800; group size for tournament selection: 5; parameter s for biasing initialization toward cheaper edges: 2.5; mutation probability 0.7. Each run was terminated when no new best solution could be identified during the last $10,000$ iterations.

We compare the memetic algorithm, called MA, to the heuristics from Khuller and Thurimella [5] (KT), Zhu et al. [10] (ZKR), and the hybrid genetic algorithm from Ljubić and Kratica [6] (LK). For smaller instances, we were able to derive optimum solution values by a not yet published branch-and-cut approach. Table 2 shows in column C^* these optimum values marked by '*' or otherwise best-known solution values. For the heuristic approaches, the qualities of final solutions are reported as percentage gaps with respect to C^*: $gap = (cost(S) - C^*)/C^* \cdot 100\%$.

KT was run once for each leaf-node becoming the root of branching, and the best obtained gaps are shown. ZKR could only by applied to smaller instances owing to its high computational effort. The same is true for LK, for which the shown gaps are adopted from [6]; they represent best values obtained from 10 runs per instance. MA's gaps are averaged over 30 runs for all instances, and σ shows the gaps' standard deviations. t gives the CPU-times on a PentiumIII/800MHz PC and *evals* the number of evaluated solutions until the finally best solutions had been identified. The success rate SR is the percentage of MA's runs that yielded optimum or best-known solutions.

It can be seen that KT performed generally worst. For the smaller instances, where results of ZKR and LK are available, MA found optimum solutions in any run. Furthermore, MA scaled well to larger instances. In all our test cases, it could identify solutions with gaps less than 3% with high reliability, as shown by the small deviations. The running times and needed numbers of evaluations increase only moderately with the problem size. Figure 3 shows two exemplary solutions.

6   Conclusions

The main features of the proposed memetic algorithm for the vertex-biconnectivity augmentation problem are: The effective deterministic preprocessing which reduces the search space in many cases dramatically, the local improvement procedure which guarantees local optimality of any created candidate solution, and the strong heritability and locality of the proposed recombination, respectively mutation. Furthermore, the local cost-based heuristics in the edge-selections of initialization, recombination, and mutation play a significant role.

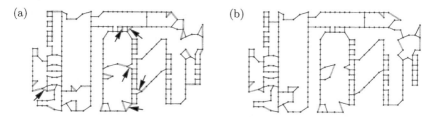

Fig. 3. Solutions to the Euclidean problem instance a280 (100) found by (a) Khuller and Thurimella's heuristic (gap: 29.4%, $|S| = 53$) and (b) the memetic algorithm (optimal, $|S| = 40$). The solutions' augmentation edges are shown in gray. In (a), the arrows mark obviously redundant edges.

Empirical tests indicate that the algorithm calculates solutions of high quality which are optimal in many cases. In particular the approach scales well to large problem instances owing to its relatively low computational effort for the creation and local improvement of one candidate solution.

References

1. K. P. Eswaran and R. E. Tarjan. Augmentation problems. *SIAM Journal on Computing*, 5(4):653–665, 1976.
2. G. N. Frederickson and J. Jájá. Approximation algorithms for several graph augmentation problems. *SIAM Journal on Computing*, 10(2):270–283, 1981.
3. H. N. Gabow, Z. Galil, T. Spencer, and R. E. Tarjan. Efficient algorithms for finding minimum spanning trees in undirected and directed graphs. *Combinatorica*, 6(2):109–122, 1986.
4. T.-S. Hsu and V. Ramachandran. On finding a minimum augmentation to biconnect a graph. *SIAM Journal on Computing*, pages 889–912, 1993.
5. S. Khuller and R. Thurimella. Approximation algorithms for graph augmentation. *Journal of Algorithms*, 14(2):214–225, 1993.
6. I. Ljubić and J. Kratica. A genetic algorithm for the biconnectivity augmentation problem. In C. Fonseca, J.-H. Kim, and A. Smith, editors, *Proceedings of the 2000 IEEE Congress on Evolutionary Computation*, pages 89–96. IEEE Press, 2000.
7. I. Ljubić and G. R. Raidl. An evolutionary algorithm with hill-climbing for the edge-biconnectivity augmentation problem. In E. J. Boers, S. Cagnoni, J. Gottlieb, E. Hart, P. L. Lanzi, G. R. Raidl, R. E. Smith, and H. Tijink, editors, *Applications of Evolutionary Computation*, volume 2037 of *LNCS*, pages 20–29. Springer, 2001.
8. P. Moscato. Memetic algorithms: A short introduction. In D. Corne et al., editors, *New Ideas in Optimization*, pages 219–234. McGraw Hill, 1999.
9. G. R. Raidl and I. Ljubić. Evolutionary local search for the edge-biconnectivity augmentation problem. *to appear in Information Processing Letters*, 2001.
10. A. Zhu, S. Khuller, and B. Raghavachari. A uniform framework for approximating weighted connectivity problems. In *Proceedings of the 10th ACM-SIAM Symposium on Discrete Algorithms*, pages 937–938, 1999.

Genetic, Iterated and Multistart Local Search for the Maximum Clique Problem

Elena Marchiori

Free University Amsterdam, Department of Computer Science
De Boelelaan 1081a, 1081 HV Amsterdam, The Netherlands
elena@cs.vu.nl

Abstract. This paper compares experimentally three heuristic algorithms for the maximum clique problem obtained as instances of an evolutionary algorithm scheme. The algorithms use three popular heuristic methods for combinatorial optimization problems, known as genetic, iterated and multistart local search, respectively.

1 Introduction

A clique of an undirected graph is a subgraph in which all pairs of distinct nodes are connected by an edge. A maximum clique of a graph is a clique with maximum number of nodes. Computing the maximum clique of a graph is a paradigmatic combinatorial optimization problem which is encountered in many different real life applications, such as cluster analysis, information retrieval, mobile networks, and computer vision (see, e.g., the survey in [1]). MC is highly intractable: it is one of the first problems which has been proven to be NP-hard [2]. Moreover, even its approximations within a constant factor are NP-hard [3,4]. Due to these strong negative results on the computational complexity of MC, many researchers have concentrated their effort on designing efficient heuristics yielding sub-optimal solutions of satisfactory quality (e.g., [5,1,6]).

Genetic algorithms have been applied with success to various hard combinatorial optimization problems. On the MC problem pure genetic algorithms have poor performance when compared with other local search techniques [7,8]. It is not yet clear which graph properties can be used to measure the hardness of this problem for GAs. Experimental investigations [9,10] indicate that neither graph density, size, relative maximum clique size, relative number of cliques, nor epistasis variance [11] are appropriate hardness measures.

The situation improves when problem knowledge is incorporated in GAs in the form of ad-hoc genetic operators and/or local optimization techniques [12,13,14,15,16]. In particular, in [17] we have introduced a hybrid GA for the maximum clique yielding results competitive with those obtained by state-of-the-art heuristic algorithms. The hybrid GA combines a simple GA with a local search procedure which generates maximal cliques.

In this paper we compare experimentally (a slightly different version of) this hybrid GA with an iterated local search algorithm and a multistart local

S. Cagnoni et al. (Eds.): EvoWorkshops 2002, LNCS 2279, pp. 112–121, 2002.
© Springer-Verlag Berlin Heidelberg 2002

search algorithm, which are obtained by choosing ad-hoc parameter settings of the hybrid GA (population size, termination condition, mutation and crossover rate). In this way we obtain three heuristic algorithms based on genetic local search, iterated local search, and multistart local search, respectively, which use an equal local search procedure as core of the optimization process, and explore a similar number of search points. These similarities allows one to compare fairly the three approaches the algorithms are based on.

The genetic local search algorithm (called GENE) consists of the application of genetic operators to a population of local optima produced by a local search procedure. The process is iterated until either a solution is found or a maximal number of generations is reached. The final output is the best solution found in all iterations. The iterated local search algorithm (ITER) acts repeatedly on just one candidate solution and outputs the best solution found in all iterations. The multistart local search algorithm (MULT) applies the local search procedure to each element of a large set of candidate solutions and outputs the best solution contained in the set. This has not to be confused with the restart method used in local search algorithms where execution can be periodically restarted in case past events indicate the search could be stuck in an attraction basin of some local optima.

The effectiveness of these three algorithms is tested on standard benchmark instances for MC collected at the DIMACS Center. The genetic and iterated local search algorithms exhibit similar performance, with results comparable to those of the best heuristic algorithms tested at the DIMACS Implementation Challenge for Maximum Clique, Graph Coloring, and Satisfiability [6]. Instead, the results of the multistart algorithm are of inferior quality.

The following notation and terminology is used throughout the paper. A graph is denoted by G, its nodes by $m, n, \ldots$. Nodes of a graph of size N are supposed to be indexed with integers from 1 to N. A subgraph C_G of G is a *clique* if every two distinct nodes n, n' in C_G are connected with an edge. A *maximal clique* C_G of G is a clique which is not properly contained in any other clique of G. A *maximum clique* C_G of G is a clique of maximum size (i.e./ number of maximum number of nodes). Note that a maximum clique is maximal but not vice versa.

2   The Algorithms

We use an approach called genetic local search (see e.g., [18]), which amounts to the repeated application of genetic operators to selected individuals of a population of local optima. The scheme algorithm we use is called GLS and is summarized in pseudo code below, where $P(t)$ denotes the population P at iteration t, $|P(t)|$ its size, and LMC is a local search procedure for finding maximal cliques that will be described later on. At each iteration, a new population is generated from the actual one as follows. Two fit individuals called parents are selected, crossover and mutation are applied to produce two offsprings which are then optimized by applying LMC to each of them. The best two individuals amongst parents and offsprings are selected and added to the new population (*keep-two-best* replacement mechanism, see [19]). This process is repeated until the new population reaches the size of the actual one.

In our implementation a chromosome represents a subgraph S_G by means of a bit string x of length N, where $x_i = 1$ (resp. $x_i = 0$) means that node i is (resp. is not) in S_G. Set-theoretic operators on graphs are translated to operators on binary strings in the expected way.

We employ a generational model with fitness proportional selection rule (roulette-wheel, cf. [20]) and elitism [21], where the two best individuals of a population are copied to the population of the next generation.

The fitness function $f : Chrm \rightarrow [0, N]$ is defined by $f(x) = |Nodes(x)|$ (the number of nodes of x) if x is a clique; $f(x) = 0$ otherwise.

Finally, we use classical blind genetic operators: uniform crossover [22], and swap mutation which swaps the values of two randomly selected genes. Note these operators describe meaningful operations on graphs, where crossover corresponds to a merge of graphs and mutation corresponds to exchange of nodes between graphs.

```
PROCEDURE GLS
   BEGIN
    t := 0;
    initialize P(t);
   apply LMC to each element of P(t);
    evaluate P(t);
    WHILE (NOT termination-condition) DO
     BEGIN
      t := t+1;
      WHILE (|P(t)| < |P(t-1)|) DO
       BEGIN
        select parents from P(t-1);
        recombine parents;
        mutate children;
        apply LMC to each of the children;
        evaluate children;
        insert in P(t) best two of parents and children;
       END
     END
   END
```

The local search algorithm LMC used in GLS is described below.

```
PROCEDURE LMC
BEGIN
  S_G := Perturb(S_G);
  S_G := Repair(S_G);
  S_G := Extend(S_G);
END
```

This procedure transforms a subgraph into a maximal clique: first, the subgraph is perturbed by deleting and adding some nodes randomly selected (Perturb); next, it is reduced to a clique (Repair); finally it is extended to a maximal clique (Extend) using a sequential greedy heuristic. The three steps are described in detail below.

Perturb(S_G):

1. For every node n, $1 \leq n \leq N/2$, if n is in S_G then with small probability (typical value 0.1) remove n from S_G. (In this step nodes are supposed to be sorted in increasing order with respect to their degree. The upper bound $N/2$ has been chosen empirically).

2. Add the sequence $s, s + 1, \ldots, s + e$ of nodes to S_G, with s and e randomly chosen, where $3 \leq e \leq BK/2$ and s in $[1, N - e]$ (BK is the size of the largest known clique G. The upper bound $BK/2$ has been chosen empirically).

Repair(S_G): Let V be the set of nodes of S_G. Repeat the following steps until V becomes empty.

1. Choose randomly a node n in V.

2. With low probability (typical value 0.01) remove n from S_G; otherwise delete from S_G and from V each node of S_G that is not connected with n, except n itself.

3. Remove n from V.

Extend(S_G): Let V be the set of nodes of $G \setminus S_G$. Repeat the following steps until V becomes empty.

1. Choose a random node n in V:

2. if $\{n\} \cup S_G$ is a clique then add n to S_G.

3. remove n from V.

We consider the following three instances of GLS, obtained by setting the parameters crossover rate, mutation rate, population size, and termination criterion to specific values. The parameter settings are such that all algorithms process approximately the same number (equal to 20,000) of individuals.

- Genetic algorithm (GENE). It is obtained from GLS by setting population size to 10, mutation rate to 0.1, crossover rate to 0.9 and termination-condition to 2,000 generations. This algorithm performs genetic local search.
- Iterated local search (ITER). It is obtained from GLS by setting population size to 1, mutation rate to 0 (hence no genetic operators are used), and termination-condition to 20,000 generations. This algorithm iterates local search (using LMC) starting from one random point in the search space.
- Multistart local search (MULT). This is obtained from GLS by setting population size to 20,000 and termination-condition to 0 generations (hence no genetic operators are used). This algorithm performs a local search (using LMC) from several points randomly distributed over the entire search space.

3   Experimental Comparison

In order to test and compare the three algorithms we consider the benchmark instances employed in the International Implementation Challenge on Maximum Clique, Graph Coloring, and Satisfiability organized by the Center for Discrete

Mathematics and Theoretical Computer Science (DIMACS) [6]. The algorithms described in [6] are based on various approaches, like tabu search, simulated annealing, and neural networks. These instances are available from the DIMACS archive using ftp to `dimacs.rutgers.edu` directory `pub/challenge` or at URL `http://dimacs.rutgers.edu/` `Challenge/`. The 37 instances here considered include random graphs (`Cx.y` and `DSJCx.y` of size x and density 0.y), Steiner Triple Graphs (`MANNx` with up to 3321 nodes and 5506380 edges), Brockington Graphs (`brockx_2` and `brockx_4` of size x), Sanchis graphs (`genx_p0.9_z` of size x), Hamming graphs (`hamming8-4` and `hamming10-4`), Keller graphs (`keller4`, `keller5`, `keller6` with up to 3361 nodes and 4619898 edges), and P-hat graphs (`p_hatx-z` of size x).

GLS has been implemented in C++ and run on a Sun Ultra 250, UltraSPARC-II 400MHz. The initial population is generated randomly, where each gene of a chromosome is set to 1 with probability 0.2, otherwise it is set to 0. These and the other GA parameters have been set to values experimentally determined after a small number of trials. `MULT`, `ITER` and `GENE` are run 10 times on each graph instance using different random seeds. The results of the experiments are summarized in Table 1 which reports the best, average and standard deviation of clique size, and the best result obtained by the fifteen heuristic algorithms for MC presented at the DIMACS Challenge (column labeled DIMACS best). Table 2 which contains average and standard deviation of time to find the best solution (in seconds).

The multistart variant `MULT` has worst performance: `MULT` outperforms `ITER` and `GENE` only on one instance (`brock200_4`), while on all other instances it yields in general results of poor quality. `ITER` and `GENE` find maximal cliques of comparable quality: `ITER` outperforms `GENE` on five instances while `GENE` outperforms `ITER` on four instances (entries in bold style). The variance of the results obtained by `ITER` and `GENE` is also comparable. On 23 of the 37 instances `ITER` is able to find the best value found by the DIMACS algorithms.

The experiments on the DIMACS snapshot indicate that `ITER` and `GENE` are the most effective of the three variants. It seems that most of the work in guiding the search towards promising regions is performed by the local search procedure `LMC`, while the genetic operators are useful for escaping from local optima, due to their disruptive effect (e.g. on random graphs `C500.9`, `C1000.9` and `C2000.9`).

The best heuristic algorithm for the MC problem we are aware of is based on reactive local search (RLS) [5]. This algorithm is roughly ten times faster than `ITER`, and yields the best known results in almost all the DIMACS benchmark instances, with very small standard deviations. RLS employs a sophisticated search strategy that exploits information about past events in two ways: memorization of past events as done in tabu search, and a reactive strategy which regulates the search diversification. Moreover an explicit memory-influenced restart is activated periodically. In contrast, GLS is a Markov (i.e., memory-less) process, where the next generation depends only on the current one.

3.1   Comparison with Hybrid Genetic Algorithms

We are aware of two hybrid genetic algorithms for MC which have being tested on some benchmarks from the DIMACS repository.

The algorithm by Bui and Eppley [12], called GMCA, is a hybrid GA with local optimization for improving the chromosomes. Moreover pre-processing is used for reordering the nodes of the graph in such a way that nodes which are likely to belong to a clique (e.g. with high degree) occur near each other. The fitness function is the weighted sum of density and size of the subgraph represented by the chromosome.

In [16] Sakamoto et al. introduce a GA for the maximum independent set problem, here called SLS. An independent set of a graph G is a clique of the complement of G (it has the nodes of G but only those edges which are not in G). The maximum clique and the maximum independent problem are equivalent since a maximum clique of G is a maximum independent set of the complement of G. In SLS a chromosome is a permutation of the nodes of the graph. A greedy decoding method is used which constructs a independent set using the nodes in the order in which they appear in the chromosome. The chromosome is then replaced with the sequence of nodes of the independent set followed by the remaining nodes of the graph. The fitness function is the size of the independent set minus the minimum of the sizes of the independent sets in the actual population.

In Table 3 GMCA, SLS, and GENE are compared by reporting the results contained in [12] and [16] concerning experiments on a set of DIMACS instances. Both GMCA and SLS use a population of size 50 and are run for 50 generations. In order to compare running times - GMCA and SLS are run on a Sun SPARC LX - we converted the running time of GENE to Sun SPARC LX time using the Dhrystone score. It turns out that our computer is about 17 times faster than a Sun SPARC LX.

On 12 instances (in bold style) GENE outperforms GMCA and SLS; on the other instances SLS and GENE yield equal best results, and they outperform GMCA on 10 instances. Concerning the running time, GENE is faster than GMCA and SLS, except on the MANN* instances: in particular on MANN_a45 and MANN_a81 GENE is about ten times slower than SLS, but is able to find the best known result.

The results also indicate that instances from the Cfat, Johnson and Hamming classes can be regarded as GA easy.

References

1. I.M. Bomze, M. Budinich, P.M. Pardalos, and M. Pelillo. The maximum clique problem. *Handbook of Combinatorial Optimization*, 4, 1999.
2. R.M. Karp. Reducibility among combinatorial problems. In *Complexity of Computer Computations*, pages 85–103. Plenum Press, NY, 1972.
3. U. Feige, S. Goldwasser, S. Safra, L. Lovász, and M. Szegedy. Approximating clique is almost NP-complete. In *Proc. 32nd Annual IEEE Symposium on the Foundations of Computer Science (FOCS)*, pages 2–12, 1991.

4. J. Hastad. Clique is hard to approximate within $n^{1-\epsilon}$. In *Proc. 37th Annual IEEE Symposium on the Foundations of Computer Science (FOCS)*, pages 627–636, 1996.
5. R. Battiti and M. Protasi. Reactive local search for the maximum clique problem. *Algorithmica*, 29(4):610–637, 2001.
6. D. Johnson and M. Trick (Eds.). *Cliques, Coloring, and Satisfiability*. AMS, DIMACS Series in Discrete Mathematics and Theoretical Computer Science, vol 26, 1996.
7. B. Carter and K. Park. How good are genetic algorithms at finding large cliques: an experimental study. Technical report, Boston University, Computer Science Department, MA, October 1993.
8. K. Park and B. Carter. On the effectiveness of genetic search in combinatorial optimization. In *Proceedings of the 10th ACM Symposium on Applied Computing*. ACM Press, 1995.
9. T. Haynes. Clique detection as a royal road function. In *Genetic Programming*, 1998.
10. T. Soule and J.A. Foster. Genetic algorithm hardness measures applied to the maximum clique problem. In T. Bäck, editor, *Seventh International Conference on Genetic Algorithms*, pages 81–88. Morgan Kaufmann, 1997.
11. Y. Davidor. Epistasis variance: a viewpoint on GA-hardness. In G.J.E. Rawlins, editor, *Foundations of Genetic Algorithms*, pages 23–35. Morgan Kaufmann, 1991.
12. T.N. Bui and P.H. Eppley. A hybrid genetic algorithm for the maximum clique problem. In L.J. Eshelman, editor, *Proceedings of the 6th International Conference on Genetic Algorithms (ICGA)*, pages 478–484. Morgan Kaufmann, 1995.
13. C. Fleurent and J.A. Ferland. Object-Oriented imlementation of heuristic search methods for graph coloring, maximum clique, and satisfiability. In D. Johnson and M. Trick, editors, *Cliques, Coloring and Satisfiability*. AMS, DIMACS Series in Discrete Mathematics and Theoretical Computer Science, vol 26, 1996.
14. J.A. Foster and T. Soule. Using genetic algorithms to find maximum cliques. Technical report, Dept. of Computer Science, Univ. Idaho, 12 1995.
15. A.S. Murthy, G. Parthasarathy, and V.U.K. Sastry. Clique finding - a genetic approach. In *Proceedings of the 1st IEEE Conference on Evolutionary Computation*, pages 18–21. IEEE Press, 1994.
16. A. Sakamoto, X. Liu, and T. Shimamoto. A genetic approach for maximum independent set problems. *IEICE Trans. Fundamentals*, E80-A(3):551–556, 1997.
17. E. Marchiori. A simple heuristic based genetic algorithm for the maximum clique problem. In J. Carroll et al., editor, *ACM Symposium on Applied Computing*, pages 366–373. ACM Press, 1998.
18. P. Merz and B. Freisleben. Genetic local search for the TSP: New results. In *IEEE International Conference on Evolutionary Computation*, pages 159–164. IEEE Press, 1997.
19. D. Thierens and D. Goldberg. Elitist recombination: an integrated selection recombination GA. In *Proceedings of the 1st IEEE Conference on Evolutionary Computation*, pages 508–518. IEEE Press, 1994.
20. Z. Michalewicz. *Genetic Algorithms + Data Structures = Evolution Programs*. Springer-Verlag, Berlin, 1994.
21. K.A. De Jong. An analysis of the behaviour of a class of genetic adaptive systems. Doctoral Dissertation, University of Michigan, Dissertation Abstract International 36(10), 5140B, 1975.
22. G. Syswerda. Uniform crossover in genetic algorithms. In J. Schaffer, editor, *Third International Conference on Genetic Algorithms*, pages 2–9. Morgan Kaufmann, 1989.

Table 1. Results for DIMACS 'snapshot': clique size

Graph	MULT		GENE		ITER		DIMACS best
	Avg(Stdv)	Best	Avg(Stdv)	Best	Avg(Stdv)	Best	
C125.9	32.6(0.5)	33	33.8(0.4)	34	34(0.0)	34	34
C250.9	39.1(0.7)	40	42.8(0.7)	44	43.0(0.6)	44	44
C500.9	46.7(0.6)	48	52.2(1.6)	**56**	52.7(1.4)	55	57
C1000.9	53.5(1.2)	56	61.6(2.1)	**66**	61.6(1.6)	64	67
C2000.9	59.7(0.9)	62	68.2(2.4)	**72**	68.7(1.2)	70	75
DSJC500.5	12.0(0.0)	12	12.2(0.4)	13	12.1(0.3)	13	15
DSJC1000.5	13.1(0.3)	14	13.3(0.5)	14	13.5(0.5)	14	15
C2000.5	14.1(0.3)	15	14.2(0.4)	15	14.2(0.4)	15	16
C4000.5	15.2(0.4)	16	15.4(0.5)	16	15.6(0.5)	16	18
MANN_a27	124.8(0.4)	125	125.6(0.5)	126	126.0(0.0)	126	126
MANN_a45	339.7(0.5)	340	342.4(0.5)	343	343.1(0.8)	**345**	345
MANN_a81	1091.2(0.7)	1092	1096.3(0.6)	1097	1097.0(0.4)	**1098**	1098
brock200_2	12(0.0)	12	10.5(0.7)	12	10.5(0.8)	12	12
brock200_4	15.7(0.9)	**17**	15.4(0.5)	16	15.5(0.5)	16	17
brock400_2	21.7(0.6)	23	22.5(0.7)	24	23.2(0.7)	**25**	25
brock400_4	21.8(0.4)	22	23.6(0.8)	**25**	23.1(0.5)	24	24
brock800_2	18.0(0.6)	19	19.3(0.6)	20	19.1(0.8)	**21**	21
brock800_4	18.0(0.0)	18	18.9(0.5)	20	19.0(0.4)	20	21
gen200_P0.9_44	36.3(0.5)	37	39.7(1.6)	44	39.5(1.6)	44	44
gen200_P0.9_55	43.4(2.0)	46	50.8(6.4)	55	48.8(7.6)	55	55
gen400_P0.9_55	43.7(0.6)	45	49.7(1.2)	55	49.1(1.0)	51	55
gen400_P0.9_65	44.6(0.9)	47	53.7(7.4)	65	51.2(4.7)	65	65
gen400_P0.9_75	47.7(1.7)	52	60.2(12.1)	75	62.7(12.3)	75	75
hamming8-4	15.7(0.9)	16	16.0(0.0)	16	16.0(0.0)	16	16
hamming10-4	32.0(0.4)	33	37.7(1.9)	40	38.8(1.2)	40	40
keller4	11.0(0.0)	11	11.0(0.0)	11	11.0(0.0)	11	11
keller5	23.9(0.9)	25	26.0(0.8)	27	26.3(0.6)	27	27
keller6	45.3(1.1)	48	51.8(1.5)	55	52.7(1.8)	**56**	59
p_hat300-1	8.0(0.0)	8	8.0 (0.0)	8.0	8.0 (0.0)	8.0	8
p_hat300-2	22.9(0.9)	25	25(0.0)	25	25(0.0)	25	25
p_hat300-3	31.0(0.4)	32	34.6(0.9)	36	35.1(0.8)	36	36
p_hat700-1	9.1(0.3)	10	9.8(0.9)	11	9.9(0.7)	11	11
p_hat700-2	35.5(0.9)	37	43.5(0.8)	44	43.6(0.7)	44	44
p_hat700-3	49.5(1.4)	52	60.4(1.0)	62	61.8(0.6)	62	62
p_hat1500-1	10.2(0.4)	11	10.8(0.4)	11	10.4(0.5)	11	12
p_hat1500-2	46.9(1.0)	48	63.8(1.0)	65	63.9(2.0)	65	65
p_hat1500-3	64.3(1.3)	67	92.4(1.3)	94	93.0(0.8)	94	94

Table 2. Results for DIMACS 'snapshot': time till best (in seconds)

Graph	MULT	ITER	GENE
	Avg(Stdv)	Avg(Stdv)	Avg(Stdv)
C125.9	2.4(0.1)	0.5(0.6)	0.1(0.1)
C250.9	4.4(0.1)	2.4(2.0)	3.7(3.5)
C500.9	8.8(0.1)	2.7(2.6)	5.7(5.5)
C1000.9	18.2(0.1)	8.6(6.5)	12.4(13.4)
C2000.9	46.3(0.3)	24.8(23.8)	33.8(34.2)
DSJC500.5	6.6(0.1)	0.4(0.5)	2.1(4.9)
DSJC1000.5	13.4(0.1)	2.3(3.8)	2.2(1.9)
C2000.5	29.3(0.1)	2.3(2.0)	7.2(12)
C4000.5	63.2(0.3)	15.7(8.0)	13.5(14.0)
MANN_a27	14.4(0.1)	15.6(10.1)	3.7(4.0)
MANN_a45	92.7(0.4)	54.4(44.3)	135.2(106.4)
MANN_a81	1203.3(39.5)	693.9(922.5)	2773.8(1158.3)
brock200_2	2.7(0.0)	0.2(0.2)	1.3(1.8)
brock200_4	2.8(0.1)	0.5(0.6)	0.9(1.6)
brock400_2	5.7(0.1)	2.0(2.3)	1.9(3.3)
brock400_4	5.7(0.1)	1.3(1.4)	1.3(1.4)
brock800_2	11.0(0.1)	3.9(4.9)	5.2(7.4)
brock800_4	11.0(0.1)	4.1(3.4)	7.8(7.7)
gen200_P0.9_44	3.5(0.0)	1.7(1.5)	1.3(1.3)
gen200_P0.9_55	3.5(0.0)	1.3(1.5)	1.4(2.9)
gen400_P0.9_55	6.8(0.0)	2.8(3.0)	3.8(6.5)
gen400_P0.9_65	6.8(0.0)	2.7(3.3)	4.3(4.3)
gen400_P0.9_75	6.8(0.1)	4.6(3.2)	3.7(6.3)
hamming8-4	3.5(0.0)	0.0(0.0)	0.0(0.0)
hamming10-4	15.5(0.1)	5.3(4.4)	5.8(5.6)
keller4	2.4(0.0)	0.0(0.0)	0.0(0.0)
keller5	10.9(0.1)	4.0(3.9)	9.1(10.2)
keller6	69.2(0.1)	36.2(23.8)	60.2(55.0)
p_hat300-1	3.7(0.0)	0.9(0.8)	0.4(0.5)
p_hat300-2	4.1(0.0)	0.5(0.5)	0.5(0.6)
p_hat300-3	4.4(0.0)	1.5(1.7)	3.6(4.5)
p_hat700-1	8.7(0.1)	2.6(3.8)	5.8(6.8)
p_hat700-2	9.5(0.0)	1.2(1.2)	1.6(1.9)
p_hat700-3	10.5(0.0)	4.5(4.4)	5.6(6.3)
p_hat1500-1	19.5(0.1)	2.1(3.1)	14.2(14.7)
p_hat1500-2	21.6(0.1)	12.2(9.2)	9.1(7.8)
p_hat1500-3	24.5(0.0)	7.1(11.4)	14.6(18.2)

Table 3. Results for DIMACS benchmark graphs: GMCA, SLS, GENE

Graph	GMCA		SLS		GENE	
	Avg Time	Best	Avg Time	Best	Avg Time	Best
c-fat200-1	8.2	12	12.3	12	0.0	12
c-fat500-1	33.2	14	60.7	14	0.0	14
johnson16-2-4	6.0	8	4.5	8	0.0	8
johnson32-2-4	187.4	16	63.2	16	1.7	16
keller4	13.3	11	9.1	11	0.0	11
keller5	438.1	18	256.7	27	63.7	27
keller6	-	-	4798.6	50	1023.4	**55**
hamming10-2	886.6	512	351.3	512	37.2	512
hamming8-2	53.0	128	21.2	128	1.7	128
san200_0.7_1	51.7	30	16.6	30	10.2	30
san400_0.5_1	411.2	7	43.1	13	42.5	13
san400_0.9_1	128.6	50	70.5	100	47.6	100
sanr200_0.7	21.5	17	14.6	18	10.5	18
sanr400_0.5	69.6	12	45.8	13	42.5	13
san1000	704.3	8	242.8	10	15.3	10
brock200_1	27.9	20	14.9	20	15.3	20
brock200_2	-	-	12.1	10	22.1	**12**
brock200_4	-	-	14.0	16	15.3	16
brock400_1	118.8	20	50.9	23	44.2	**24**
brock400_2	-	-	52.5	24	32.3	24
brock400_4	-	-	56.8	23	22.1	**25**
brock800_1	460.8	18	172.1	18	27.2	**19**
brock800_2	-	-	188.8	19	66.3	**20**
brock800_4	-	-	171.8	20	69.7	20
p_hat300_1	20.0	8	24.1	8	6.8	8
p_hat300_2	-	-	30.4	25	8.5	25
p_hat300_3	-	-	36.1	35	61.2	**36**
p_hat500_1	49.1	9	62.0	9	3.5	9
p_hat700_1	310.1	8	117.7	11	105.4	11
p_hat700_2	-	-	165.0	44	27.2	44
p_hat700_3	-	-	192.7	61	95.2	**62**
p_hat1000_1	671.0	8	247.6	10	54.4	10
p_hat1500_1	1580.3	10	573.2	11	249.9	11
p_hat1500_2	-	-	952	64	154.7	**65**
p_hat1500_3	-	-	1079.5	92	248.2	**94**
MANN_a27	121.8	125	52.4	126	62.9	126
MANN_a45	916.9	337	339.5	341	2298.4	**345**
MANN_a81	-	-	6249.0	1094	47154.6	**1098**

An Experimental Investigation of Iterated Local Search for Coloring Graphs

Luis Paquete and Thomas Stützle

Darmstadt University of Technology, Computer Science Department, Intellectics Group
Alexanderstr. 10, 64283 Darmstadt

Abstract. Graph coloring is a well known problem from graph theory that, when attacking it with local search algorithms, is typically treated as a series of constraint satisfaction problems: for a given number of colors k one has to find a feasible coloring; once such a coloring is found, the number of colors is decreased and the local search starts again. Here we explore the application of Iterated Local Search on the graph coloring problem. Iterated Local Search is a simple and powerful metaheuristic that has shown very good results for a variety of optimization problems. In our research we investigated several perturbation schemes and present computational results on a widely used set of benchmarks problems, a sub-set of those available from the DIMACS benchmark suite. Our results suggest that Iterated Local Search is particularly promising on hard, structured graphs.

1 Introduction

The Graph Coloring Problem (GCP) is a well known combinatorial problem defined as follows: Given a directed graph $G = (V, E)$, where V is the set of $|V| = n$ vertices and $E \subseteq V \times V$ is the set of edges, and an integer k (number of colors), find a mapping $\Psi : V \mapsto 1, 2, ..., k$ such that for each $[u, v] \in E$ we have $\Psi(v) \neq \Psi(u)$. In fact, this problem statement corresponds to the decision version, where we try to find a solution, satisfying an additional constraint on the value of the objective function. In the optimization counterpart, the GCP can be defined as to find the minimum k, that is to find the chromatic number χ_G of G.

The GCP is an interesting problem for theory and practice. In fact, several classes of real-life problems such as examination timetabling [1] and frequency assignment [2] can be modelled as GCPs. Yet, it cannot be expected that an algorithm can find, in polynomial time, a solution to any GCP instance, because it is $\mathcal{NP}$-hard [3]. In fact, exact algorithms can solve only small size instances [4]. For larger instances approximate algorithms have to be used and a large number of such algorithms has been proposed [5,6,7,8,9,10].

This article explores the application of Iterated Local Search (ILS) [11] to the GCP. ILS consists in the iterative application of a local search procedure to starting solutions that are obtained by the previous local optimum through a solution perturbation. So far, ILS has only been applied to optimization problems. In this article we apply ILS to the decision variant of the GCP. This is no limitation, because the optimization variant can be stated as a sequence of constraint satisfaction problems, where k is being decremented sequentially by one until no admissible mapping exists, meaning that $\chi_G = k+1$. In this case, the problem is attacked as a constraint satisfaction problem. The good performance

S. Cagnoni et al. (Eds.): EvoWorkshops 2002, LNCS 2279, pp. 122–131, 2002.

of our ILS especially on structured graphs suggests that it may be worthwhile to also consider applications of ILS to other types of constraint satisfaction problems or to the well known satisfiability problem in propositional logic.

The article is structured as follows. Section 2 presents a review of approximation algorithms for the GCP and Section 3 introduces available benchmark sets. Section 4 introduces ILS and describes some details of the ILS implementation for the GCP. Section 5 presents the experimental results and we conclude in Section 6.

2   Approximate Algorithms for Graph Coloring

Approximate algorithms for the GCP fall into two main classes, construction heuristics and local search algorithms.

Construction heuristics start from an empty solution and successively augment a partial coloring until the full graph is colored. During the solution construction these algorithms maintain feasibility that is, they return a conflict free coloring. Well known construction heuristics are the Brelaz heuristic [5], the Recursive Largest First (RLF) heuristic [10] or iterated, randomized construction heuristics like the Iterated Greedy algorithm [12].

Local search for the GCP starts at some initial, inconsistent color assignment and iteratively moves to neighboring solutions, trying to reduce the number of conflicts. A pair of vertices (i, j) is in conflict, if both are assigned the same color and $(i, j) \in E$. In fact, local search applied to the GCP iteratively tries to *repair* the current color assignment guided by an evaluation function that counts the total number of conflicts. In case a candidate solution with zero conflicts is encountered, this candidate solution corresponds to a feasible coloring of the graph.

When treating the GCP as a decision problem, a commonly used neighborhood is the 1-opt neighborhood that in each step changes the color assignment of exactly one vertex. For searching the 1-opt neighborhood, two different local search architectures exist. The first is based on the *min-conflicts* heuristic (A1): In each local search step, a vertex that is in conflict is chosen at random. To this vertex a color is assigned that minimizes the number of conflicts [13]. The second scheme (A2) examines all pairs of vertices and colors (v_i, j) and performs the move with the maximal reduction of the number of conflicts; if several such moves exist, one is chosen randomly [14,6]. This latter neighborhood is often further reduced by considering moves that only affect vertices that are currently involved in a conflict (A2'). Note that A2 and A2' architectures are greedier than the min-conflicts architecture, because at each step one among a larger set of candidate moves is choosen.

In the simplest case, local search algorithms accept only improving moves and they terminate in local optima. The most successful technique to avoid this problem is a family of algorithm schemata that is often called metaheuristics. Among the first meta-heuristic approaches to the GCP were Simulated Annealing implementations. Simulated Annealing was first applied to the GCP by Chams et al. [15] and was intensively tested by Johnson et al. [9] on random graphs. Among the most widely applied metaheuristics to the GCP are Tabu Search implementations. The first implementations of Tabu Search were due to Hertz and de Werra [8]. More recently, Hao, Dorne, and Galinier

[14,7] presented the most performing Tabu Search implementations, based on the A2' local search. The solution of the GCP by Evolutionary Algorithms was proposed by Davis [16], who reported several crossover operators combined with several ordering of vertices. Eiben et al. [17] applied an Adaptive Evolutionary Algorithm to the GCP; this algorithm changes periodically the evaluation function to avoid local optima. More recently, Laguna and Martí [18] proposed an application of GRASP to the GCP and presented good results for sparse graphs. Finally, several hybrid approaches were proposed. These typically combine Evolutionary Algorithms with Tabu Search implementations. The first such approach for the GCP was proposed by Fleurent and Ferland [19].

The best computational results so far have been obtained by the Hybrid Evolutionary Algorithm (HEA) by Galinier and Hao [7]. In their algorithm, the initial population is generated by a greedy saturation algorithm. After initialization, two solutions are chosen randomly and a specific crossover operator generates a new solution based on the information of the latter. This operator builds a partial solution by exchanging subsets of the color class sets of the two chosen solutions and fills the unassigned vertices in a random fashion to obtain a complete solution. The new solution is then improved by applying the Tabu Search algorithm for a certain number of L iterations and then reinserted into the set of solutions. Galinier and Hao have reported very good performance and outperformed the Tabu Search, which previously was reported to be among the most effective local search algorithms for the GCP, on a number of hard benchmark instances.

3   Benchmark Problems

Many different benchmark problems and instance sets for the GCP are available at Joseph Culberson's Graph Coloring Page (http://www.cs.ualberta.ca/~joe/Coloring/) and Michael Trick's Graph Coloring Page (http://mat.gsia.cmu.edu/COLOR/color.html). The most prominent benchmark set is that of the Second DIMACS Implementation Challenge on Cliques, Coloring, and Satisfiability, which is available on the web site http://mat.gsia.cmu.edu/challenge.html. Many of these instances were generated in way such that the chromatic number is known, like the Leighton and the Flat graphs. The Leighton graphs were generated by a procedure that uses the number of vertices, the desired chromatic number, the average vertex degree and a random vector of integers to generate a certain number of cliques. The Flat graphs were generated in a way such that the set of vertices is partitioned into k almost equal sized sets. The number of edges is close to the expected number of edges given a certain probability p and partitioning. A flatness parameter controls the variation of the vertex degree. The DIMACS instances have a flatness parameter equal to 0, which means that the graph is uniform.

Another prominent class of benchmark instances available from the DIMACS site are random graphs, where each edge is included into the graph with a probability p (the chromatic number of these graphs is not known). These graphs were initially proposed by Johnson et al. in their experimental evaluation of Simulated Annealing [9] and were used since then in a large number of studies.

4   Iterated Local Search for Coloring Graphs

ILS [11] is based on the simple idea of improving a local search procedure by providing new starting solutions which are obtained from perturbations of the current solution. This perturbation must be sufficiently strong to allow the local search to explore different solutions, but also weak enough to prevent random restart. ILS is appealing both due to its simplicity and due to the very good results obtained, for example, the Traveling Salesman Problem [20], Graph-Partitioning [21], and Scheduling Problems [11].

To apply an ILS algorithm, four components have to be specified. These are a procedure GenerateInitialSolution() that generates an initial solution s_0, a procedure Perturbation, that modifies the current solution s leading to some intermediate solution s', a procedure LocalSearch that returns an improved solution s'', and an AcceptanceCriterion that decides to which solution the next perturbation is applied. An algorithmic scheme for ILS is given in Figure 1.

procedure *Iterated Local Search*
    s_0 = GenerateInitialSolution()
    s = LocalSearch(s_0)
    repeat
        s' = Perturbation($s, history$)
        s'' = LocalSearch(s')
        s = AcceptanceCriterion($s, s'', history$)
    until termination condition met
end

Fig. 1. Pseudocode of an iterated local search procedure (ILS)

In principle, any local search algorithm can be used, but the performance of the ILS algorithm with respect to solution quality and computation speed depends strongly on the one chosen. Very often an iterated descent algorithm is taken, but it is also possible to apply more sophisticated local search algorithms like Tabu Search, as done in our case.

The perturbation mechanism should be chosen *strong enough* to allow to leave the current local minimum and to allow the local search to explore different solutions. At the same time, the modification should be *weak enough* to keep enough characteristics of the current local minimum.

The procedure AcceptanceCriterion is used to decide from which solution the search is continued by applying the next pertubation. One important aspect of the acceptance criterion and the perturbation is to introduce a bias between intensification and diversification of the search. Intensification of the search around the best found solution is achieved, for example, by applying the perturbation always to the best found solution and using small perturbations. Diversification is achieved, in the extreme case, by accepting every new solution s'' and applying large perturbations.

4.1   Iterated Local Search Operators

In our application of ILS to the GCP we focused mainly on the choice of the local search and the perturbation operator.

Local Search. For the local search we considered to apply one based on local search architecture A1 (see Section 2) and one based on A2 and A2', respectively. In addition to a plain iterative improvement local search, we implemented Tabu Search versions for both local search architectures. In fact, after some initial experiments it was found that our ILS with the Tabu Search implementation based on A2' performed significantly better than the other variants. Yet, this improved performance is only possible because of the use of speed-up techniques for the neighborhood evaluation [19]: The implementation uses a two-dimensional table of size $n \cdot k$ where each entry $\gamma(i, j)$ stores the effect on the evaluation function incurred by changing the color of vertex i to color j. Each time a move is performed, only the part of the table that is affected by the move is updated. This table has to be initialised in $O(n^2 \cdot k)$, but each update of the matrix then only takes $O(n \cdot k)$ in the worst case (for sparse graphs even faster); in fact the complexity of each iteration is the same as for implementations of architecture A1. For the setting of the tabu list lenght we followed the scheme in [7]: the length of tabu list was taken as $Random(A) + \alpha \times |n_c|$, where n_c is the set of conflicting vertices.

Perturbation. For the perturbation we considered four different possibilities.

P1, **random moves**: Each perturbation consists of assigning to some vertices randomly chosen colors.

P2, **adding edges**: P2 consists of adding a certain number of edges during some number of iterations, leading to a modification of the instance definition. After the perturbation, the additional edges are removed again.

P3, **conflict vertices**: Assignments of randomly chosen colors to conflicting vertices.

P4, **directed diversification**: We perform p_{iter} moves using the Tabu Search procedure with long tabu list settings and mix this strategy with random moves. In fact, at each step a random choice is made whether we execute a random search step (such a step is done with a probability w_p) or a Tabu Search step (with probability $1-w_p$).

One particularity of all the perturbations is that they also use the move table from the local search in the perturbation. The reason is that, depending on the length of the perturbation, this can save a significant amount of computation time, because the re-initialisation of the move table is avoided in this case.

Preliminary experiments showed that P4 was the most promising perturbation. An investigation of the tradeoffs regarding the parameter settings for P4 showed that the larger is p_{iter}, the lower should be w_p to avoid that the final solution becomes too close to a random initial solution. Two more details are note-worthy: First, the tabu list is maintained during the perturbation as during the standard Tabu Search procedure. This has the effect that the perturbation moves (including random moves) cannot directly be undone after the perturbation. Second, the perturbation is triggered, if the local search is deemed to be stuck as indicated by the fact that for ls_{iter} iterations the Tabu Search could not find improved solutions.

Initial Solution. For the initial solution we considered random initial solutions and greedy solutions. Since for the hardest instances we could not observe a significant difference in performance, we used random initial solutions for the following experiments.

Acceptance Criteria. The Acceptance criteria is a straightforward procedure that uses one of the following rules: accept every new solution or always apply the perturbation to best solution found so far in the search. Some preliminary experiments were done to compare both acceptance criteria, and no significant difference was detected. Thus, we choose random walk which is less expensive from a computational point of view.

5   Experimental Results

5.1   Peak Performance

In this section we give some performance results of the ILS obtained by a limited parameter tuning effort. Here we report peak performance, that is, the best performance we obtained for the studied parameter settings. Observing peak performance is important, because one hint on the usefulness of a technique is that at least for some instance classes the state-of-the-art performance can be matched or even be improved upon. In fact, many other articles report peak performance after significant parameter tuning efforts like it is the case for the Hybrid Evolutionary Algorithm (HEA) and the Tabu Search presented in [7]. Since in that article the best results so far were reported for several hard benchmark problems, we compare our results to HEA as far as possible. We report results on some Leighton and Flat graphs, which are known to be very hard, and few Random graphs.

In some preliminary experiments, we first identified good parameter settings for the Tabu Search algorithm that is embedded into our ILS. We found that best performance was obtained by the setting of $A = 10$ and $\alpha = 3.5$ with a A2' architecture. Then, at a next step, the Tabu Search procedure was kept fixed (as a black box) and we performed experiments with different parameter settings for the perturbation. Regarding the perturbation, we found that P4 gave the apparently best performance and we decided to run detailed experiments with all possible combinations of parameters settings governing perturbation P4 taken from $ls_{iter} = \{10|V|, 20|V|, 30|V|\}$, $p_{iter} = \{\frac{|V|}{5}, \frac{|V|}{2}, |V|, 2|V|\}$, $w_p = \{0.25, 0.5, 0.75\}$. This results in a total of 36 experiments.

Table 1 presents the results obtained in 50 runs of the Tabu Search (as said above, these results were obtained after fine-tuning parameters A and α) and the best results found in 25 runs of each combination of parameter settings for the ILS. Each trial was given a maximum of ten million local search iterations. For a given number of colors, the results indicated are the average number of iterations in the successful runs and the fraction of successful runs. The ILS presents an additional column with the best parameter settings. The HEA results are taken from [7].

When comparing the peak performance obtained by ILS with Tabu Search, we observe that ILS obtains complete colorings in less iterations than Tabu Search in almost all instances. The most marked difference in favor of ILS is observed for the instance le450_15c.col which is solved in all 25 runs by ILS, while the Tabu Search in a large fraction of the runs stopped unsuccessfully. However, on the runs on instances DSJC250.5.col and flat300_28_0.col, ILS obtained an inferior number of colorings than Tabu Search. One possible reason could be the efficiency of the Tabu Search, giving no opportunity to ILS to outperform its underlying local search.

For some of the hardest instances, as le450_15c.col and le450_25c.col, the ILS was able to obtain colorings in less iterations than HEA. This is remarkable, since HEA was

Table 1. Experimental results. Given are the instance name, the number of colors allowed, and for each algorithm the average number of iterations in the successful runs and the percentage of successful runs.

Instances	k	Tabu Search Average	% succ	ILS Average	% succ	ls_{iter}, p_{iter}, w_p	HEA Average	% succ
DSJC125.5.col	17	244382	100	186185	100	$(20\lvert V\rvert, \frac{\lvert V\rvert}{5}, 0.25)$	-	-
DSJC250.5.col	28	4179466	88	3550914	76	$(30\lvert V\rvert, \lvert V\rvert, 0.50)$	490000	90
flat_26_0.col	26	957811	100	649107	100	$(10\lvert V\rvert, \frac{\lvert V\rvert}{2}, 0.25)$	-	-
flat_28_0.col	31	5375378	26	5354077	20	$(30\lvert V\rvert, \frac{\lvert V\rvert}{5}, 0.75)$	637000	60
le450_15a.col	15	113165	100	95246	100	$(30\lvert V\rvert, \frac{\lvert V\rvert}{2}, 0.50)$	-	-
le450_15b.col	15	66165	100	63920	100	$(30\lvert V\rvert, \frac{\lvert V\rvert}{5}, 0.50)$	-	-
le450_15c.col	15	3881149	22	451714	100	$(10\lvert V\rvert, 2\lvert V\rvert, 0.25)$	194000	60
le450_15d.col	15	6399520	8	2128483	100	$(10\lvert V\rvert, 2\lvert V\rvert, 0.25)$	-	-
le450_25c.col	26	120720	100	108312	100	$(20\lvert V\rvert, \frac{\lvert V\rvert}{5}, 0.50)$	800000	100
le450_25d.col	26	108685	100	99785	100	$(30\lvert V\rvert, \frac{\lvert V\rvert}{5}, 0.25)$	-	-

reported the best results among the large number of approximation algorithms for the GCP. However, on the instances DSJC250.5.col and flat_28_0.col, HEA shows better average behavior.

Finally, it should be remarked that, when compared to the Tabu Search results of [7] (not reported here), our Tabu Search was able to find complete coloring with an inferior chromatic number on the instances flat_28_0.col and le450_15c.col, with 31 and 15 colors, respectively. However, on the instance DSJC250.5.col our Tabu Search performed worse.

The peak performance results clearly indicate that the ILS algorithm is able to compete with state-of-art algorithms. Yet, this approach does not use any operator that uses the knowledge of the problem to guide the search, as the crossover operator in [7].

5.2 Perturbation Parameters

We analyzed the dependence of the ILS performance from the parameters that determine the perturbation. Tables 2, 3, and 4 show the average results obtained from 25 runs on the instance le450_25c.col and the percentage of successful runs. Each table is grouped by the ls_{iter} values.

The computational results suggest that higher values for ls_{iter} could be a good choice. Additionally, low values for the parameters w_p and p_{iter} seem to help. Hence, a good perturbation strength for this instance is rather different from random restart, which confirms our conjecture that a perturbation, as the one reported here, brings more advantages than a random restart. This observation holds for most instances. One exception is the instance le450_15c.col, for which the results obtained indicate that the perturbations should be disruptive enough to move the search to different regions of the search space.

Table 2. Average values for ILS with $ls_{iter} = 10|V|$ on le450_25c.col

w_p	$\frac{|V|}{5}$	$\frac{|V|}{2}$	$|V|$	$2|V|$
0.25	157298 (100%)	254323 (100%)	376345 (100%)	2498161 (100%)
0.50	153709 (100%)	192482 (100%)	278668 (100%)	- (0%)
0.75	131303 (100%)	320049 (100%)	- (0%)	- (0%)

Table 3. Average values for ILS with $ls_{iter} = 20|V|$ on le450_25c.col

w_p	$\frac{|V|}{5}$	$\frac{|V|}{2}$	$|V|$	$2|V|$
0.25	117866 (100%)	163488 (100%)	179836 (100%)	709729 (100%)
0.50	108312 (100%)	144848 (100%)	196596 (100%)	872861 (100%)
0.75	127268 (100%)	155794 (100%)	1051493 (100%)	1973531 (100%)

Table 4. Average values for ILS with $ls_{iter} = 30|V|$ on le450_25c.col

w_p	$\frac{|V|}{5}$	$\frac{|V|}{2}$	$|V|$	$2|V|$
0.25	116881 (100%)	167439 (100%)	163045 (100%)	300516 (100%)
0.50	165227 (100%)	128923 (100%)	153511 (100%)	597886 (100%)
0.75	130700 (100%)	184779 (100%)	411852 (100%)	737571 (100%)

5.3 Run-Time Distributions

Our ILS algorithm makes strong use of randomized decisions during the search and therefore the run-time required to find a coloring is a random variable; this is notable by the fact that the time needed to find such a solution varies between runs of the algorithm. We estimate the distribution the solution times by analyzing empirical run-time distributions (RTDs). RTDs give the empirical probability of finding a solution as a function of the run-time [22,23]. RTDs can be used to ease the comparison of local search algorithms, characterize the run-time behavior of local search algorithms on specific problem classes and to give valuable hints in which situations local search algorithms can be improved [22].

Here we analyze the run-time behavior of ILS and compare it with Tabu Search. We ran 100 times the Tabu Search and the ILS using the best parameter settings found on the instances le450_15b.col, le450_15c.col. Figures 2 and 3 plot the RTDs of ILS and Tabu Search on these instances.

Observing the RTDs on instance le450_15b.col on the Figure 2 (left), it is possible to conclude that both algorithms perform similarly, as also reported in Table 1. However, the behavior of the two algorithms is different beyond 100000 iterations. The curvature of Tabu Search's RTD indicates worse performance than ILS on long search paths. This different behavior could mean that ILS is capable of getting out of some local optimum in a more successful way than Tabu Search and this may be the main reason for improved performance. Figure 2 (right) also demonstrates the higher performance of ILS on the instances le450_15c.col (the RTDs on le450_15d.col look similar). The RTDs obtained

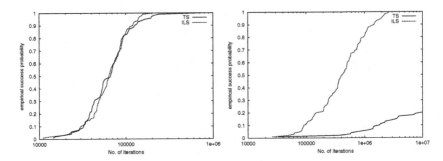

Fig. 2. Run-time distributions of Tabu Search and ILS on instance le450_15b.col (left) and le450_15c.col (right).

from these two instances prove their hardness, since Tabu Search was not able to reach more than a success rate of 20%.

6   Conclusion

In this paper we have presented an initial study of the application of ILS to the graph coloring problem. When looking at peak performance, some very promising results were obtained compared to state-of-the-art algorithms. There are a number of ways how this research will be extended in the future. First, we will extend this approach to larger instances, random and structured, to verify if the overall behaviour is similar to the one reported here. Second, we will to do a more detailed analysis of the parameter space for the perturbation. Such a study may identify even better parameter settings. Additionally, a refined study using methods from experimental design may help to better understand the interaction between the perturbation parameters. Third, additional ways of perturbing solutions will be tested, because this seems to be the key to further enhance ILS performance for the graph coloring problem. Future work will also include ways of how to automatize the adjustment of the parameters for the instance under solution.

Acknowledgments. This work was supported by the "Metaheuristics Network", a Research Training Network funded by the Improving Human Potential programme of the CEC, grant HPRN-CT-1999-00106. The information provided is the sole responsibility of the authors and does not reflect the Community's opinion. The Community is not responsible for any use that might be made of data appearing in this publication.

References

1. M.W. Carter. A survey of pratical applications of examination timetabling algorithms. *Operations Research*, 34(2):193–202, 1986.
2. D.J. Castelino, S. Hurley, and N.M. Stephens. A tabu search algorithm for frequency assignment. *Annals of Operations Research*, 63:301–320, 1996.
3. M.R. Garey and D.S. Johnson. *Computers and Intractability: A Guide to the Theory of $\mathcal{NP}$-Completeness*. Freeman, San Francisco, CA, USA, 1979.

4. A. Mehrotra and M. Trick. A column generation approach for graph coloring. *INFORMS Journal On Computing*, 8(4):344–354, 1996.

5. D. Brélaz. New methods to color the vertices of a graph. *Communications of the ACM*, 22(4):251–256, 1979.

6. C. Fleurent and J. Ferland. Genetic and hybrid algorithms for graph coloring. *Annals of Operations Research*, 63:437–464, 1996.

7. P. Galinier and J.K. Hao. Hybrid evolutionary algorithms for graph coloring. *Journal of Combinatorial Optimization*, 3(4):379–397, 1999.

8. A. Hertz and D. de Werra. Using tabu search techniques for graph coloring. *Computing*, 39:345–351, 1987.

9. D.S. Johnson, C.R. Aragon, L.A. McGeoch, and C. Schevon. Optimization by simulated annealing: An experimental evaluation: Part II, graph coloring and number partitioning. *Operations Research*, 39(3):378–406, 1991.

10. F.T. Leighton. A graph coloring algorithm for large scheduling problems. *Journal of Research of the National Bureau of Standards*, 85:489–506, 1979.

11. H.R. Lourenço, O. Martin, and T. Stützle. Iterated local search. In F. Glover and G. Kochenberger, editors, *Handbook of Metaheuristics*. Kluwer Academic Publishers, Boston, MA, USA, 2002. to appear.

12. J.C. Culberson. Iterated greedy graph coloring and the difficulty landscape. Technical Report 92-07, Department of Computing Science, The University of Alberta, Edmonton, Alberta, Canada, June 1992.

13. S. Minton, M.D. Johnston, A.B. Philips, and P. Laird. Minimizing conflicts: A heuristic repair method for constraint satisfaction and scheduling problems. *Artificial Intelligence*, 52:161–205, 1992.

14. R. Dorne and J.K. Hao. Tabu search for graph coloring, t-colorings and set t-colorings. In I.H. Osman S. Voss, S. Martello and C. Roucairol, editors, *Meta-heuristics: Advances and Trends in Local Search Paradigms for Optimization*, pages 77–92. Kluwer Academic Publishers, Boston, MA, USA, 1999.

15. M. Chams, A. Hertz, and D. De Werra. Some experiments with simulated annealing for coloring graphs. *European Journal of Operational Research*, 32:260–266, 1987.

16. L. Davis. Order-based genetic algorithms and the graph coloring problem. In *Handbook of Genetic Algorithms*, pages 72–90. Van Nostrand Reinhold; New York, 1991.

17. A.E. Eiben, J.K. Hauw, and J.I. Van Hemert. Graph coloring with adaptive evolutionary algorithms. *Journal of Heuristics*, 4:25–46, 1998.

18. M. Laguna and R. Martí. A GRASP for coloring sparse graphs. *Computational Optimization and Applications*, 19(2):165–178, 2001.

19. C. Fleurent and J. Ferland. Object-oriented implementation of heuristic search methods for graph coloring, maximum clique and satisfiability. In D.S. Johnson and M.A. Trick, editors, *Cliques, Coloring, and Satisfiability: Second DIMACS Implementation Challenge*, volume 26, pages 619–652. American Mathematical Society, 1996.

20. D.S. Johnson and L.A. McGeoch. The travelling salesman problem: A case study in local optimization. In E.H.L. Aarts and J.K. Lenstra, editors, *Local Search in Combinatorial Optimization*, pages 215–310. John Wiley & Sons, Chichester, UK, 1997.

21. O. Martin and S. W. Otto. Partitoning of unstructured meshes for load balancing. *Concurrency: Practice and Experience*, 7:303–314, 1995.

22. H.H. Hoos and T. Stützle. Evaluating Las Vegas algorithms — pitfalls and remedies. In *Proceedings of the Fourteenth Conference on Uncertainty in Artificial Intelligence (UAI-98)*, pages 238–245. Morgan Kaufmann, San Francisco, 1998.

23. H.H. Hoos and T. Stützle. Characterising the behaviour of stochastic local search. *Artificial Intelligence*, 112:213–232, 1999.

Solving Car Sequencing Problems
by Local Optimization

Markus Puchta and Jens Gottlieb

SAP AG
Neurottstr. 16, 69190 Walldorf, Germany
markus.puchta@sap.com
jens.gottlieb@sap.com

Abstract. Real-world car sequencing problems deal with lots of constraints, which differ in their types and priorities. We evaluate three permutation-based local search algorithms that use different acceptance criteria for moves. The algorithms meet industrial requirements to obtain acceptable solutions in a rather short time. It is essential to employ move operators which can be evaluated quite fast. Further, using different move types enlarges the neighbourhood, thereby decreasing the total number of local optima in the search space. The comparison of the acceptance criteria shows that the greedy approach is inferior to two variants of threshold accepting that allow escaping from local optima.

1   Introduction

Production of cars is typically divided into several steps that are performed in sequence. The manufacturing process is organized by assembly lines, where single working steps are assigned to predefined positions of the line. Each single car moves from one position to the next until the end of the line is reached. The time spent at each position is the same. Besides these working steps there are some special features that are not integrated in each car. The workers at the positions assigned to these features should be even burdened, which can be achieved by an even distribution of the cars needing those features. Due to restricted availability of material, other constraints may arise for the special features. As an example, a restricted delivery of material constrains the number of cars with the corresponding feature to a maximum in a certain period. Not all constraints have the same priority: While it may be acceptable to overload a certain position for a short time, a restricted delivery of material is much more important, which is modelled by a higher priority.

Car sequencing problems are constraint satisfaction problems [1]. They can be tackled by local optimization, which proved quite successful for combinatorial optimization problems in general [2]. We consider local search algorithms that rely on a permutation representation and evaluating constraint violations by penalty functions. The acceptance criterion inside local optimization has a strong impact on the overall results, because it determines whether search can

S. Cagnoni et al. (Eds.): EvoWorkshops 2002, LNCS 2279, pp. 132–142, 2002.
© Springer-Verlag Berlin Heidelberg 2002

escape from local optima. Therefore, we compare several acceptance criteria on benchmarks resembling real-world car sequencing problems.

Our paper presents a formal model of car sequencing problems in section 2. Three local search algorithms are proposed in section 3 and evaluated empirically in section 4. Conclusions and ideas for further research are given in section 5.

2   Modelling Car Sequencing Problems

2.1   Basic Concepts

The car sequencing problem deals with determining a sequence of cars on an assembly line, such that predefined constraints are met. Each car directly corresponds to an order that demands a specific configuration. Formally, there is a set of orders $O = \{1, \ldots, n\}$ and a set of characteristics $C = \{1, \ldots, m\}$ that each car $i \in O$ may have. The characteristics of the orders are represented by a configuration matrix $A = (a_{ij}) \in \{0, 1\}^{m \times n}$. Here, the entry a_{ij} denotes whether order $j \in O$ has the characteristic property $i \in C$. A solution candidate is a permutation $\pi : O \to O$ of the orders, where $\pi(i) = k$ means that order k is assigned to position i in the sequence. A permutation π is described by the assignments to its sequentially listed positions, i.e. $\pi = (\pi(1)\, \pi(2)\, \cdots\, \pi(n))$.

A constraint refers to a specific interval I of the complete sequence and to a specific characteristic $c \in C$ of the orders assigned to that interval. Stated formally, an interval $I = \{i \in P \mid l \leq i \leq r\}$ is a subset of consecutive positions of all possible positions $P = \{1, \ldots, n\}$ in the sequence. Given a permutation π of the orders, a constraint on the interval I restricts the characteristic vector $(a_{c\pi(l)}\, a_{c\pi(l+1)}\, \cdots\, a_{c\pi(r)})$ that indicates which order in the interval has the given characteristic property c.

We illustrate the concepts by an example with $n = 20$ orders and $m = 1$ properties, i.e. $C = \{c\}$. Let $a_{cj} = 1$ for $j \in \{2, 3, 5, 9, 11, 12, 13, 18, 20\}$ and $a_{cj} = 0$ otherwise. Given the interval $I = \{4, 5, \ldots, 13\}$, the permutation

$$\pi = (6\ \underline{9}\ 16\ \mid\ \underline{5}\ \underline{20}\ 1\ 4\ \underline{3}\ \underline{18}\ \underline{11}\ 17\ \underline{13}\ 14\ \mid\ 7\ 10\ 15\ 8\ 19\ \underline{12}\ \underline{2})$$

has the characteristic vector

$$(1\ 1\ 0\ 0\ 1\ 1\ 1\ 0\ 1\ 0)\,.$$

For the sake of clarity we marked in π the interval borders and orders with c.

2.2   Constraint Types and Penalty Functions

All specific types of constraints refer to the characteristic vector of an interval I and a property c. Some of the constraints (block, distance, subinterval quantity) are in fact conjunctions of atomic constraints. We present a penalty function for each atomic constraint, which is evaluated by 0 if the constraint is satisfied. The total penalty of a constraint that consists of several atomic constraints is the sum of the separate penalties for its atomic constraints. While linear penalty functions have been studied for car sequencing problems in [3], here we use quadratic penalty functions in order to give preference to solutions with many small violations over solutions with a few but large violations. Most constraints state a lower and an upper bound on some property of an interval. We use infinite bounds to model that there is no upper or lower bound, respectively.

Block constraint: Given $b_{min} \in \{-\infty, 2, \ldots, |I|\}$ and $b_{max} \in \{1, \ldots, |I| - 1, +\infty\}$, each block in I must have size s with $b_{min} \leq s \leq b_{max}$. A block is a subinterval containing only ones, which is surrounded by zeroes. Each constraint violation is evaluated by

$$p_{block} = \begin{cases} \left(\frac{s-b_{min}}{b_{min}}\right)^2 & \text{if } s < b_{min} \\ \left(\frac{s-b_{max}}{b_{max}}\right)^2 & \text{if } s > b_{max}. \end{cases}$$

In above example we have three blocks with sizes 2, 3, 1, respectively. Given $b_{min} = 2$ and $b_{max} = 4$, there is one constraint violation as the third block has size 1. This violation is penalized by $1/4$, which is also the total penalty.

Distance constraint: Given the values $d_{min} \in \{-\infty, 1, \ldots, |I| - 2\}$ and $d_{max} \in \{0, \ldots, |I| - 2, +\infty\}$, each distance d between two consecutive ones must satisfy $d_{min} \leq d \leq d_{max}$. Here, two ones are called consecutive if there is no other entry one between them, and their distance is the number of zeroes between them. Each constraint violation is punished by

$$p_{dist} = \begin{cases} \left(\frac{d-d_{min}}{d_{min}}\right)^2 & \text{if } d < d_{min} \\ \left(\frac{d-d_{max}}{d_{max}}\right)^2 & \text{if } d > d_{max}. \end{cases}$$

In above example five pairs of consecutive ones exist, having the distances 0, 2, 0, 0, 1 respectively. Given $d_{min} = 1$ and $d_{max} = 2$, there are three constraint violations, each of which is punished by $p_{dist} = 1$. Thus, the total penalty is 3.

Quantity constraint: Given the values $q_{min} \in \{-\infty, 1, \ldots, |I|\}$ and $q_{max} \in \{0, \ldots, |I| - 1, +\infty\}$, the quantity q of ones must meet $q_{min} \leq q \leq q_{max}$. In case of a violation, the total penalty is

$$p_{quantity} = \begin{cases} \left(\frac{q-q_{min}}{q_{min}}\right)^2 & \text{if } q < q_{min} \\ \left(\frac{q-q_{max}}{q_{max}}\right)^2 & \text{if } q > q_{max}. \end{cases}$$

Given the example, $q_{min} = 3$ and $q_{max} = 4$, there is a constraint violation yielding a total penalty of $4/9$.

Subinterval quantity constraint: Given the values $s \in \{1, \ldots, |I|\}$, $q_{min} \in \{-\infty, 1, \ldots, s\}$ and $q_{max} \in \{0, \ldots, s - 1, +\infty\}$, the quantity q of ones in each subinterval of size s must meet $q_{min} \leq q \leq q_{max}$.[1] A constraint violation is

---

[1] Thus, the quantity constraint is equivalent to the corresponding subinterval quantity constraint with $s = |I|$.

punished by

$$p_{subquantity} = \begin{cases} \left(\frac{q-q_{min}}{q_{min}}\right)^2 & \text{if } q < q_{min} \\ \left(\frac{q-q_{max}}{q_{max}}\right)^2 & \text{if } q > q_{max}. \end{cases}$$

In the example there are three constraint violations for $s = 7$, $q_{min} = 2$ and $q_{max} = 3$, which are punished by 4/9, 1/9 and 1/9, causing a total penalty 2/3.

Even distribution constraint: There are $|I|$ subintervals that cover the first position of I. Let s be the size of such a subinterval, and t the number of ones in it. Supposed there are T ones in I, then t should be as close to $s \cdot T/|I|$ as possible. The penalty for each subinterval is

$$p_{distribution} = \left(\frac{s}{|I|} - \frac{t}{T}\right)^2 .$$

Minimizing the sum of all subinterval's penalties causes an even distribution of all ones among the positions of the interval I. In the example, the penalties for the subintervals are 1/225, 4/225, 1/900, 1/225, 0, 1/225, 4/225, 1/900, 1/100 and 0, which sum up to the total penalty 11/180.

2.3   The Fitness Function

The concepts introduced so far allow to define a feasible solution as a permutation that yields penalty 0 for each constraint. However, many practical problems are over-constrained, i.e. no feasible solution exists at all. Further, not all constraints may have the same priority: some may be very important and should be satisfied if possible, while others represent preferences that are not as critical and for which violations may be acceptable. The manufacturer's priorities are given a priori by constant weights $w_k > 0$, one for each constraint k. The resulting fitness function is

$$f(\pi) = \sum_k w_k \cdot p_k(\pi)$$

where $p_k(\pi)$ is the total penalty assigned to π with respect to the kth constraint. The fitness function is to be minimized and hence high weights represent high priorities. Assuming our example and the weights 3, 1, 5, 3, 2 for the five constraints, the objective value of the permutation π is

$$f(\pi) = 3 \cdot 1/4 + 1 \cdot 3 + 5 \cdot 4/9 + 3 \cdot 2/3 + 2 \cdot 11/180 = 1097/180 .$$

3   Local Optimization of Car Sequencing Problems

3.1   Move Operators

The search space of the car sequencing problem is $\Pi = \{\pi \mid \pi : O \rightarrow O\}$, the set of all permutations of the orders O. Local search requires a neighbourhood relation defined on the search space. We define the neighbourhood implicitly by move operators that transform a permutation $\pi \in \Pi$ into a neighbour $\pi' \in$

Π. Given a set of move operators, the neighbourhood of π is the set of all permutations π' that can be generated by applying a move to π. Here, we consider the following move operators.

- **Insert.** One order is moved from its current position to another, thereby also changing other orders by one position. In our example of the previous section, an Insert of order 20 at position 10 in π yields

$$\pi' = (\ 6\ 9\ 16\ 5\ \underline{1\ 4\ 3\ 18\ 11}\ \underline{20}\ 17\ 13\ 14\ 7\ 10\ 15\ 8\ 19\ 12\ 2)\,.$$

- **Swap.** The positions of two orders are exchanged. A Swap of the orders 20 and 14 in π results in

$$\pi' = (\ 6\ 9\ 16\ 5\ \underline{14}\ 1\ 4\ 3\ 18\ 11\ 17\ 13\ \underline{20}\ 7\ 10\ 15\ 8\ 19\ 12\ 2)\,.$$

- **Transposition.** Being a special case of Swap, this move exchanges two orders having neighbouring positions in the sequence. In the example, the orders 5 and 20 have neighbouring positions and are swapped yielding

$$\pi' = (\ 6\ 9\ 16\ \underline{20\ 5}\ 1\ 4\ 3\ 18\ 11\ 17\ 13\ 14\ 7\ 10\ 15\ 8\ 19\ 12\ 2)\,.$$

- **SwapS.** This move swaps two orders that are similar but not identical. Defining the difference of orders $k, l \in O$ as $d(k,l) = |\{j \in C \mid a_{jk} \neq a_{jl}\}|$, orders k, l are identical if $d(k, l) = 0$, and similar if $d(k, l) \leq 2$. In our example the orders 20 and 4 are similar but not identical, and hence may be swapped[2], yielding

$$\pi' = (\ 6\ 9\ 16\ 5\ \underline{4}\ 1\ \underline{20}\ 3\ 18\ 11\ 17\ 13\ 14\ 7\ 10\ 15\ 8\ 19\ 12\ 2)\,.$$

- **Lin2Opt.** A part of the sequence is inverted, which yields for the subsequence from position 12 to 16 in π the permutation

$$\pi' = (\ 6\ 9\ 16\ 5\ 20\ 1\ 4\ 3\ 18\ 11\ 17\ \underline{15\ 10\ 7\ 14\ 13}\ 8\ 19\ 12\ 2)\,.$$

- **Random.** This move randomly re-arranges a part of the sequence. Supposed the selected part ranges from position 6 to 11, this operator may produce

$$\pi' = (\ 6\ 9\ 16\ 5\ 20\ \underline{18\ 3\ 11\ 17\ 4\ 1}\ 13\ 14\ 7\ 10\ 15\ 8\ 19\ 12\ 2)\,.$$

These moves affect only parts of the total sequence — which are underlined in above examples — and hence it is faster to evaluate only the difference in the fitness due to the changed positions, instead of completely evaluating the whole sequence after each move. The smaller the changed parts, the smaller is the time needed to evaluate the changed fitness. The two special cases of Swap, Transposition and SwapS, are evaluated very quickly: Transposition changes only two neighboring positions, and SwapS modifies only one or two characteristics in

---

[2] The orders $\{2, 3, 5, 9, 11, 12, 13, 18, 20\}$ are identical and cannot be swapped by this move operator; the same holds for the orders $\{1, 4, 6, 7, 8, 10, 14, 15, 16, 17, 19\}$.

the sequence, so that only constraints for these changed characteristics must be evaluated. As a consequence, most changes in the fitness are very small, which makes the landscape rather smooth and thus enables effective local search.

Move operators affect different aspects of permutations, like relative ordering, absolute positions, or adjacency of the items in a permutation [4]. It is known for some combinatorial optimization problems that not all information types are important: Consider e.g. the symmetric traveling salesman problem, where solutions are directly represented by edges and thus, only the adjacency information coded in a permutation is relevant. However, in our case, several information types are relevant since the notion of blocks correponds to adjacency and the definition of an even distribution relies on absolute positions. Therefore, we use several operators, each of which focuses on another information type. While Lin2Opt maintains most adjacency relations (all but two edges are preserved), Swap changes only two absolute positions, and Insert changes only the relative ordering between the inserted item and the other changed items.

3.2   Acceptance Criteria of Moves

Local optimization starts with an initial solution and then generates a sequence of moves, each of which is applied to the current solution if the move meets some acceptance criterion. If only improving moves are accepted, the search process terminates in a local optimum. Supposed this happens in early stages of the search, then the remaining time is wasted if non-improving moves are rejected. To make better use of the total time available for determining a solution, different strategies have been proposed to escape from local optima, like e.g. simulated annealing [5] or threshold accepting [6]. We consider three different strategies for our empirical comparison:

- **Greedy.** This strategy tries to improve the fitness step by step. A move is only accepted if the resulting fitness is not worse than the current solution. Moves that are not changing the fitness are called trivial moves; these moves are helpful to escape from plateaus in the fitness landscape. As we use several alternative move types, the neighbourhood is large enough to prevent freezing in a local optimum after short running times.
- **Threshold accepting (TA).** In contrast to the greedy strategy, TA also accepts moves that lead to a worse fitness if the fitness deterioration Δf is not greater than a certain parameter th called threshold. The probability for accepting a move is defined by the Heaviside function

$$\Theta(th - \Delta f) = \begin{cases} 1 & \text{if } \Delta f \leq th \\ 0 & \text{otherwise.} \end{cases}$$

During search the threshold is lowered by $th_{i+1} = th_i \cdot \alpha$ after certain time intervals, given some control parameter $\alpha \in [0.95, 0.99]$, an initial (high) value th_{init} and a final (low) value th_{freeze}. Thus, the threshold sequence is $(th_0, th_1, \ldots, th_t)$ where $th_0 = th_{init}$ and $th_t = th_{freeze}$. This scheme resembles the geometric cooling schedule of the temperature in Simulated

Annealing [5]. Note that TA with a threshold $th = 0$ is identical to the Greedy strategy.

– **Threshold accepting with a bouncing strategy.** If TA is stuck in a local optimum at a low threshold it has no chance to leave it. A possibility to leave this local optimum is to increase the threshold, so that TA can accept worse solutions. Once the local optimum has been left, the threshold is decreased as usual, which may lead to a new local optimum with even better fitness. The new threshold th_{bounce} should not be too high in order to avoid the solution being completely re-organized and random. Unlike a new restart with a completely randomized solution, this "bouncing" initiates local optimization starting from a relatively good solution. If th_{bounce} is too low, the system remains frozen in its local optimum. In addition to the parameters of classical TA — α, th_{init} and th_{freeze} — there is the value th_{bounce} and the number of bouncing-steps bs. We assume th_{bounce} to be among the threshold sequence of TA, i.e. $th_{bounce} = th_s$ for some $s \in \{0, \ldots, t\}$. This yields a new threshold sequence $((th_0, th_1, \ldots, th_t), (th_s, \ldots, th_t), \ldots, (th_s, \ldots, th_t))$.

Figure 1 illustrates a classical threshold sequence and a bouncing sequence derived from it. We refer the reader to [7,8] for the statistical background of the bouncing strategy. Note that similar ideas have also been applied within simulated annealing, like e.g. feedback temperature schedules [9].

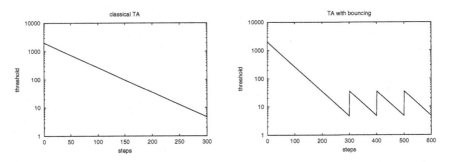

Fig. 1. Threshold sequences for $\alpha = 0.98$, the classical variant with $t = 300$ (left) and a bouncing strategy with $bs = 3$ and $s = 200$ (right)

4   Empirical Analysis

4.1   Benchmarks

We consider six benchmarks representing each a realistic scenario of the daily production of a German car manufacturer. Each instance consists of 1000 orders and 25 to 30 characteristics. Constraints of types block, distance, quantity and subinterval quantity are defined for several characteristics and intervals. Further,

80 % of the characteristics should be evenly distributed. The priority for the even distribution is always lower than other constraints' priorities for the same characteristic. The benchmarks mainly differ in the priorities.

As block constraints have higher priorities than the even distribution, defining both for one characteristic should yield an even distribution of the blocks. No solutions without constraint violations do exist, because the instances are over-constrained. An analysis reveals that e.g. for one instance there are two characteristics 1 and 2, for which a block constraint with $b_{min} = 5$ is defined for 1 and a distance constraint with $d_{min} = 3$ is defined for 2. As 50 % of the orders with characteristic 1 also have characteristic 2 it is not possible to meet both constraints. Note that there might be a feasible solution if only 40 % of the orders with characteristic 1 had characteristic 2.

4.2   Experimental Setup

The initial solution is a random permutation. Each single move is drawn randomly from the six move types, i.e. each type has probability 1/6 to be selected. Although most operators — except for Random — provide high locality, we further increase locality by restricting the affected subsequence to a maximum length of $n/20$. Thus, at most 5 % of all positions are modified by a move. This guarantees small fitness changes and short evaluation times, see table 1.

The parameters of classical threshold accepting are adjusted according to a procedure suggested in [7]. Before starting the search process, 0.5 % of the total CPU time is spent in sampling moves. Then, the initial threshold th_{init} is set such that 75 % of the moves would have been accepted. This prevents the search process to start with a random walk. The final threshold th_{freeze} is set to the minimal fitness difference produced among all deteriorating moves. Such value is expected to enforce convergence towards a local optimum. As we want the time spent at each threshold to be identical, this time is determined by α, for which we use a standard value 0.98. Note that we also tried other values between 0.95 and 0.99, but without observing significant differences.

The bouncing strategy relies on the same parameters as selected for classical threshold accepting. In addition, we have to determine the number of bouncing steps bs and th_{bounce}. After experimenting with some values, we chose $bs = 5$ and $s = 0.6 \cdot t$, which determines th_{bounce} as described in section 3.2.

Table 1. Efficiency of move types on the first benchmark

Operator	Insert	Swap	Transposition	SwapS	Lin2Opt	Random
Moves / 100 seconds	128 542	127 121	514 532	1 971 348	127 652	126 592

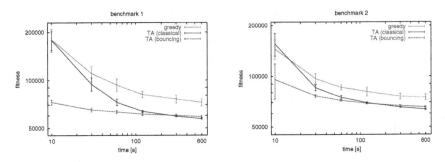

Fig. 2. Results on two representative benchmarks

4.3   Results

The three algorithms are compared on a Pentium III 600 MHz for different CPU time limits of 10, 30, 60, 120, 300 and 600 seconds. We selected these limits for our comparison of the algorithms since a realistic upper bound for the optimization time in practice appears to be five minutes. 30 runs per benchmark are performed for each algorithm and run time limit. Since the benchmarks are over-constrained, no feasible solution is found throughout all runs.

The outcomes of the algorithms for different run times are shown in figure 2 for two representative benchmarks. The result of one run is the fitness of the best solution found. For each combination of algorithm and CPU time we present the average result of 30 runs and its standard deviation.

Although the greedy strategy is often expected to converge in a local optimum in an early stage of the search, this is not the case here. Due to the large neighbourhood induced by four structurally different moves, the fitness landscape contains fewer local optima than when using one single move operator only. Investing more time in exploring the current neighbourhood increases the probability still to find an improving move, and hence better results can be expected for higher CPU time limits. In general, we observe larger deviations of the results compared to the deviations of the other algorithms. This is caused by the dependance of the final solution from the initial solution; a greedy strategy cannot leave the basin of attraction of a local optimum.

Threshold accepting is superior to the greedy strategy for CPU times greater than 10 seconds. This holds not only for the average result, even the worst results of TA are better than the best obtained by greedy. Increasing the time limit improves the average result and decreases the corresponding deviation. Threshold accepting is much more robust than greedy in finding good solutions, since it yields better solutions with smaller deviations. This is explained by its ability to escape from local optima and their surrounding basin of attraction, which makes the search more independent of the initial solution.

For CPU times less than 120 seconds the bouncing strategy leads to much better results than TA. The deviation for the short CPU times is much smaller than the deviations of the other two strategies. This is due to the bouncing strategy's enhanced capability — with respect to TA — of leaving local optima. This positive effect is reduced for larger running times, because long TA runs have enough time to escape from bad local optima in an early stage of the search. Basically, TA and the bouncing strategy lead to comparable results (note the logarithmic fitness scale).

5   Conclusion

The car sequencing problem is a constraint satisfaction problem with various different types of constraints. Having defined appropriate penalty functions, we perceive it as a permutation problem which can be tackled by local search using typical permutation operators. The results are satisfactory and meet the industrial requirements to obtain good solutions in a rather short time. It is essential to use (i) move operators that can be evaluated quite fast, and (ii) different types of moves, which enlarges the neighbourhood and hence reduces the total number of local optima.

Our comparison of three acceptance criteria shows that the greedy approach is inferior to both threshold accepting variants concerning solution quality and robustness. The bouncing strategy yields good solutions in a very short time, but it needs a lot of additional parameter tuning.

The promising results obtained for the benchmarks, which resemble real-world customers' problems, raise the question whether even more complex problems can be tackled by our algorithms. Preliminary results demonstrate that our approach can be extended in a straightforward way. Among the extensions are new constraint types like due dates or earliest starting times, and a more complex manufacturing structure involving a network of assembly lines and buffers.

There is a potential to increase our algorithms' performance. All move types have the same probability to be applied to the current solution. Fine-tuning these probabilities — e.g. such that fast moves are favoured — may reduce the computational efforts without deteriorating the final outcome. The parameters could be tuned by a priori experiments, or they could be controlled by an adaptive scheme based on the history of the search process. We are also comparing evolutionary algorithms to the local search strategies. First results show that in particular the use of specific crossover operators can be beneficial.

References

1. E. Tsang. *Foundations of Constraint Satisfaction.* Academic Press, 1993
2. E. H. L. Aarts and J. K. Lenstra (eds.). *Local Search in Combinatorial Optimization.* John Wiley & Sons, 1997
3. J. Schneider, J. Britze, A. Ebersbach, I. Morgenstern and M. Puchta. Optimization of Production Planning Problems — A Case Study for Assembly Lines. *International Journal of Modern Physics C*, Volume 11, No. 5, 949 – 972, 2000
4. J. Gottlieb. *Evolutionary Algorithms for Constrained Optimization Problems.* PhD Thesis, Technical University of Clausthal, 1999. Published by Shaker, Aachen, 2000

5. S. Kirkpatrick, C. D. Gelatt Jr. and M. P. Vecchi. Optimization by Simulated Annealing. *Science*, Volume 220, 671 – 680, 1983
6. G. Dueck and T. Scheuer. Threshold Accepting: A General Purpose Optimization Algorithm Appearing Superior to Simulated Annealing. *Journal of Computational Physics*, Volume 90, 161 – 175, 1990
7. M. Puchta. *Physikalische Optimierungsverfahren angewendet auf Produktionsplanungsprobleme*. Diplomarbeit, Universität Regensburg, 1998
8. J. Schneider, I. Morgenstern and J. M. Singer. Bouncing towards the optimum: Improving the results of Monte Carlo optimization algorithms. *Physical Review E*, Volume 58, Number 4, 5085 – 5095, 1998
9. M. Kolonko and M. T. Tran. Convergence of Simulated Annealing with Feedback Temperature Schedules. *Probability in the Engineering and Informational Sciences*. Volume 11, 279 – 304, 1997

Evolution Strategies, Network Random Keys, and the One-Max Tree Problem

Barbara Schindler[1], Franz Rothlauf[1], and Hans-Josef Pesch[2]

[1] Department of Information Systems, University of Bayreuth/Germany
[2] Department of Applied Mathematics, University of Bayreuth/Germany
barbara.schindler@stud.uni-bayreuth.de rothlauf@uni-bayreuth.de
hans-josef.pesch@uni-bayreuth.de

Abstract. Evolution strategies (ES) are efficient optimization methods for continuous problems. However, many combinatorial optimization methods can not be represented by using continuous representations. The development of the network random key representation which represents trees by using real numbers allows one to use ES for combinatorial tree problems.

In this paper we apply ES to tree problems using the network random key representation. We examine whether existing recommendations regarding optimal parameter settings for ES, which were developed for the easy sphere and corridor model, are also valid for the easy one-max tree problem.

The results show that the $\frac{1}{5}$-success rule for the $(1+1)$-ES results in low performance because the standard deviation is continuously reduced and we get early convergence. However, for the $(\mu + \lambda)$-ES and the (μ, λ)-ES the recommendations from the literature are confirmed for the parameters of mutation τ_1 and τ_2 and the ratio μ/λ. This paper illustrates how existing theory about ES is helpful in finding good parameter settings for new problems like the one-max tree problem.

1 Introduction

Evolution strategies [1,2,3] are a class of direct, probabilistic search and optimization methods gleaned from the model of organic evolution. In contrast to genetic algorithms (GAs) [4], which work on binary strings and process schemata, ES have been dedicated to continuous optimization problems. The main operator of ES is mutation, whereas recombination is only important for the self-adaption of the strategy parameters. Random network keys (NetKeys) have been proposed by [5] as a way to represent trees with continuous variables. This work was based on [6] and allows to represent a permutation by a sequence of continuous variables.

In this work we investigate the performance of ES for tree problems when using the continuous NetKey representation. Because ES have been designed for solving continuous problems and have shown good performance therein, we expect ESs to perform well for network problems when using NetKeys. Furthermore, we want to examine whether the recommendations for the setting of ES

S. Cagnoni et al. (Eds.): EvoWorkshops 2002, LNCS 2279, pp. 143–152, 2002.

parameters, that are derived for the sphere and the corridor model, are also valid for the easy 8 node one-max tree problem. In analogy to the sphere and corridor models, the one-max tree problem [5,7] is also easy and ES are expected to perform well. Finally, we compare the performance of ES and GAs for the one-max tree problem. We wanted to know which of the two search approaches, mutation versus crossover, performs better for this specific test problem.

The paper is structured as follows. In section 2 we present the NetKey encoding and present its major characteristics. This is followed by a short review of the one-max tree problem. In section 4, after taking a closer look at the different types of ES (subsection 4.1), we perform an analysis of the adjustment of ES parameters for the one-max problem (subsection 4.2), and finally compare the performance of ESs to GAs (subsection 4.3). The paper ends with concluding remarks.

2   Network Random Keys

This section gives a short overview about the NetKey encoding.

Network random keys are adapted random keys (RKs) for the representation of trees. RKs allow us to represent permutations and were first presented in [6]. Like the LNB encoding [8] NetKeys belong to the class of weighted representations. Other tree representations are Prüfer numbers [9], direct encodings [10], or the determinant encoding [11].

When using NetKeys, a key sequence of l random numbers $r_i \in [0, 1]$, where $i \in \{0, \ldots l - 1\}$, represents a permutation r^s of length l. From the permutation r^s of length $l = n(n - 1)/2$ a tree with n nodes and $n - 1$ links is constructed using the following algorithm:

(1) Let $i = 0$, G be an empty graph with n nodes, and r^s the permutation of length $l = n(n - 1)/2$ that can be constructed from the key sequence r. All possible links of G are numbered from 1 to l.
(2) Let j be the number at the ith position of the permutation r^s.
(3) If the insertion of the link with number j in G would not create a cycle, then insert the link with number j in G.
(4) Stop, if there are $n - 1$ links in G.
(5) Increment i and continue with step 2.

With this calculation rule, a unique, valid tree can be constructed from every possible key sequence. We give some properties of the encoding:

- Standard crossover and mutation operators work properly and the encoding has high locality and heritability.
- NetKeys allow a distinction between important and unimportant links.
- There is no over- or underspecification of a tree possible.
- The decoding process goes with $O(l \log(l))$, where $l = n(n - 1)/2$.

Examining NetKeys reveals that the mutation of one key results either in the same tree, or in a tree with no more than two different links. Therefore, NetKeys have high locality. Furthermore, standard recombination operators, like x-point or uniform crossover, create offspring that inherit the properties of their parents that means they have the same links like their parents. If a link exists in a parent, the value of the corresponding key is high in comparison to the other keys. After recombination, the corresponding key in the offspring has the same, high value and is therefore also used with high probability for the construction of the tree.

A benefit of the NetKey encoding is that genetic and evolutionary algorithms (GEAs) are able to distinguish between important and unimportant links. The algorithm which constructs a tree from the key sequence uses high-quality links with high key values and ensures that they are not lost during the GEA run.

NetKeys always encode valid trees. No over- or underspecification of a tree is possible because the construction rule ensures that only valid solutions are decoded. Thus, no additional repair mechanism are needed.

The key sequence that is used for representing a tree has length $l = n(n-1)/2$. Constructing a a tree results in sorting the l keys that goes with $O(l \log(l))$. Therefore, in comparison to other representations like Prüfer numbers the decoding process is more complex and demanding. For a more detailed description of the NetKey encoding the reader is referred to [5].

3 The One-Max Tree Problem

This section gives a short overview of the one-max tree problem. For further information please refer to [5].

For the one-max tree problem an optimal solution (tree) is chosen either randomly or by hand. The structure of this tree can be determined: It can be a star, a list, or an arbitrary tree with n nodes. In this work we only consider the optimal solution to be an arbitrary tree.

For the calculation of the fitness of the individuals, the distance d_{ab} between two trees G_a and G_b is used. It is defined as

$$d_{ab} = \frac{1}{2} \sum_{i=1}^{n-1} \sum_{j=0}^{i-1} |l_{ij}^a - l_{ij}^b|,$$

where l_{ij}^a is 1 if the link from node i to node j exists in tree G_a and 0 if it does not exist in G_a. n denotes the number of nodes. This definition of distance between two trees is based on the Hamming distance [12] and $d_{ab} \in \{0, 1, \ldots, n-2\}$.

When using this distance metric for a minimization problem the fitness of an individual G_i is defined as the distance $d_{i,opt}$ to the optimal solution G_{opt}. Therefore, $f_i = d_{i,opt}$, and $f_i \in \{0, 1, \ldots, n-2\}$. An individual has fitness (cost) of $n-2$ if it has only one link in common with the best solution. If the two individuals do not differ ($G_i = G_{opt}$), the fitness (cost) of G_i is $f_i = 0$. In this work we only want to use a minimization problem. Because this test problem is similar to the standard one-max-problem it is easy to solve for mutation-based GEAs, but somewhat harder for recombination-based GAs [13].

4   Performance of Evolution Strategies and Adjustment of Parameters

In this section, after a short introduction into the functionality of evolution strategies, we present an investigation into the adjustment of ES parameters for the one-max tree problem when using the NetKey encoding. The section ends with a short comparison of ES and GA for this specific problem.

4.1   A Short Introduction into Evolution Strategies

ESs were developed by Rechenberg and Schwefel in the 1960s at the Technical University of Berlin in Germany [14]. First applications were experimental and dealt with hydrodynamical problems like shape optimization of a bended pipe, drag minimization of a joint plate [2] and a structure optimization of a two-phase flashing nozzle [3].

The simple $(1 + 1)$-ES uses n-dimensional real valued vectors and creates one offspring $x' = \{x'_1, \ldots x'_n\}$ from one parent $x = \{x_1, \ldots x_n\}$ by applying mutation with identical standard deviations σ to each parental allele x_i.

$$x'_i = x_i + \sigma \cdot N_i(0,1) \quad \forall\, i = 1, \ldots, n.$$

$N(0,1)$ denotes a normal distributed one-dimensional random variable with expectation zero and standard deviation one. $N_i(0,1)$ indicates that the random variable is sampled anew for each possible value of the counter i. The resulting individual is evaluated and compared to its parent, and the better one survives to become the parent of the next generation. For the $(1 + 1)$-ES a theoretical convergence model for two specific problems, the sphere model and the corridor model, exists. The $\frac{1}{5}$-success rule reflects the theoretical result that, to achieve fast convergence, on average one out of five mutations should result in higher fitness values [14, p. 123].

To incorporate the principle of a population, [15] introduced the $(\mu + \lambda)$-ES and the (μ, λ)-ES. This notation considers the selection mechanism and the number of parents μ and offspring λ. For the $(\mu + \lambda)$-ES, the μ best individuals survive out of the union of the μ parents and the λ offspring. In the case of the (μ, λ)-ES, only the best μ offspring form the next parent generation. Both population-based ES start with a parent population of μ individuals. Each individual a consists of an n-dimensional vector $x \in \mathbb{R}^n$ and l standard deviations $\sigma \in \mathbb{R}_+$. One individual is described as $a = (x, \sigma)$ [16].

For both, $(\mu + \lambda)$-ES and (μ, λ)-ES, recombination is used for the creation of the offspring. Mostly, discrete recombination is used for the decision variables x'_i and intermediate recombination is used for the standard deviations σ'_i. Discrete recombination means that x'_i is randomly taken from one parent, whereas intermediate recombination creates σ'_i as the arithmetic mean of the parents standard deviations.

However, the main operator in ES is mutation. It is applied to every individual after recombination:

$$\sigma'_k = \sigma_k \cdot \exp(\tau_1 \cdot N(0,1) + \tau_2 \cdot N_k(0,1)) \ \forall\, k = 1, 2, \ldots, l,$$
$$x'_i = x_i + \sigma'_i \cdot N_i(0,1) \qquad\qquad \forall\, i = 1, 2, \ldots, n.$$

The standard deviations σ_k are mutated using a multiplicative, logarithmic normally distributed process with the factors τ_1 and τ_2. Then, the decision variables x_i are mutated by using the modified σ'_k. This mutation mechanism enables the ES to evolve its own strategy parameters during the search, exploiting an implicit link between appropriate internal model and good fitness values. One of the major advantages of ES is seen in its ability to incorporate the most important parameters of the strategy, e.g standard deviations, into the search process. Therefore, optimization not only takes place on object variables, but also on strategy parameters according to the actual local topology of the object function. This capability is called self-adaption.

4.2 Adjustment of Parameters

Over time, many recommendations for choosing ES parameters have been developed mainly for the simple sphere and the corridor model. We want to investigate if these recommendations also hold true for simple one-max tree problems represented using the real-valued NetKey encoding.

When using the $(1+1)$-ES there are two possibilities for choosing the standard deviation σ. It can be either fixed to some value or adapted according to the $\frac{1}{5}$-success rule. For sphere and corridor models this rule results in fastest convergence. However, sometimes the probability of success cannot exceed $\frac{1}{5}$. For problems, where the objective function has discontinuous first partial derivatives, or at the edge of the allowed search space, the $\frac{1}{5}$-success rule does not work properly. Especially in the latter case, the success rule progressively forces the sequence of iteration points nearer to the boundary and the step lengths are continuously reduced without the optimum being approached with comparable accuracy [17].

Figure 3(a) and Figure 1 illustrate the problems of the $\frac{1}{5}$-success rule when using ES for solving an 8 node one-max tree problem. The plots show the fitness and the standard deviation σ over the number of fitness calls. The initial standard deviation $\sigma_0 = 2$ and we performed 200 runs. Due to the success rule the standard deviation is continuously reduced and we get early convergence. The same results have been obtained for larger 16 and 32 node problem instances. This behavior of $(1+1)$-ES can be explained when examining the NetKey encoding. In section 2 we saw that only $n-1$ out of $n(n-1)/2$ links are used for constructing the tree. Therefore, a mutation of one allele often does not result in a change of the represented tree. Many mutations do not result in a different phenotype but only change the genotype. However, for the $\frac{1}{5}$-rule we assume that every mutation results in a different phenotype and about every fifth new phenotype is superior to its parent. Therefore, the one-max tree problem is more difficult than the fully easy sphere and corridor models, and the $\frac{1}{5}$-rule can not be used.

Instead, we can use a fixed standard deviation σ. In Figure 3(b) we show the fitness after 10 000 iterations over the standard deviation σ for the 8 and 16 node one-max tree problem. The results indicate that the $(1 + 1)$-ES shows the best performance for a fixed standard deviation of $\sigma \approx 0.2 - 0.4$. Larger standard deviations do not result in a faster convergence but the search becomes random. With smaller standard deviations we also do not get better solutions, because with small standard deviations of mutation we only make slow progress.

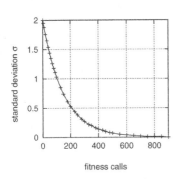

Fig. 1. Standard deviation σ over number of fitness calls. We use the $\frac{1}{5}$-rule and a $(1+1)$-ES for the 8 node one-max tree problem.

To overcome the problem of the $(1 + 1)$-ES getting stuck in local optima, population-based ES approaches like $(\mu + \lambda)$-ES and (μ, λ)-ES have been proposed. We want to examine how one can adjust the strategy parameters τ_1 and τ_2. The standard deviations are mutated using a multiplicative, logarithmic normally distributed process. The logarithmic normal distribution is motivated as follows. A multiplicative modification process for the standard deviations guarantees positive values for σ and smaller modifications must occur more often than larger ones [18].

Because the factors τ_1 and τ_2 are robust parameters, [18] suggests setting them as follows: $\tau_1 \propto (\sqrt{2\sqrt{n}})^{-1}$ and $\tau_2 \propto (\sqrt{2n})^{-1}$. Newer investigations indicate that optimal adjustments are in the interval $[0.1, 0.2]$ [19,20]. τ_1 and τ_2 can be interpreted in the sense of "learning rates" as in artificial neural networks, and preliminary experiments with proportionality factors indicate that the search process can be tuned for particular objective functions by modifying these factors. We investigated in Figure 2 for the $(\mu + \lambda)$-ES whether the recommendations of Schwefel or Kursawe are also valid for the 8 node one-max tree problem. The plots show how the best fitness after 100 generations depends on τ_1 and τ_2. The results confirm the recommendations from Kursawe to initialize the parameters in the interval $[0.1, 0.2]$, where $\tau_2 > \tau_1$. The best solutions are $\tau_1 = 0.1$ and $\tau_2 = 0.15$ for the $(\mu + \lambda)$-ES and $\tau_1 = 0.15$ and $\tau_2 = 0.2$ for the (μ, λ)-ES.

The next part of our investigation focuses on the optimal proportion of μ parents to λ offspring to maximize the convergence velocity. [17] proposed a $(1, 5)$-ES or a $(1, 6)$-ES that is nearly optimal for sphere and corridor models. Figure 3(c) $((\mu + \lambda)$-ES$)$ and Figure 3(d) $((\mu, \lambda)$-ES$)$ show the fitness over the number of fitness calls for the 8 node one-max tree problem. We used a population size of $N = \mu + \lambda = 200$, $\tau_1 = 0.13$, $\tau_2 = 0.16$, $\sigma_0 = 0.5$, and performed 1000 runs for every parameter setting. The results show that a ratio of $\frac{\mu}{\lambda} \in \{\frac{1}{4} \ldots \frac{1}{7}\}$ results in good performance for the $(\mu + \lambda)$-ES and the (μ, λ)-ES. These results confirm the recommendations from [21] and [16]. The investigations for the sphere model

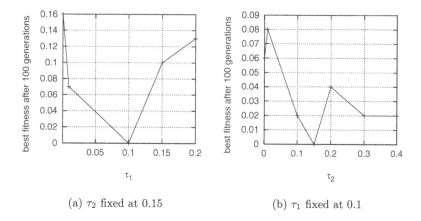

(a) τ_2 fixed at 0.15   (b) τ_1 fixed at 0.1

Fig. 2. Best fitness at end of the run over the strategy parameters τ_1 and τ_2 for the $(\mu + \lambda)$-ES. In Figure 2(a) we fixed τ_2 at 0.15 and varied τ_1, and in Figure 2(b) we fixed τ_1 at 0.1 and varied τ_2. We used $N = 200$, $\sigma_0 = 0.5$, $\mu/\lambda = 0.25$, 100 generations, and 1000 runs per plot.

indicated that the ratio of $\frac{\mu}{\lambda} \approx \frac{1}{7}$ is optimal concerning the accelerating effect of self-adaption. This ratio also provides the basic parameterization instrument for controlling the character of the search. Decreasing μ/λ emphasizes on path-oriented search and convergence velocity, while increasing μ/λ leads to a more volume-oriented search.

Finally, we want to examine the population size N which provides optimal convergence velocity. The population size mainly depends on the representation of the individuals and the optimization problem. Guidelines for choosing proper population sizes N when using NetKeys for the one-max tree problem and using selectorecombinative GAs were shown in [5]. In Figure 3(e) we compare the fitness over the number of fitness calls for the $(\mu + \lambda)$-ES for different population sizes $N = \mu + \lambda$. We used $\tau_1 = 0.13$, $\tau_2 = 0.16$, $\sigma_0 = 0.5$, $\mu/\lambda = 0.25$ and performed 1000 runs for every parameter setting. The results reveal that for a 8 node problem, a population size $N = 200$ is enough to allow ES to find the optimal solution reliably and fast.

Our investigations indicate that the simple $\frac{1}{5}$-rule for the $(1 + 1)$-ES from [14] does not work when using NetKeys. However, when using $(\mu + \lambda)$-ES or (μ, λ)-ES the recommendations for the simple sphere and corridor models from [18], [19], and [17] can also be used for the one-max tree problem using NetKeys. The existing guidelines help us to choose proper strategy parameters τ_1, τ_2, and the ratio $\frac{\mu}{\lambda}$. For further information about the use of ES for tree problems using the NetKey encoding the reader is referred to [22].

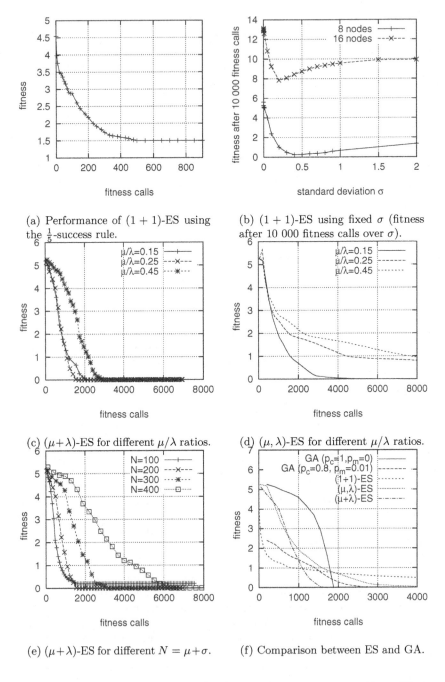

(a) Performance of $(1 + 1)$-ES using the $\frac{1}{5}$-success rule.

(b) $(1 + 1)$-ES using fixed σ (fitness after 10 000 fitness calls over σ).

(c) $(\mu + \lambda)$-ES for different μ/λ ratios.

(d) (μ, λ)-ES for different μ/λ ratios.

(e) $(\mu + \lambda)$-ES for different $N = \mu + \sigma$.

(f) Comparison between ES and GA.

Fig. 3. Adjustment of ES parameters for the 8 node one-max tree problem. All plots show the fitness over the number of fitness calls (except Figure 3(a) which shows best fitness after 10 000 fitness calls over σ.)

4.3   A Comparison to Genetic Algorithms for the One-Max Tree Problem

After identifying optimal strategy parameters for ES we want to compare the performance of ES with GAs for the one-max tree problem using NetKeys.

For both optimization methods, ES and GA, we use uniform crossover. For the GA we implemented a roulette-wheel selection scheme and used $N = 200$. Mutation in the context of GA means that the value of one key is randomly changed. As before, we used for the ES $\tau_1 = 0.13$, $\tau_2 = 0.16$, $\sigma_0 = 0.5$, $N = \mu + \lambda = 200$, and performed 1000 runs for every parameter setting.

Figure 3(f) compares the performance of ES and GAs for the one-max tree problem. We plot the fitness over the number of fitness calls. The results show that a $(\mu + \lambda)$-ES has the highest performance. The $(1 + 1)$-ES gets stuck and is not able to find the optimal solution.

5   Conclusions

In this paper we extended the use of evolution strategies to combinatorial tree problems. Evolution strategies are designed for continuous optimization problems and can be applied to trees when using the continuous network random key (NetKey) representation. We examined for the small 8 node one-max tree problem how to adjust the parameters of the $(1 + 1)$-, $(\mu + \lambda)$-, and (μ, λ)-ES and compared their performance to a simple GA.

The results showed that the recommendations regarding the adjustment of the ES parameters (τ_1, τ_2, and μ/λ) for simple sphere and corridor models can also be used for the easy one-max tree problem when using the NetKey encoding. Only the $\frac{1}{5}$-success rule for the $(1 + 1)$-ES does not hold true for the one-max tree problem because most of the mutations do not change the represented tree. Therefore, the strategy parameter σ is continously reduced and the algorithm gets stuck.

The results indicate that existing theory about ES can often help in finding good parameter settings for new types of problems. We want to encourage researchers when developing new representations or techniques to first look at existing theory, to check if they can be used advantageously, and not to reinvent the wheel.

References

1. H.-P. Schwefel. Kybernetische Evolution als Strategie der experimentellen Forschung in der Strömungstechnik. Master's thesis, Technische Universität Berlin, 1965.
2. I. Rechenberg. Cybernetic solution path of an experimental problem. Technical Report 1122, Royal Aircraft Establishment, Library Translation, Farnborough, Hants., UK, 1965.
3. H.-P. Schwefel. Experimentelle Optimierung einer Zweiphasendüse. Bericht 35, AEG Forschungsinstitut Berlin, Projekt MHD-Staustahlrohr, 1968.

4. J. H. Holland. *Adaptation in natural and artificial systems*. University of Michigan Press, Ann Arbor, MI, 1975.
5. F. Rothlauf, D. E. Goldberg, and A. Heinzl. Network random keys – a tree network representation scheme for genetic and evolutionary algorithms. Technical Report No. 8/2000, University of Bayreuth, Germany, 2000. to be published in Evolutionary Computation.
6. J. C. Bean. Genetics and random keys for sequencing and optimization. Technical Report 92-43, Department of Industrial and Operations Engineering, University of Michigan, Ann Arbor, MI, June 1992.
7. Franz Rothlauf. *Towards a Theory of Representations for Genetic and Evolutionary Algorithms: Development of Basic Concepts and their Application to Binary and Tree Representations*. PhD thesis, University of Bayreuth/Germany, 2001.
8. C. C. Palmer. *An approach to a problem in network design using genetic algorithms*. unpublished PhD thesis, Polytechnic University, Troy, NY, 1994.
9. H. Prüfer. Neuer Beweis eines Satzes über Permutationen. *Archiv für Mathematik und Physik*, 27:742–744, 1918.
10. Günther R. Raidl. An efficient evolutionary algorithm for the degree-constrained minimum spanning tree problem. In *Proceedings of 2000 IEEE International Conference on Evolutionary Computation*, pages 43–48, Piscataway, NJ, 2000. IEEE.
11. F. Abuali, R. Wainwright, and D. Schoenefeld. Determinant factorization and cycle basis: Encoding schemes for the representation of spanning trees on incomplete graphs. In *Proceedings of the 1995 ACM/SIGAPP Symposium on Applied Comuting*, pages 305–312, Nashville, TN, February 1995. ACM Press.
12. R. Hamming. *Coding and Information Theory*. Prentice-Hall, 1980.
13. D. E. Goldberg, K. Deb, and D. Thierens. Toward a better understanding of mixing in genetic algorithms. *Journal of the Society of Instrument and Control Engineers*, 32(1):10–16, 1993.
14. I. Rechenberg. Bionik, Evolution und Optimierung. *Naturwissenschaftliche Rundschau*, 26(11):465–472, 1973.
15. H.-P. Schwefel. *Evolutionsstrategie und numerische Optimierung*. PhD thesis, Technical University of Berlin, 1975.
16. T. Bäck. *Evolutionary Algorithms in Theory and Practice*. Oxford University Press, New York, 1996.
17. H.-P. Schwefel. *Evolution and Optimum Seeking*. Wisley & Sons, New York, 1995.
18. H.-P. Schwefel. *Numerische Optimierung von Computer-Modellen mittels der Evolutionsstrategie*. Birkhäuser, Basel, 1977. from Interdisciplinary Systems Research, volume 26.
19. F. Kursawe. *Grundlegende empirische Untersuchungen der Parameter von Evolutionsstrategien - Metastrategien*. PhD thesis, University of Dortmund, 1999.
20. V. Nissen. *Einführung in evolutionäre Algorithmen: Optimierung nach dem Vorbild der Evolution*. Vieweg, Wiesbaden, 1997.
21. H.-P. Schwefel. Collective phenomena in evolutionary systems. In P. Checkland and I. Kiss, editors, *Problems of Constancy and Change - The Complementarity of Systems Approaches to Complexity*, volume 2, pages 1025–1033, Budapest, 1987. Papers presented at the 31st Annual Meeting of the International Society for General System Research.
22. Barbara Schindler. Einsatz von Evolutionären Stratgien zur Optimierung baumförmiger Kommunikationsnetzwerke. Master's thesis, Universität Bayreuth, Lehrstuhl für Wirtschaftsinformatik, Mai 2001.

Evolutionary Computational Approaches
to Solving the Multiple Traveling Salesman Problem
Using a Neighborhood Attractor Schema

Donald Sofge[1], Alan Schultz[1], and Kenneth De Jong[2]

[1] Navy Center for Applied Research in Artificial Intelligence,
Naval Research Laboratory
{Sofge, Schultz}@aic.nrl.navy.mil
[2] Department of Computer Science,
George Mason University
kdejong@cs.gmu.edu

Abstract. This paper presents a variation of the Euclidean Traveling Salesman Problem (TSP), the Multiple Traveling Salesman Problem (MTSP), and compares a variety of evolutionary computation algorithms and paradigms for solving it. Techniques implemented, analyzed, and discussed herein with regard to MTSP include use of a neighborhood attractor schema (a variation on k-means clustering), the "shrink-wrap" algorithm for local neighborhood optimization, particle swarm optimization, Monte-Carlo optimization, and a range of genetic algorithms and evolutionary strategies.

1 Introduction

The Euclidean Traveling Salesman Problem (ETSP), or simply TSP, is a well-known and extensively studied benchmark for many new developments in combinatorial optimization [1],[2],[3],[4],[5],[6], including techniques in evolutionary computation [7],[8],[9]. A very simple yet highly practical extension of TSP is a TSP with multiple sales agents. This problem is defined as follows:

Def. MTSP(n, k) --- given n cities and k sales agents, find the k closed circuit paths which minimize the sum of the squares of the path lengths.

This paper will henceforth refer to this problem as the multiple traveling salesman problem, or MTSP. The MTSP is actually a two-level optimization problem. On the first level we wish to determine the optimal subdivision of cities into k groups. The second level of optimization is to find the minimum length circuit for each of the groups of cities. This paper will be primarily concerned with the first level of optimization, determining the optimal subdivision of cities into groups, since this area has received less attention in the literature than solving the Euclidean TSP.

S. Cagnoni et al. (Eds.): EvoWorkshops 2002, LNCS 2279, pp. 153–162, 2002.

The MTSP is similar to the k-TSP problem discussed in [5], but excludes the requirement that each circuit include a common base city. The cluster optimization task is also related to the minimum sum-of-squares clustering problem (MSSC) [10]. However, in MSSC the distance measurement for each cluster is the sum of squared distances between each point and the cluster centroid, whereas in MTSP the relevant distance measurement for each cluster is the squared circuit length.

A note of explanation about the definition of MTSP given above is in order. Why do we want to find the k closed-circuit-paths which minimize the sum of the squares of the circuit lengths, rather than simply minimizing the sum of the circuit lengths? The reason is in order to distribute the work load as equitably as possible amongst the k sales agents. Thus, in addition to the constraint typically found in TSP of finding the minimum length circuit, we have imposed an additional constraint for distributing the cities amongst the sales agents. Minimization of the sum of the squares of the circuit lengths satisfies these constraints.

2   Approach to Solving MTSP

MTSP is a two-level optimization problem, with one problem being to determine how to subdivide the list of n cities into k groups or neighborhoods (i.e. clustering), and the second problem being to optimize the circuits for each cluster or neighborhood. While it may be possible to route first then cluster [5], we chose to cluster first then route since this seemed to be a more logical decomposition of the problem.

Since our cities lie in a Euclidean space and we wish to minimize local path lengths, it is reasonable to use a distance metric between each city and each cluster center to determine neighborhood affiliation. However, with MTSP we have an additional constraint of distributing the circuit lengths roughly evenly amongst the k neighborhoods. Since we cannot assume homogeneous distribution of the cities in Euclidean space, and it is quite reasonable to expect that given a randomly generated set of city coordinates (or those found on a real map) some areas will be significantly more densely populated with cities than others, we need another term or condition to better define our clusters.

2.1   Neighborhood Attractor Schema

The central idea behind the neighborhood attractor schema is that a number of moveable attractors are added to the map which have the effect of "capturing" cities. Each attractor has a position (x, y) and an attraction constant (a). The attraction between each city and each attractor is calculated as the attraction constant of the cluster attractor (a) divided by the distance (or squared distance) between the city and attractor. This feature allows the neighborhood attractors to then partition the non-homogeneously distributed cities into clusters such that the second constraint of MTSP (equitable distribution of work) is more easily satisfied. This approach is referred to as the neighborhood attractor schema. A result of applying the neighborhood attractor schema to MTSP(30,5) is shown in **Figure 1**.

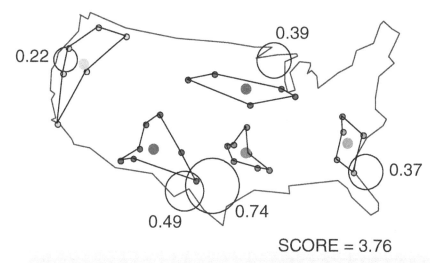

SCORE = 3.76

Fig. 1. MTSP(30,5) Optimal Solution Using Neighborhood Attractor Schema

This approach is roughly analogous to a gravitational attractor model where each planetoid or moon is "affiliated" with a planet based upon the gravitational attraction between the planetoid and the planet. In this case the "planetoids" are the cities (each with unit mass) and the "planets" are the basins of attractions, with each having a position (x,y) and mass parameter. The positions and mass parameters for the attractors are optimized using an evolutionary algorithm, particle swarm or other technique.

In **Figure 1** the large circles represent the neighborhood attractors, with their diameters determined by their respective attractor constants (numbers shown next to circles). The smaller solid circles inside each cluster show the unweighted centroid for each cluster, which is used in the shrink-wrap algorithm discussed in the next section.

One advantage this schema has over k-means-clustering models is that the neighborhood attractors need only compete with other neighborhood attractors to determine city affiliations and corresponding path lengths. They do not need to move to the centers of the clusters of cities they attract. This feature gives them more flexibility for partitioning the cities into equitable subgroups, especially important where the distribution of cities is highly non-homogeneous.

2.2 Optimizing Neighborhood Closed Circuit Paths

Given a MTSP(n,k) and a "solution" of neighborhood attractors (and their corresponding city cluster groups), we still need a means of optimizing closed-circuit paths for each of the neighborhoods in order to score or determine the fitness of our solution. Finding such an optimum for each neighborhood becomes a standard Euclidean TSP. Since we do not wish to address the full computational burden of solving k neighborhood TSPs before we can even evaluate our solution of k neighborhood attractors, a shortcut for estimating a good (if not always optimal) closed circuit path for each neighborhood is desirable.

A rather simple and fairly elegant technique for accomplishing this is the shrink-wrap algorithm (though certainly not the only one, see [5] for others). The mean for each cluster of cities is calculated, and then the cities' coordinates are translated in Euclidean space such that the cluster mean is at the origin. The (x, y) coordinates are then converted to polar coordinates (θ, ρ), and the order of cities is sorted by the polar coordinates (θ first, then ρ). The sorted order list is then linked in ascending order, with an additional link added from the last city in the list back to the first. This process is illustrated in **Figure 2**. The cities are then restored to their original Euclidean coordinates, but the polar sorted linked list order is retained.

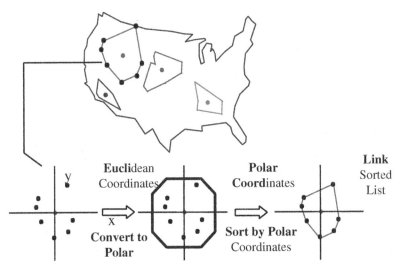

Fig. 2. The Shrink-Wrap Algorithm

The shrink-wrap algorithm thus gives a good first estimate of the node order for an optimum closed-circuit path for each neighborhood. From this we calculate a path length for each neighborhood, and by summing the squares of the k closed-circuit paths, we can generate a score for each solution set of k neighborhood attractors.

2.3 Chromosome Representation

A solution to MTSP(n, k) consists of k closed circuit paths, and may be completely determined for a particular group of n cities by a vector of k neighborhood attractors.

The city affiliations for each attractor are determined by the relative parameters of the solution set of attractors, and similarly the closed circuit path lengths are determined using the shrink-wrap algorithm. Since the attractor schema requires three parameters for each neighborhood attractor (**x,y,a**) with position (**x,y**) and attraction constant (**a**), then a solution to MTSP(n, k) can be specified as a vector of length 3*k (each solution consists of k triplets (**x,y,a**)), where each parameter is a real-valued number between zero and one.

Each neighborhood attractor has an associated linked list of cities (of varying length, possibly including zero length) which define the neighborhood circuit. Genetic algorithms which use mutation operators generally specify a fixed mutation rate, often between 0.001 and 0.1, which is applied to each gene in a chromosome independently of whether any mutations have occurred in other parts of the same chromosome.

2.4 A Menu of EA Designs

A wide variety of evolutionary algorithms have been developed which can be applied to solving optimization problems such as MTSP [11]. In this effort a broad representative sampling of these methods was implemented and applied to MTSP. These techniques may be roughly subdivided into three classes of evolutionary computation techniques: genetic algorithms, evolutionary strategies, and other methods.
The techniques implemented and tested were:

- Genetic Algorithm (μ, λ) using Mutation and Crossover
- Genetic Algorithm (μ, λ) using Mutation Only
- Genetic Algorithm (μ, λ) using Mutation and Crossover, plus Elitism
- Evolutionary Strategy ($\mu + \lambda$) using Mutation and Crossover
- Evolutionary Strategy ($\mu + \lambda$) using Mutation Only
- Generational Monte-Carlo Optimization
- Particle Swarm Optimization

Note that the terms "genetic algorithm" and "evolutionary strategy" are used to distinguish (μ, λ) strategies from ($\mu + \lambda$) strategies, with real-valued representations used for both. The "genetic algorithms" use a strategy with the property that each generation the parent population is replaced by their offspring, commonly known as (μ, λ). When the elitism operator is added, the single fittest member from the parent population is retained in the population the following generation.

The "evolutionary strategies" generate offspring from the parent population, but then the offspring must compete with the parents, hence a ($\mu + \lambda$) strategy. The populations are concatenated and sorted by score, and only the fittest members survive with the remainder discarded. Since the fittest member is already propagated forward from the parent generation, there is no need to use an elitism operator.

The Generational Monte-Carlo Optimization method used was implemented as a ($\mu + \lambda$) evolutionary strategy, but with mutation rate of (1.0). During each generation a number of new individuals equal to the population size are randomly generated, and then they are added to the existing population. The population is then sorted and truncated. During the trials the new individuals were randomly and independently produced, but the evaluation and sorting of them was generational for ease of comparison.

The particle swarm optimization method (described below) defines a population of particles in n-space (where n is the number of genes to be represented) which have position and velocity. The particle positions and velocities are updated based on a combination of parameters including previous states, the global best yet found (analogous to use of an elitism policy), pre-assigned constants, and random factors.

2.5 Particle Swarm Optimization

Particle swarm optimization was first proposed by Kennedy and Eberhart [12] and was inspired by natural biological processes based on the notion of swarm behavior in animal or insect populations. A swarm is defined as a population of interacting elements which is able to achieve some global objective through collaborative search. Swarm optimization relies on the tendency of a swarm population to converge on a center of mass along critical dimensions, resulting in achievement of an optima [13].

In particle swarm optimization (PSO) each individual or chromosome is considered to be a particle in hyperspace having both position and velocity. The particles are then "flown" through hyperspace (a chromosome with n genes would exist in an n-dimensional space), with its velocity and position information influenced by its previous state, its previous best (fittest) state, and the global best for the entire population.

Each particle i is represented as
$$\mathbf{x}_i = (x_{i1}, x_{i2}, x_{i3}, \ldots, x_{in}) \tag{1}$$
and has velocity
$$\mathbf{v}_i = (v_{i1}, v_{i2}, v_{i3}, \ldots, v_{in}) . \tag{2}$$

The previous best position for each particle is stored, and given by
$$\mathbf{pbest}_i = (p_{i1}, p_{i2}, p_{i3}, \ldots, p_{in}) . \tag{3}$$

The global best for all particles yet seen,
$$\mathbf{gbest} = (g_1, g_2, g_3, \ldots, g_n) . \tag{4}$$

The particle update equations are then given by:
$$\mathbf{v}_{i+1} = w*\mathbf{v}_i + c_1*\text{rand1}*(\mathbf{pbest}_i - \mathbf{x}_i) + c_2*\text{rand2}*(\mathbf{gbest} - \mathbf{x}_i) , \tag{5}$$
$$\mathbf{x}_{i+1} = \mathbf{x}_i + \mathbf{v}_{i+1} \tag{6}$$

where w is inertial weight, c_1 and c_2 are constants, and rand1 and rand2 are random values between 0.0 and 1.0.

When implementing PSO it is important that great care is taken in choosing w, c_1 and c_2, as well as the initial magnitudes of the velocity vectors. A heuristic for annealing w is given by [13], that w should start at 0.9 and be decreased linearly to 0.4 over 1000 generations. In this effort the population generally converged within 100 generations, even using the cited heuristic. Instead, the following parameter values were used:

initial velocity = 0.1 * rand,   where   0.0 < rand < 1.0
$0.4 \leq w \leq 0.9$,   decreased linearly over 100 generations
$c_1 = 1.0, \quad c_2 = 0.1$

The results of PSO on MTSP are discussed below under **Results**.

2.6   Mutation Operator with Parameter Space Search

The MTSP(n,k) landscape is characterized by plateaus for which mutations often result in offspring with scores equal to those of the parents. This feature inhibits continued progress to the optimum. One approach to address this is to augment the mutation operator with the capability to perform a search in parameter space in order to find a mutation for which the offspring fitness is not identical to that of the parent.

Once a decision is made to generate a mutation (using a generated random number compared with the mutation rate), a mutation delta is "suggested" to the mutation operator. The operator adds the mutation delta to the parent individual and generates an offspring individual. The fitness of the offspring is measured, and if the fitness is different from that of the parent, the offspring is kept. However, if the fitness of the offspring is the same as that of the parent, then the mutation operator begins incrementing the mutation delta, generating and testing offspring until an offspring is generated which has a different fitness from the parent. The parameter search must recognize when it has reached the edge of the parameter space (e.g. edge of the map), after which it may change the direction of delta. The parameter search must also recognize when it is not possible to find an acceptable mutation using that parent, for which it has the ability to select a new parent. The parameter search rule is given:

Parameter Search Rule:
  if **fitness(offspring)==fitness(parent)**
  then   **increment mutation delta by 10%**
  (if limit of parameter space reached, change sign of delta)
  generate new offspring from parent
  (if selected parameter set fully searched and no new)
  (fitness is found, then select another parameter to mutate)
  repeat until **fitness(offspring)<>fitness(parent)**

3   Results

Results were generated for three different problems: MTSP(30,5), MTSP(20,3) and MTSP(50,3). The different algorithms applied are listed below. Each algorithm was run for 100 generations, with 10 independent runs performed using different random seeds.

- Genetic Algorithm (μ, λ) with Mutation & Crossover (GA-MC)
- Genetic Algorithm (μ, λ) with Mutation, Crossover and Elitism (GA-MCE)
- Genetic Algorithm (μ, λ) with Mutation Only (GA-M)
- Evolutionary Strategy $(\mu + \lambda)$ with Mutation & Crossover (ES-MC)
- Evolutionary Strategy $(\mu + \lambda)$ with Mutation Only (ES-M)
- Particle Swarm Optimization (PSO)
- Generational Monte-Carlo Optimization (GMCO)

All of the algorithms were at least fairly successful at finding good solutions to each problem. The normalized results of all of the algorithms are shown following in **Figure 3**. The genetic algorithms that did not use elitism performed worst. This is apparently due to the need for high exploration and the loss of good solutions when the parent generations are completely replaced by their offspring. Preserving the best parent and adding it to the offspring generation (as shown with GA-MCE) substantially improves performance, and yields one of the best algorithms overall.

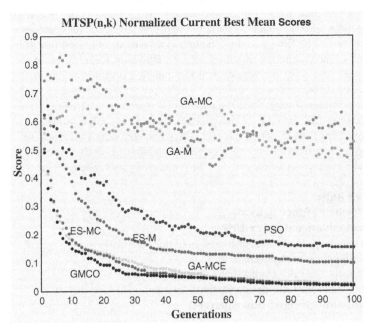

Fig. 3. Normalized Results of All EAs on MTSPs

Particle swarm optimization performed reasonably well, but often failed to locate the global optimum solution. Since this algorithm is quite sensitive to its parameter settings, and comparatively little effort was made to fine-tune PSO for this application, it is quite possible and reasonable that better results could be obtained with PSO with different parameter settings. One obvious target is a different annealing schedule for the inertial weight **w**.

The evolutionary strategies performed better than PSO and the GAs without elitism. The algorithm ES-MC, employing both mutation and crossover, performed significantly better than ES-M, which used mutation only. Addition of the crossover operator seems to provide an effective mechanism for finding better solutions. In fact, the algorithm ES-MC was one of the top performing techniques.

Perhaps the most interesting result is the excellent performance of the GMCO strategy. This algorithm essentially operates as an evolutionary strategy where a number of random individuals are generated each generation, and then the offspring are added to the parents, they are sorted and the fittest members are kept. Thus, this algorithm combines strong elitism along with high exploration. Overall, this technique seemed to offer the best results. An interesting question is whether it would work as well on more complex problem sets with larger values for k, the number of agents.

For each algorithm the chromosome population size was kept at 10. A number of runs (though not the full matrix) were done with higher values of k and larger population sizes. A strictly subjective observation was that for best performance population size should scale with k. Thus, for 3-5 agents a population of 10 was sufficient, but for $k=10$, a more suitable chromosome population size was 20-30. This suggests that the complexity of the problem increases not so much with the number of cities, but with the number of sales agents. However, the issues of complexity and scalability need to be studied more thoroughly using larger numbers of cities and various combinations on the number of sales agents and chromosome population sizes.

4 Conclusions and Future Research

In this paper a variation of the Euclidean Traveling Salesman Problem (TSP), the Multiple Traveling Salesman Problem (MTSP), was presented and discussed. This problem is representative of many real-world multi-level optimization problems, and thus offers a platform for development and testing of evolutionary computation paradigms for solving such problems. An approach to solving MTSPs was presented using the Neighborhood Attractor Schema, the Shrink-Wrap algorithm, and a variety of evolutionary computation algorithms and paradigms. The results were encouraging, but also showed that much more could be done to understand these problems and evolutionary computation approaches to solving them.

References

1. Bellman, R.E., "Dynamic programming treatment of the traveling salesman problem", Journal of the ACM, 9:61—63 (1962)
2. Bellmore, M. and G.L. Nemhauser, "The traveling salesman problem: a survey", Operations Res. 16, 538-558 (1968)
3. Karp, R.M., "Probabilistic analysis of partitioning algorithms for the traveling-salesman problem in the plane", Math. of Operations Research, 2:209—224 (1977)
4. Papadamitriou, C.H., "The Euclidean traveling salesman problem is NP-complete", Theoretical Computer Science, 4:237—244 (1977)
5. Lawler, E.L., Lenstra, J.K., Rinnooy Kan, A.H.G and D.B. Shmoys. The traveling salesman problem. Wiley, New York (1985)
6. Arora, S., "Nearly linear time approximation schemes for Euclidean TSP and other geometric problems", In 38th Annual Symposium on Foundations of Computer Science, pages 554-563, Miami Beach, Florida, 20-22 October (1997)
7. Goldberg, D.E., "Messy genetic algorithms: Motivation, Analysis, and First results", Complex Systems, Vol. 3, pp. 493-530 (1989)
8. Jog, P., Suh, J.Y., and D. Van Gucht, "The Effects of Population Size, Heuristic Crossover and Local Improvement on a Genetic Algorithm for the Traveling Salesman Problem", Proc. of the 3rd Intl. Conference on Genetic Algorithms, pp. 110-115, Morgan Kaufmann, (1989)
9. Julstrom, B.A., "Insertion Decoding Algorithms and Initial Tours in a Weight-Coded GA for TSP", Genetic Programming 1998: Proc. of the Third Annual Conference, pp. 528-534, Morgan Kaufmann, 22-25 (1998)
10. Hansen, P., and N. Mladenoviæ, "J-Means: A new local search heuristic for minimum sum-of-squares clustering", Pattern Recognition, Vol. 34 (2) 405-413, 2001.
11. De Jong, K., Course on Evolutionary Computation, George Mason University, (1998)
12. Kennedy, J. and R.C. Eberhart, "Particle Swarm Optimization", Proc. IEEE Intl. Conf. on Neural Networks, pp. IV:1942-1948 (1995)
13. Eberhart, R.C., and Y. Shi, "Evolving Artificial Neural Networks", Proc. Intl. Conf. on Neural Networks and the Brain - Beijing (1998)

Boosting ACO with a Preprocessing Step

Christine Solnon

LISI - University Lyon 1 - 43 bd du 11 novembre, 69 622 Villeurbanne cedex, France

Abstract. When solving a combinatorial optimization problem with the Ant Colony Optimization (ACO) metaheuristic, one usually has to find a compromise between guiding or diversifying the search. Indeed, ACO uses pheromone to attract ants. When increasing the sensibility of ants to pheromone, they converge quicker towards a solution but, as a counterpart, they usually find worse solutions. In this paper, we first study the influence of ACO parameters on the exploratory ability of ants. We then study the evolution of the impact of pheromone during the solution process with respect to its cost's management. We finally propose to introduce a preprocessing step that actually favors a larger exploration of the search space at the beginning of the search at low cost. We illustrate our approach on Ant-Solver, an ACO algorithm that has been designed to solve Constraint Satisfaction Problems, and we show on random binary problems that it allows to find better solutions more than twice quicker.

1 Introduction

When solving a combinatorial optimization problem with the ACO metaheuristic [1], one usually has to find a compromise between two dual goals, i.e., guiding or diversifying the search.

Guiding the search with ACO: ACO aims at guiding the search towards "promising" states, that are close to the best solutions found so far. Indeed, it has been shown for a large number of combinatorial optimization problems that higher quality local minima tend to be closer to global minima than lower quality local minima. Actually, this property underlies the motivation of evolutionary approaches which are of little interest on problems for which the correlation between solution fitness and distance to optimal solutions is too low [2,3].

In order to reach this guidance goal, ACO uses a stigmergetic communication mechanism: ants lay pheromone on graph edges in order to attract other ants, and the quantity of pheromone laid is proportional to the path quality so that edges participating to the best paths become more attractive. Improved performances can be obtained when introducing an elitist strategy where only the best ants deposit pheromone, as proposed in the $\mathcal{MAX} - \mathcal{MIN}$ Ant System [3].

Diversifying the search with ACO: While guiding the search, one must also favor exploration to discover new, and hopefully more successful, areas of the search space. Diversification is particularly important at the beginning of the search in order to avoid premature convergence towards local optima.

S. Cagnoni et al. (Eds.): EvoWorkshops 2002, LNCS 2279, pp. 163–172, 2002.

In order to reach this diversification goal, ants choose their way to go in a stochastic way —with respect to transition probabilities— so that an edge with a small quantity of pheromone can be chosen —with a small probability— even though there are other edges with higher quantities of pheromone. To emphasize exploration, [3] imposes lower and upper bounds τ_{min} and τ_{max} on pheromone trails so that for all pheromone trails τ_{ij}, $\tau_{min} \leq \tau_{ij} \leq \tau_{max}$. These bounds prevent the relative differences between pheromone trails from becoming too extreme so that the probability of choosing an edge never becomes null. Also, pheromone trails are initialized at τ_{max} at the beginning of the algorithm, thus achieving a higher exploration of the search space during the first cycles.

Influence of parameters on guidance/diversification: The behaviour of ants can be influenced by modifying parameter values [4]. In particular, diversification can be emphasized both by decreasing the value of the pheromone factor weight α —so that ants become less sensitive to pheromone trails— and by increasing the value of the pheromone persistence rate ρ —so that pheromone evaporates more slowly. When increasing the exploratory ability of ants in this way, one usually finds better solutions, but as a counterpart it takes longer time to find them. Different approaches have been proposed for finding "good" parameter values, such as [5]. However, one always has to choose between values that give priority to solution quality —by favoring diversification — or values that give priority to computation time —by favoring guidance.

Overview of the paper: The goal of the preprocessing step presented in this paper is to increase diversification at low cost, so that ants can find better solutions in much less time. The basic idea is to collect a significant number of paths without using pheromone. These collected paths constitute a kind of sampling of the search space and they are used to initialize pheromone trails before running the ACO algorithm. The paper is organized as follows: we first study the impact of ACO parameters with respect to guidance and diversification; we then study the evolution of the impact of pheromone trails during solution process, and we show that during the first cycles pheromone does not influence much on ants behaviour whereas its management is tedious and expensive; we finally describe the preprocessing step that actually favors a larger exploration of the search space at the beginning of the search at low cost. This approach is illustrated with Ant-Solver, an ACO algorithm designed to solve Constraint Satisfaction Problems (CSPs). The next section recalls some definitions and terminology on CSPs and random binary CSPs and briefly describes Ant-Solver.

2   Background

CSPs: A *CSP* [6] is defined by a triple (X, D, C) such that $X = \{X_1, ..., X_n\}$ is a finite set of n variables, D is a function which maps every variable $X_i \in X$ to its domain $D(X_i)$, i.e., the set of values that can be assigned to X_i, and C is a set of constraints, i.e., relations between variables that restrict the set of values that can be assigned simultaneously to the variables. An *assignment*, denoted by $\mathcal{A} = \{<X_1, v_1>, <X_2, v_2>, \ldots, <X_k, v_k>\}$, is a set of variable-value pairs and corresponds to the simultaneous assignment of values $v_1, v_2, \ldots, v_k$ to variables $X_1, X_2, \ldots, X_k$ respectively. The *cost* of an assignment $\mathcal{A}$ is defined by the number of violated constraints in $\mathcal{A}$. A *solution of a CSP* (X, D, C) is a complete assignment for all the variables in X that satisfies all the constraints in C, i.e., the cost of which is 0.

Random binary CSPs: Binary CSPs only have binary constraints, i.e., each constraint involves exactly two variables. Binary CSPs can be generated at random. A class of randomly generated CSPs is characterized by 4 components $< n, m, p_1, p_2 >$ where n is the number of variables, m is the uniform domain size, p_1 is a measure of the connectivity and p_2 is a measure of the tightness of the constraints. Experiments reported in this paper have been obtained with random binary CSPs generated according to model A as described in [7], i.e., p_1 is the probability of adding a constraint between two different variables, and p_2 is the probability of ruling out a pair of values between two constrained variables. As incomplete approaches cannot detect inconsistency, we report experiments performed on feasible instances only, i.e., CSPs that do have at least one solution. Instances are forced to be feasible by first randomly generating a solution.

Description of Ant-Solver: The algorithm for solving CSPs, called Ant-Solver, is based on the ACO metaheuristic [1] and is sketched below.

> **procedure** Ant-Solver(X, D, C)
> $\quad \tau \leftarrow$ InitializePheromoneTrails$()$
> $\quad$ **repeat**
> $\quad\quad$ **for** k in $1..nbAnts$ **do**
> $\quad\quad\quad \mathcal{A}_k \leftarrow \emptyset$
> $\quad\quad\quad$ **while** $\mid \mathcal{A}_k \mid < \mid X \mid$ **do**
> $\quad\quad\quad\quad X_j \leftarrow$ SelectVariable$(X, \mathcal{A}_k)$
> $\quad\quad\quad\quad v \leftarrow$ ChooseValue$(\tau, X_j, D(X_j), \mathcal{A}_k)$
> $\quad\quad\quad\quad \mathcal{A}_k \leftarrow \mathcal{A}_k \cup \{<X_j, v>\}$
> $\quad\quad\quad \mathcal{A}_k \leftarrow$ ApplyLocalSearch$(\mathcal{A}_k)$
> $\quad\quad \tau \leftarrow$ UpdatePheromoneTrails$(\tau, \{\mathcal{A}_1, \ldots, \mathcal{A}_{nbAnts}\})$
> $\quad$ **until** $cost(\mathcal{A}_i) = 0$ for some $i \in \{1..nbAnts\}$ **or** max cycles reached

We now describe the pheromone graph on which artificial ants lay pheromone trails, and the different used functions. More details can be found in [8].

The **pheromone graph** associates a vertex with each variable-value pair $<X_i, v>$ such that $X_i \in X$ and $v \in D(X_i)$. There is an edge between any pair of vertices corresponding to two different variables. The amount of pheromone laying on an edge $(<X_i, v>, <X_j, w>)$ is noted $\tau(<X_i, v>, <X_j, w>)$. Intuitively, this trail represents the learned desirability of assigning simultaneously value v to variable X_i and value w to variable X_j. Lower and upper bounds τ_{min} and τ_{max}, such that $0 < \tau_{min} \leq \tau_{max}$, are explicitly imposed on pheromone trails.

The function "**InitializePheromoneTrails**$()$" initializes the amount of pheromone laying on each edge of the pheromone graph to τ_{max}.

The function "**SelectVariable**$(X, \mathcal{A}_k)$" returns a variable $X_j \in X$ that is not yet assigned in $\mathcal{A}_k$. This choice can be performed randomly, or with respect to some commonly used variable ordering. Experiments reported in this paper have been performed with the smallest-domain ordering, i.e., the function returns a variable that has the smallest number of consistant values with respect to the already assigned variables (ties are broken randomly).

The function "**ChooseValue**$(\tau, X_j, D(X_j), \mathcal{A}_k)$" returns a value $v \in D(X_j)$ to be assigned to X_j. The choice of v is done according to the ACO metaheuristic, i.e., with respect to a probability $p(v, \tau, X_j, D(X_j), \mathcal{A}_k)$ which depends on a pheromone factor $\mathcal{P}$ and a quality factor $\mathcal{Q}$:

$$p(v, \tau, X_j, D(X_j), \mathcal{A}_k) = \frac{[\mathcal{P}(\tau, \mathcal{A}_k, X_j, v)]^\alpha [\mathcal{Q}(\mathcal{A}_k, X_j, v)]^\beta}{\sum_{w \in D(X_j)} [\mathcal{P}(\tau, \mathcal{A}_k, X_j, w)]^\alpha [\mathcal{Q}(\mathcal{A}_k, X_j, w)]^\beta}$$

where α and β are two parameters which determine the relative importance of pheromone and quality factors. The pheromone factor $\mathcal{P}(\tau, \mathcal{A}_k, X_j, v)$ corresponds to the sum of all pheromone trails laid on all edges between $<X_j, v>$ and the assignments in $\mathcal{A}_k$, i.e.,

$$\mathcal{P}(\tau, \mathcal{A}_k, X_j, v) = \sum_{<X_l, m> \in \mathcal{A}_k} \tau(<X_l, m>, <X_j, v>)$$

and the quality factor $\mathcal{Q}(\mathcal{A}_k, X_j, v)$ is inversely proportional to the number of new violated constraints when assigning value v to variable X_j, i.e.,

$$\mathcal{Q}(\mathcal{A}_k, X_j, v) = 1/(1 + cost(\{<X_j, v>\} \cup \mathcal{A}_k) - cost(\mathcal{A}_k))$$

The function "**ApplyLocalSearch**$(\mathcal{A}_k)$" improves the constructed assignment $\mathcal{A}_k$ by performing some local search, i.e., by iteratively changing some variable-value assignments. Different heuristics can be used to choose the variable to be repaired and the new value to be assigned to this variable. Experiments reported in this paper have been performed with the min-conflict heuristic [9], i.e., for each repair, we first randomly select a conflicting variable —involved in some violated constraints— and then choose a value for this variable which minimizes the number of conflicts (ties are broken randomly).

The function "**UpdatePheromoneTrails**$(\tau, \{\mathcal{A}_1, \ldots, \mathcal{A}_{nbAnts}\})$" updates pheromone trails according to the $\mathcal{MAX} - \mathcal{MIN}$ Ant System: all pheromone trails are uniformly decreased and then pheromone is added on edges participating to the construction of the best assignment. Hence, at the end of each cycle, the quantity of pheromone laying on each edge (i, j) is updated as follows:

$\tau(i, j) \leftarrow \rho * \tau(i, j)$
if $i \in \mathcal{A}_{Best}$ and $j \in \mathcal{A}_{Best}$ then $\tau(i, j) \leftarrow \tau(i, j) + 1/cost(\mathcal{A}_{Best})$
if $\tau(i, j) < \tau_{min}$ then $\tau(i, j) \leftarrow \tau_{min}$
if $\tau(i, j) > \tau_{max}$ then $\tau(i, j) \leftarrow \tau_{max}$

where ρ is the trail persistence parameter such that $0 \leq \rho \leq 1$ and $\mathcal{A}_{Best}$ is the best assignment —the cost of which is minimal— of $\{\mathcal{A}_1, \ldots, \mathcal{A}_{nbAnts}\}$. One should remark that all possible pairs $(i, j) \in \mathcal{A}_{Best}^2$ are rewarded.

3   Impact of α and ρ on the Ants Behaviour

α determines the weight of pheromone factors when computing transition probabilities: when increasing α, ants become more sensitive to pheromone. ρ determines trail persistence: a value close to 0 implies a high evaporation so that ants have a "shorter-term" memory and recent experiments are emphasized with respect to older ones. The influence of these two parameters is illustrated below on $< 100, 8, 0.14, p_2 >$ random binary CSPs (for each tightness value p_2, we display results obtained on 300 different feasible problem instances; the other parameters have been setted to $\beta = 10$ and $nbAnts = 8$).

- Influence on the success rate (within a limit of 1000 cycles):

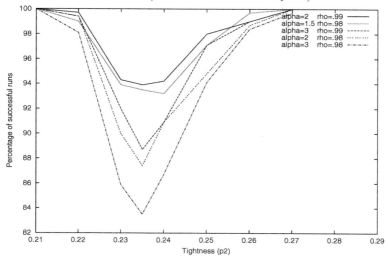

- Influence on the CPU time spent to find a solution (average results on the successful runs only):

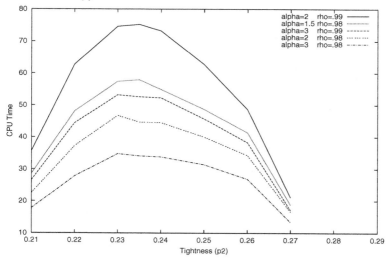

Hence, when considering the success rate criterion, the best results are obtained with low values of α and high values of ρ such as $\alpha = 2$ and $\rho = 0.99$. However, in this case the convergence is rather slow so that the CPU time spent to find a solution is more than twice more than when $\alpha = 3$ and $\rho = 0.98$. Moreover, with even smaller values of α and greater values of ρ —i.e., when $\alpha \leq 1.5$ and $\rho \geq 0.99$— ants converge so slowly that the success rate, within a limit of 1000 cycles decreases —e.g., less than 80% for $< 100, 8, 0.14, 0.23 >$ instances when $\alpha = 1.5$ and $\rho = 0.99$.

As a conclusion, when setting α and ρ one can either choose values that favor a quicker convergence —such as $\alpha = 3$ and $\rho = 0.98$— or values that slow down the convergence —such as $\alpha = 2$ and $\rho = 0.99$. In the first case, Ant-Solver will quickly find rather good assignments, eventhough it may not find the best one. In the last case, Ant-Solver will need more time to find good assignments, but it will more often succeed in actually finding the best one. In the rest of the paper, we have set α to 2 and ρ to 0.99 in order to give priority to the success rate. The next section shows that with such values pheromone only starts to influence ants after a significant number of cycles so that the convergence is rather slow.

4   Impact versus Cost of Pheromone

In order to favor a larger exploration, and more particularly at the beginning of the search, Ant-Solver derives its features from the $\mathcal{MAX} - \mathcal{MIN}$ Ant System, i.e., pheromone trails are limited to an interval $[\tau_{min}, \tau_{max}]$ and are initialized to τ_{max}. Hence, after n cycles of Ant-Solver, the quantity of pheromone laying on an edge (i, j) is such that $\rho^n * \tau_{max} \leq \tau(i, j) \leq \tau_{max}$ where ρ is the pheromone persistence parameter and usually has a value very close to 1. As a consequence, during the first cycles, pheromone factors nearly all have the same values and do not significantly influence the computation of transition probabilities. This is illustrated below on a typical example of execution of Ant-Solver on a $< 100, 8, 0.14, 0.23 >$ random binary instance (with $\alpha=2$ and $\rho=0.99$).

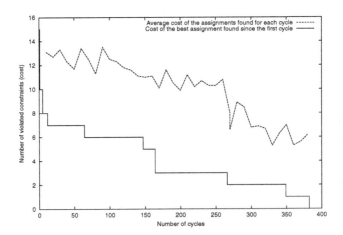

During a first step, roughly corresponding to the first 100 cycles, the average cost of the constructed assignments is rather stable: in average, around 12 constraints are violated in each constructed assignment. This first step corresponds to a learning step, where the search space is widely explored and pheromone information is collected. Then, in a second step, pheromone information gathered during the first 100 cycles starts to influence the collective behaviour of ants, so that they become able to find better paths: the average number of violated constraints decreases from 12 at cycle 100 to 6 at cycle 350 while the cost of the best assignment found decreases from 6 at cycle 100 to 0 at cycle 382.

However, if the influence of pheromone is deliberately reduced at the beginning of the search so that pheromone information only becomes effective after some 100 or so cycles, managing pheromone is tedious and expensive in time. Indeed, let us consider a CSP (X, D, C), and let $n = |X|$ be the number of variables and $p = \sum_{X_i \in X} |D(X_i)|$ be the number of vertices of the associated graph of pheromone. To construct a complete assignment, the computation of pheromone factors of all transition probabilities will require $\mathcal{O}(n * p)$ operations. Then, at the end of each cycle, pheromone laying will require $\mathcal{O}(n^2)$ operations for each rewarded assignment and pheromone evaporation will require $\mathcal{O}(p^2)$ operations. The other operations, that do not deal with pheromone, are the selection of variables, the computation of quality factors and the local repair phase. The complexity of these operations mainly depends on the kind of constraints considered. However, when choosing appropriate data structures that allow incremental computations, the complexity of these operations is usually lower than $\mathcal{O}(p^2)$. Actually, when n and p increase, most of the computation time is used to manage pheromone. For example, on $< 100, 8, 0.14, 0.23 >$ CSPs, assignments are constructed nearly 3 times faster when pheromone is not used.

One should remark that the cost of managing pheromone is more important for Ant-Solver than for many other ACO algorithms. Let us consider for example Ant System described in [10] and designed to solve Traveling Salesman Problems. If the pheromone graph has n vertices, then the computation of all pheromone factors for constructing a complete path will require $\mathcal{O}(n)$ operations; pheromone laying will also require $\mathcal{O}(n)$ operations for each rewarded path and pheromone evaporation will require $\mathcal{O}(n^2)$ operations. However, for large instances with thousands of cities, pheromone management can also become expansive. In this case, a solution can be to deliberately reduce the set of candidate vertices to the closest cities.

5   Description of the Preprocessing Step

The introduction of a preprocessing step is motivated by the fact that pheromone is rather expensive to manage whereas it actually allows to improve the collective behaviour only after some 100 or so cycles. The idea is to collect a significant number of local minima by performing "classical" local search, i.e., by iteratively constructing complete assignments —without using pheromone— and repairing them. These assignments constitute a kind of sampling of the search space. Then, we select from this sample set the n_{Best} best local minima and use them to initialize pheromone trails. Finally, we continue the search guided by pheromone trails with Ant-Solver.

We shall call *SampleSet* the set of all local minima collected during the preprocessing step, and $BestOf(n_{Best}, SampleSet)$ the n_{Best} best local minima in *SampleSet*.

To determine the number of local minima that should be collected in *SampleSet*, one has to find a tradeoff between computing a large number of local minima, that are more representative of the search space, and a smaller number of local minima, that are more quickly computed. Actually, experiments showed us that the harder a problem is, the more local minima should be computed. Therefore, instead of computing a fixed number of local minima within *SampleSet*, we only fix n_{Best} —the number of local minima that will be selected from *SampleSet* to initialize pheromone trails— and we introduce a limit rate ϵ on the quality improvement of these best local minima, i.e., we stop collecting new local minima in *SampleSet* when the average cost of $BestOf(n_{Best}, SampleSet)$ has not been improved by more than ϵ% for a while. The preprocessing procedure is more precisely described below:

> **procedure** preprocessing
>> compute n_{Best} local minima and store them in *SampleSet*
>> **repeat**
>>> $OldCost \leftarrow \sum_{\mathcal{A} \in BestOf(n_{Best}, SampleSet)} cost(\mathcal{A})$
>>> compute n_{Best} more local minima and add them to *SampleSet*
>>> $NewCost \leftarrow \sum_{\mathcal{A} \in BestOf(n_{Best}, SampleSet)} cost(\mathcal{A})$
>> **until** $NewCost/OldCost > 1 - \epsilon$ **or** a solution has been found

In this procedure, local minima are computed by performing local search on complete assignments that are constructed like in Ant-Solver, i.e., by iteratively selecting a variable to be assigned and then choosing a value for this variable. However, the function "ChooseValue" is modified so that values are chosen with respect to quality factors only, i.e., the probability of choosing a value v for a variable X_j is $p(v, \tau, X_j, D(X_j), \mathcal{A}_k) = \mathcal{Q}(\mathcal{A}_k, X_j, v)^\beta / \sum_{w \in D(X_j)} \mathcal{Q}(\mathcal{A}_k, X_j, w)^\beta$

Finally, at the end of the preprocessing step, if no solution has been found, the set $BestOf(n_{Best}, SampleSet)$ —that contains the n_{Best} best local minima of *SampleSet*— is used to initialize pheromone trails: the quantity of pheromone laying on each edge (i, j) of the pheromone graph is set to

$$\tau(i, j) \leftarrow \sum_{\mathcal{A}_k \in BestOf(n_{Best}, SampleSet)} \Delta\tau(\mathcal{A}_k, i, j)$$

where $\Delta\tau(\mathcal{A}_k, i, j)$ is the quantity of pheromone deposited on edge (i, j) for the complete assignment $\mathcal{A}_k$ and is defined as follows:

$$\Delta\tau(\mathcal{A}_k, i, j) = 1/cost(\mathcal{A}_k) \text{ if } i \in \mathcal{A}_k \text{ and } j \in \mathcal{A}_k$$
$$\Delta_\tau(\mathcal{A}_k, i, j) = 0 \qquad \text{otherwise}$$

One should remark that if a pheromone trail is initialized to 0 (resp. to a high value greater than τ_{max}) then the updatePheromoneTrails function will bound it at the end of the first cycle to τ_{min} (resp. τ_{max}).

6   Experimental Results

The preprocessing step is parameterized by n_{Best}, the number of assignments that are extracted from the sample set to initialize pheromone trails, and ϵ, the limit rate on the quality improvement. The influence of these two parameters has been experimentally studied on random binary CSPs. With respect to CPU times, the best results are obtained with the highest values of ϵ and the smallest values of n_{Best}. With respect to success rates, the best results are usually obtained with the smallest values of ϵ and the largest values of n_{Best}. However, we have noticed that, when $n_{Best} \geq 300$, success rates usually decrease. A good tradeoff between CPU time and success rate appears when n_{Best}=200 and ϵ=2%.

The table below displays experimental results on $< 100, 8, 0.14, p_2 >$ random binary CSPs, with n_{Best}=200, ϵ=2%, α=2, β=10, ρ=0.99 and $nbAnts$=8.

p_2	Success rate		Nb of assignments		CPU Time	
	AS with P	AS	AS with P	AS	AS with P	AS
0.21	52.7 + 47.3 = 100.0	100.0	1354 + 153 = 1507	1145	10.6 + 4.7 = 15.3	35.7
0.22	8.6 + 90.9 = 99.5	99.7	1356 + 506 = 1862	1984	11.5 + 15.4 = 26.9	62.7
0.23	0.4 + 95.2 = 95.6	94.3	1603 + 664 = 2267	2288	15.4 + 20.3 = 35.7	74.6
0.24	0.7 + 94.6 = 95.3	94.2	1610 + 746 = 2356	2237	15.2 + 20.2 = 35.4	73.2
0.25	2.9 + 95.3 = 98.2	98.0	1486 + 343 = 1829	1880	15.3 + 10.6 = 25.9	62.7
0.26	10.6 + 88.7 = 99.3	99.0	1392 + 147 = 1539	1432	14.8 + 4.9 = 19.7	48.7
0.27	69.1 + 30.9 = 100.0	100.0	747 + 9 = 756	614	8.2 + 0.4 = 8.6	21.1

For each value of p_2, we have considered 300 different feasible instances and the table successively displays results obtained by Ant-Solver with preprocessing (AS with P) and Ant-Solver without preprocessing (AS), with respect to 3 criteria: the success rate, the number of constructed assignments (average on successful runs only) and the CPU time spent to find a solution (average on successful runs only). For the results obtained with preprocessing, we successively display the part due to the preprocessing step, the part due to the solution process after preprocessing and the total (both for preprocessing and solving).

In this table, one can remark that success rates are slightly improved by preprocessing (except when p_2=0.22). One can also note that the total number of constructed assignments is comparable whereas the CPU time is always twice as small when using preprocessing. Let us consider for example $< 100, 8, 0.14, 0.23 >$ instances. In average, the number of constructed assignments is nearly the same (2267 and 2288). However, for AS with P, 1603 of these assignments are constructed during preprocessing, without using pheromone, so that they are rather quickly computed (in 15.2s). Then, after preprocessing, solutions are found in 20.3s in average, after the construction of 664 assignments, corresponding to 83 cycles of Ant-Solver. Hence, the total time spent to find a solution is 35.7s. As a comparison, when running Ant-Solver without preprocessing, solutions are found in 74.6s in average, after the construction of 2288 assignments, corresponding to 286 cycles of Ant-Solver. Finally, one can also note that if nearly half of the easiest instances are solved during preprocessing, only very few of the hardest instances have been solved during preprocessing.

This shows that random restart of local search with the min-conflict heuristic gives very low performances on hard problems and that ACO actually boosts local search in this case.

As a conclusion, these experimental results show that the preprocessing step introduced in this paper actually boosts Ant-Solver so that it can find solutions twice as fast. This acceleration of the convergence is not performed to the detriment of the exploratory behaviour of ants —as it is usually the case when one emphasizes the influence of pheromone by increasing α or decreasing ρ. Actually, it also allows Ant-Solver to find a few more solutions.

Another possibility to improve the convergence behaviour could be to reduce ρ so that pheromone evaporates quicker and convergence is boosted, but at the same time to introduce occasional pheromone re-initialization. Such an approach, that uses data mining technics to extract information from pheromone matrices, has been proposed in [11] and actually allowed us to boost ACO when solving permutation Constraint Satisfaction Problems (the goal of which is to find an hamiltonian path in a graph that satisfies a given set of constraints). Further works will investigate similar approaches on Ant-Solver, and compare them with preprocessing.

References

1. M. Dorigo and G. Di Caro. the Ant Colony Optimization Meta-Heuristic. In *New Ideas in Optimization*, McGraw Hill, pages 245–260, 1999.
2. P. Merz and B. Freisleben. Fitness landscapes and memetic algorithm design. In *New Ideas in Optimization*, McGraw Hill, pages 245–260, 1999.
3. T. Stutzle and H.H. Hoos. MAX-MIN Ant System. In *Journal of Future Generation Computer Systems*, 16:889–914, 2000.
4. A. Colorni, M. Dorigo and V. Maniezzo. An investigation of some properties of an "Ant algorithm". In *PPSN'92*, pages 509–520, Elsevier, 1992.
5. H. M. Botee and E. Bonabeau. Evolving Ant Colony Optimization. In *Advanced complex systems*, 1:149–159, 1998.
6. E.P.K. Tsang. Foundations of Constraint Satisfaction. *Academic Press*, 1993.
7. E. MacIntyre, P. Prosser, B. Smith and T. Walsh. Random Constraints Satisfaction: theory meets practice. In *CP'98*, volume 1520 of LNCS, Springer-Verlag, pages 325–339, 1998.
8. C. Solnon. Ants can solve Constraint Satisfaction Problems. *Research report*, 2001.
9. S. Minton, M.D. Johnston, A.B. Philips and P. Laird. Minimizing Conflicts: a Heuristic Repair Method for Constraint Satistaction and Scheduling Problems. In *Artificial Intelligence*, 58:161–205, 1992.
10. M. Dorigo, V. Maniezzo and A. Colorni. The Ant System: Optimization by a Colony of Cooperating Agents. In *IEEE Transactions on Systems, Man, and Cybernetics-Part B*, 26(1):29–41, 1996.
11. A. Stuber and C. Solnon. Boosting ACO algorithms with data mining technics (in french). In *JNPC'01*, pages 271–282, 2001.

A Memetic Algorithm
Guided by *Quicksort* for the Error-Correcting
Graph Isomorphism Problem

Rodolfo Torres-Velázquez and Vladimir Estivill-Castro

School of Electrical Engineering and Computer Science,
University of Newcastle,
University Drive, Callaghan, NSW 2308, Australia
{rodolfo,vlad}@cs.newcastle.edu.au

Abstract. Sorting algorithms define paths in the search space of $n!$ permutations based on the information provided by a comparison predicate. We guide a Memetic Algorithm with a new mutation operator. Our mutation operator performs local search following the path traced by the *Quicksort* mechanism. The comparison predicate and the evaluation function are made to correspond and guide the evolutionary search. Our approach improves previous results for a benchmark of experiments of the Error-Correcting Graph Isomorphism. For this case study, our new Memetic Algorithm achieves a better quality vs effort trade-off and remains highly effective even when the size of the problem grows.

1 Introduction

Memetic Algorithms (MAs) combine Genetic Algorithms (GAs) and Local Search (LS). They deal successfully with hard combinatorial optimization problems [1,2], [3,4,5,6,7,8,9,10]. Previously [9,10] we proposed to integrate classical sorting in GAs towards improved permutation search, in particular we applied *Insertion Sort*. We called our approach *MA-sorting*. Here, we build on those results and propose a new mutation operator based on *Quicksort*. Our approach is specific to problems involving permutations, but in any other sense it is domain-independent. Therefore, it is applicable to the Error-Correcting Graph Isomorphism, but also to other problems involving permutations. Domain-independent techniques are appealing because of their lower development cost. We integrate *MA-sorting* with *Quicksort* to demonstrate convincingly that our approach is general and flexible. We obtain improved results for a benchmark of experiments of the *Error-Correcting Graph Isomorphism*. This problem seems so hard that in this case, genetic algorithms do perform better than several other optimization strategies [11]. Our experiments using our *MA-sorting* with *Quicksort* show a marked improvement over previous optimization with GAs [11,9].

S. Cagnoni et al. (Eds.): EvoWorkshops 2002, LNCS 2279, pp. 173–182, 2002.
© Springer-Verlag Berlin Heidelberg 2002

2   The Error-Correcting Graph Isomorphism Problem

A benchmark problem arises from the use of graphs in pattern matching [12,11,13]. Graphs are combinatorial objects that have been widely used in applications where structured objects emerge in a natural way. Thus, the interest in finding efficient algorithms to deal with the Graph Isomorphism problem.

Graph Isomorphism (GI) is an important problem in Graph Theory whose precise computational complexity remains unknown [13]. It requires to find a bijection of the vertices so that the edge structure is the same. Labeling the vertices from the same set corresponds to finding a permutation π. However, it is not a minimization problem. The interesting aspect is that in practice a close variant is a hard minimization problem. In real world applications of pattern matching, the existence of noise, distortion, uncertainty or measurement errors, together with weights associated to nodes and edges, translates the GI problem into its inexact version: the inexact Graph Isomorphism (iGI) or Error-Correcting Graph Isomorphism (ECGI) [12]. In order to define this problem we first need the notion of attributed graph.

Definition 1 (AG). *An* attributed graph *[12] is a 4-tuple* $G_a = (V, E, \alpha, \beta)$ *where V is a finite nonempty set of vertices; $E \subset V \times V$ is a set of distinct ordered pairs of distinct elements in V called edges; $\alpha : V \to \Re$ is a function called vertex interpreter; and $\beta : E \to \Re$ is a function called edge interpreter.*

In what follows G is a graph, $V(G)$ denotes the vertices of G and $E(G)$ denotes the edges of G.

Definition 2 (ECGI). *The* Error-Correcting Graph Isomorphism *problem is that given two attributed graphs* **AGs**, $G_a = (V(G_a), E(G_a), \alpha_G, \beta_G)$ *and* $H_a = (V(H_a), E(H_a), \alpha_H, \beta_H)$, *with* $\mid V(G_a) \mid = \mid V(H_a) \mid$, *we must find a permutation of vertices* $\pi : V(G_a) \to V(H_a)$ *so that some metric of total dissimilarity between the graph* $G'_a = (\pi[V(G_a)], \pi[E(G_a)], \alpha_{G'}, \beta_{G'})$ *and the graph* H_a *is minimized.*

To illustrate the virtues of our *MA-sorting* we will describe its application to this harder combinatorial minimization problem. This is also chosen as a benchmark because GAs have previously shown to be effective [11,9].

3   *Quicksort* as Local Search Directs Mutation

LS has been shown successful to deal with hard combinatorial optimization problems [14]. In such problems the search space is the set S of feasible solutions and the purpose is to find a solution $i^* \in S$ such that given a cost function $f : S \to \Re$, $f(i^*)$ is globally optimal. LS algorithms iteratively search for a better solutions in the local neighborhood of a current solution i. The neighborhood $N \subseteq S$ of a LS algorithm is the set of solutions that can be reached from the current solution i by the application of a neighborhood operator $\nu : S \to 2^S$. LS mechanisms are approximation algorithms with the common feature of an underlying neighborhood operator $\nu(i)$ which is used to guide the search.

LS comprises four basic steps: *Initialization, Nomination, Comparison* and *Termination.* An outline of the basic LS mechanism is displayed in Fig. 1. A well known disadvantage of LS is that it can be trapped in local optima potentially far away from the global optimum.

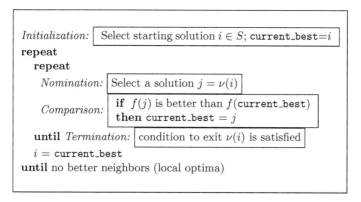

Initialization: Select starting solution $i \in S$; current_best$=i$
repeat
  repeat
    Nomination: Select a solution $j = \nu(i)$
    Comparison: **if** $f(j)$ is better than $f($current_best$)$
                   then current_best $= j$
  until *Termination:* condition to exit $\nu(i)$ is satisfied
  $i =$ current_best
until no better neighbors (local optima)

Fig. 1. Pseudo-code for Algorithm Local Search (LS).

Sorting algorithms are a well understood mechanism to deal with a class of permutation problems [15]. Although it is unusual, these algorithms can be conceptualized as LS mechanisms, where the local information provided by a pair of elements is enough to decide the search direction in a deterministic way. Sorting algorithms define paths in the search space S_n of $n!$ permutations, from any permutation to the goal permutation [9]. Using measures of presortedness it is possible to formalize sorting as an optimization problem. This problem deals with sequences of items from a total order. An *inversion* is any pair of elements that are in the wrong order. The total number of inversions in a sequence X is the measure *Inv*. This measure corresponds to the smallest number of adjacent swaps necessary to bring the sequence to sorted order. Relaxing the requirement of adjacency in swaps results in the measure *Exc*. *Exc* is the minimum number of exchanges (arbitrary swaps) required to sort the sequence. The input sequence X to a sorting problem defines a permutation $\pi(X)$ on $\{1, \ldots, |X|\}$ by $\pi(X)[i]$ is the final position of x_i when X is sorted. More than 150 years ago, Cayley realized that $Exc(X) = |X|-$ the number of cycles in $\pi(X)$. Thus, the identity permutation has X cycles, and Exc is zero, the minimum. Sorting is now formalized as follows. Minimize $Exc(X)$ or alternatively as Minimize $Inv(X)$.

We model *Quicksort* as a LS mechanism. *Quicksort* is a divide and conquer strategy for sorting invented in 1960 by C. A. R. Hoare. In *Quicksort*, we divide the sequence of items to be sorted into two subsequences, a left and right parts, and then call the *Quicksort* procedure recursively to sort the two parts. There are several ways to achieve such a partitioning. Our implementation adheres to the description of Sedgewick [16] because it is considered as the best scheme [15].

During partitioning, we choose a pivot v and arrange that all the items in the left part are less than the pivot and all those in the right part are greater than v. This process keeps a left pointer l and a right pointer r. Pointer l scans from the left and pointer r scans from the right moving inward until the left pointer finds an element greater than the pivot and the right pointer finds an element less than the pivot. Every time such two elements are found, they are swapped (and the two pointers restart scanning inward). If they cross, we exchange the pivot with the element pointed to by the left pointer completing the partition. Sedgewick [16, p 118] describes pseudo-code for *Quicksort*. We now indicate how *Quicksort* is adapted to the framework of LS in Fig. 1. We will use *Quicksort* to induce a neighborhood operator in our *MA-sorting*. The *Quicksort* neighborhood operator is dynamic and depends on the current state of its process. This state is encoded by the left pointer l and the right pointer r. Thus, we consider the neighborhood operator $\nu : S \times L \times R \to 2^S$ where $L = \{1, \ldots, l, \ldots, n-1\}$, $R = \{2, \ldots, r, \ldots, n\}$ and $r > l$. For example, consider a current permutation $\pi_i = \langle 1, 4, 5, 2, 3 \rangle$ and the state $(l = 2, r = 4)$ (the left and right pointers in *Quicksort*) after *Initialization* (and perhaps some further processing). The *Nomination* step suggests several permutations. The advancement of the left pointer suggests $\pi_j^l = \nu(\pi_i, l, r) = \langle 1, 2, 5, 4, 3 \rangle$. Also, by the advancement of the right pointer, it also suggests $\pi_j^r = \nu(\pi_i, l, r) = \langle 1, 4, 2, 5, 3 \rangle$. We make each comparison predicate $< (\pi_i, \pi_j)$ correspond with the *Comparison* step in LS. Note that in an algorithm like *Quicksort* the *Nomination* step is integrated into the *Comparison* step. There could be several strategies for the *Termination* step. One of them is to exhaust the neighborhood $\nu(i)$ while another alternative is *first improvement*. For the latter, we stop exploring $\nu(i)$ as soon as an improvement is found and we make this improvement the current solution. We adopted first improvement, and in this example, when the result of the *Comparison* step is *True*, the LS mechanism ends with $\pi_j = \langle 1, 4, 2, 5, 3 \rangle$.

Insertion Sort has the property that every swap suggested along its way to sorting a sequence monotonically reduces *Inv*. This is not true of *Quicksort*. A single swap, or an entire partitioning step (that sends all elements less than the pivot to the left and those larger to the right) may also increase the value of *Inv* or *Exc* temporarily. However, Lemma 3.2 in [17] proves that *Quicksort* monotonically reduces *Exc* on the average.

In our integration of the MA algorithm, a mutation is the result of a swap suggested by *Quicksort*. The acceptance or not of the swap is performed by evaluating the objective function in the entire permutation. It is important to notice that we do not compare the individual elements that are being swapped, but the fitness of the entire permutations as a whole, before and after the swapping. However, if the resulting permutation represents an improvement (with respect to the objective function), the message relayed to the concurrent *Quicksort* is that the comparison returned *True*, as if the two elements were swapped (from the perspective of *Quicksort* we compared the individuals). Symmetrically, if the resulting permutation does not correspond to an improvement with respect to the objective function, the message relayed back to *Quicksort* is that the comparison resulted in *False* (as if the individual items were not swapped).

The swap mutation operator used by Wang et al. [11] chooses uniformly a random pair of positions in the chromosomes and swaps the contents of the selected positions. This operation often produces a chromosome with worse performance. This is not considered a problem when a mutation operator is conceived as a diversity generator. By contrast, we apply the results of the mutation operator only if we achieve better performance. In this sense our mutation operator is a hill-climber. Other mutation operators which perform LS have been proposed in literature. Many of them [1,3,7,8] are based on variants of the 2-opt algorithm proposed by Croes [18].

In our approach [9,10], classical sorting works as a map for the search space S_n of $n!$ permutations. Thus, when a mutation is to be applied with given mutation probability p_m, *Quicksort* defines which mutation operation to perform. The mutation probability is per individual. We refer to our mutation operator as *SORT*. Note that the evaluation function, which guides the evolutionary search, and the *Comparison* function f, which guides our mutation operator, are made to correspond. The only explicit knowledge about the problem in our *MA-sorting* algorithms is that the problem involves permutations. Our *SORT* mutation operator determines facts like item i should be before item j. That is, it adopts how sorting algorithms learn the permutation to sort the input.

4   A Benchmark of Graphs

We reproduce the experimental setting of Wang et al. [11]. Their method for the construction of instances of the ECGI problem provides the optimal solution, and these instances constitute a benchmark where to test the effectiveness of GAs. While this family of graphs represents only some instances of the ECGI problem, they constitute a parameterized model with respect to the size of the graph and the separation between the graphs that constitute a problem instance. Attributed undirected graphs are encoded as an edge-weight adjacency matrices. Let $n = |G_a(V)|$ be the number of vertices of G_a and $M(G_a) = [m_{i,j}]$ be the $n \times n$ edge-weight matrix of the attributed undirected graph G_a given by $m_{i,i} = \alpha(v_i)$ and $m_{i,j} = \beta(v_i, v_j)$. Note that two graphs are isomorphic only if their edge-weight adjacency matrices differ by a permutation of rows and columns, that is, if $M(H_a) = P \cdot (G_a) \cdot P^T$, where P and P^T are the matrix representation of a permutation π and its transpose, respectively (i.e. P is a permutation matrix).

The construction of Wang et al. assigns integer values in $[0, 100]$ to each cell $m_{i,j}$ of the matrix $M(G_a)$, independently and uniformly (each integer entry $m_{i,j}$ is chosen with uniform probability $1/101$ and independently of other entries). In order to build the graph H_a (the isomorphic graph), a random permutation π^* is uniformly constructed (each permutation has probability $1/n!$). The permutation matrix P is π^* applied to the columns of the identity matrix.

The graph H'_a has edge-weight adjacency matrix $M(H'_a) = P \cdot M(G_a) \cdot P^T$. We add noise for the matching to be inexact. For each entry in the upper triangle of $M(H'_a)$ an integer is selected uniformly in the interval $[-\varepsilon, +\varepsilon]$ (that is, with probability $1/(2\varepsilon + 1)$), where $\varepsilon \in [0, 20]$ is a parameter of the experiment regulating the approximate mismatch in the pair of graphs. In this process, noise is only added to the upper triangle and changes are reflected to keep an undirected graph. The resulting matrix is the edge-weight adjacency matrix for the attributed graph H_a. The minimization algorithms will receive as input the pair G_a, H_a and we hope that they will recover π^*. Measuring to what extent the algorithms recover π^* allows to evaluate the quality of the optimization.

In Definition 2, we left open the formulation of a criterion of dissimilarity between attributed graphs. Wang et al. experimented with at least two fitness functions, and concluded that the absolute total error (**ATE**) in the entries of the edge-weight adjacency matrices is more accurate.

Definition 3 $(ATE(\pi))$. *Given two attributed graphs G_a and H_a, the Absolute Total Error of a permutation π is the 1-norm of the matrix $M(G_a) - P \cdot M(H_a) \cdot P^T$, where P is the permutation matrix of π. The explicit form of the Absolute Total Error is given by $ATE(\pi) = \sum_{i=1}^{n} \sum_{j=1}^{n} |m_{i,j} - m_{\pi(i),\pi(j)}|$.*

Wang et al. re-write this as a maximization problem: find the permutation π that maximizes $Mx\_ATE(\pi)$ given by $Mx\_ATE(\pi) = C'_{max} - ATE(\pi)$, where C'_{max} is the maximum possible value of $ATE(\pi)$.

Wang et al. genetic algorithm GAB outperforms branch-and-bound and other optimization approaches for the ECGI [11]. They use an initialization heuristic named *status matching* insert one chromosome into the initial randomly-generated population GAB. We have found that the heuristic on its own provides good approximated solutions when noise levels are low (i.e. ε is close to 0). We implemented *MA-sorting* guided by *Quicksort* and left fixed the common aspects with Wang et al. to demonstrate that hybridization as proposed here is the source of the much-improved optimization. That is, our experiments are designed to illustrate that our mutation operator SORT achieves a hybridization with sorting methods that speeds up convergence to much better solutions.

Table 1 summarizes the operators and parameters of the *MA-sorting* used in our experiments. Operators like *PMX* [19] are well known in the GA community when applied to optimization problems searching for permutations. We will skip PMX details here. Similarly, we skip description of the Inhibiting Selection [11], Terminate upon Thresh Convergence [11] and details of Ranking Scaling [20].

Table 1. *MA-sorting* Settings and Parameters.

Setting	Option		Parameter	Value
Encoding Scheme	Integer Permutation		Population Size	30
Scaling	Ranking Scaling		Crossover Probability	0.9
Selection	Inhibiting		Mutation Probability	0.1
Termination	Upon Thresh Convergence		Elitism	True
Crossover	PMX			
Mutation	SORT			

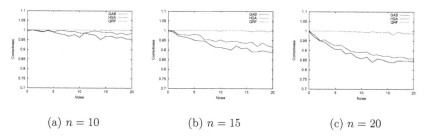

(a) $n = 10$ (b) $n = 15$ (c) $n = 20$

Fig. 2. Average correctness of ECGI algorithms. Three graph sizes.

In our experiments, the mutation operator is the result of continuing *Quicksort* (from a state encoded in the two partition pointers until first improvement) each time the mutation operator is requested. In order to carry out our experiments, we used the software *GAlib* genetic algorithm package, written by Matthew Wall at the Massachusetts Institute of Technology.

5   Experimental Results

We use the fitness value of the best chromosome π_{found} in the final population to assess the quality of solutions. We compute the ratio between $Mx\_ATE(\pi_{found})$ and the optimum value $Mx\_ATE(\pi^*)$ (recall permutation π^* was used to construct H_a from G_a). This ratio is called *correctness* and is given by

$$correctness = 1 - |Mx\_ATE(\pi_{found}) - Mx\_ATE(\pi^*)|/Mx\_ATE(\pi^*).$$

Applying the methodology described in Section 4, we constructed 50 pairs of graphs for each integer value of $\varepsilon \in [0, 20]$, and each size $n \in \{10, 15, 20\}$. For each graph pair we executed three times each GA and selected the best solution found. So, we generated 3150 test cases, and executed 9450 times each GA.

The results are displayed in Fig. 2. The plots correspond to three values for the size n of the graph. The x-axis shows the range of ε values. The y-axes is correctness scale, from 70% correct to 100% correct. The plotted lines are average *correctness* over the 50 problems. The dotted lines correspond to results for our *MA-sorting* guided by *Quicksort*, labeled in the figure as QRF. The solid lines correspond to GAB's (Wang et al. algorithm). The broken lines correspond to our previous results [9] (*MA-sorting* guided by *Insertion Sort*), labeled HGA.

This figure illustrates that our *MA-sorting* with *Quicksort* performs remarkably better. QRF is consistently superior for all the tested size of graphs and for all noise values ε. QRF is highly effective for all the tested sizes, whereas both GAB and HGA are less effective as the size of the graph grows. Moreover, QRF is more robust because it is less sensitive to noise than its counterparts. We conclude that *MA-sorting* when guided by *Quicksort* is consistently better than both GAB [11] and HGA [9] for a broad spectrum of test cases, for different size graphs, and for low, medium and high noise levels.

Table 2. CPU time for different size graphs using GAB, HGA and QRF.

Results over 50×21(noise levels) = 1050 problems (3 runs per problem)
Average CPU time in seconds.   Relative overhead.

Graph Size	μ_{GAB}	μ_{HGA}	μ_{QRF}
10	0.172	0.230	0.259
15	0.526	0.911	1.381
20	1.266	2.488	4.708

Graph Size	$\frac{\mu_{GAB}}{\mu_{GAB}}$	$\frac{\mu_{HGA}}{\mu_{GAB}}$	$\frac{\mu_{QRF}}{\mu_{GAB}}$
10	1	1.3	1.5
15	1	1.7	2.6
20	1	2.0	3.7

Table 2 shows average CPU time in seconds for different size graphs when *MA-sorting* is guided by *Quicksort*, in comparison with GAB and HGA. It is clear that hybridization adds overhead, but it is constant for the range of ε values and it also remains acceptable in the dimension of n values. Thus, the *MA-sorting* guided by *Quicksort* is effective, obtaining impressive quality solutions for reasonable requirements of CPU time.

6 Discussion

The first remarkable point derived from this work is the outstanding performance obtained when a Memetic Algorithm is guided by *Quicksort* for the Error-Correcting Graph Isomorphism problem. This illustrates that our approach is generic and can be used to incorporate other classical sorting algorithms besides *Quicksort*. One can actually expect that the hybridization would work with Mergesort, Heapsort and so on.

For this case study, the neighborhood operator suggested by *Quicksort* is superior to that suggested by *Insertion Sort*. We conjecture that the main reason for the better performance, is that, in contrast with *Insertion Sort* where the sequence of comparisons is predetermined and restricted to adjacent items, the sequence of comparisons prescribed by *Quicksort* depend on what we may have learned from previous comparisons. That is, in the *Quicksort* case, the result of each comparison determines which neighbors of the current permutation are to be compared next. Moreover, our approach reduces the danger of getting caught in a bad local optimum by the effect of a conventional crossover operator (as suggested by Grefenstette [1]).

It is important to analyze to what extent our algorithms is a general blackbox that does not make assumptions about the landscape of the search space neither about the nature of the problem, besides the fact that it involves permutations. Specially, the No Free Lunch (NFL) Theorems [21] imply that a good algorithm would pay the price by being bad in many of the other optimization problems (those problems that have search space S_n and as objective function some function $f : S_n \to \Re$). First, one has to review the assumptions of the NFL theorems. Searching algorithms for NFL have no cost for remembering where they have been before, an issue that certainly implies a space-time trade-off in any practical algorithm. Also, there are some restrictions over the apriori distribution of all optimization problems (the space of all functions $f : S_n \to \Re$).

Otherwise, exhaustive search (which meets the requirements of an algorithm that does not repeat function evaluations) would be a widely used algorithm in practice for optimization problems. Notice that exhaustive search is a variant of LS; simply let $\nu(i) = S_n$ and produce a list of all permutations in $O(1)$ amortized time (by a classical algorithms like generating all permutations lexicographicly [22]). Exhaustive search would perform better maximization than other algorithms in functions like $High_K(\pi) = 0$ for all $\pi \in S_n \setminus \{\pi_K\}$ and $High_K(\pi_K) = 1$ where π_K is a permutation with high Kolmogorov Complexity (assuming n is large enough, such permutations exist [23]). But certainly, the optimization problems on permutations we expect to arise in practice would have their objective function encoded in a (short) program; thus, these problems could not be like optimizing $High_K$. So the distribution of optimization problems we expect our algorithms to face has a very marked non-flat distribution.

Our hybridization with sorting algorithms does assume something about the nature of the optimization problem on permutations. It assumes regularity of the landscape with respect to a measure of presortedness. Problems should be such that a permutation that is very similar in sorted order with respect to the optimal solution has also a good value for the objective function. Thus, it makes sense for the optimizer to learn heuristically; that is, record information "i before j" suggested by the sorting algorithm. For the case study of the ECGI problem, a permutation that is a close order to the optimal permutation is a reasonable match between the two given graphs to reduce the error in the match. We conjecture that knowing which measure of presortedness is suitable for an optimization problem would go a long way in suggesting the sorting algorithms for the hybridization. Although we can not say which of the many measures of presortedness best suits the ECGI problem, we see that Inv usually results in gaining information at a rate that demands more sorting comparisons than Exc. We believe this is a reasonable trade-off in performing local search in Genetic Algorithms without including additional knowledge about the problem. In this sense our approach is less brittle than a problem-specific Memetic Algorithm.

Acknowledgment

The first author was supported by a fellowship from the National Council of Science and Technology (CONACYT), México, grant number 135385.

References

1. Greffenstette, J.: Incorporating problem specific knowledge into genetic algorithms. Davis, L., ed.: Genetic Algorithms and Simulated Annealing, Pitman (1987) 42–60
2. Mühlenbein, H.: Parallel genetic algorithms, population genetics and combinatorial optimization. Schaffer, J., ed.: Proc. 3rd Int. Conf. Genetic Algorithms, George Mason Univ., Morgan Kaufmann (1989) 416–421
3. Davis, L., ed.: Handbook of Genetic Algorithms. Van Nostrand Reinhold (1991)
4. Mühlenbein, H.: Evolution in time and space - the parallel genetic algorithm. Rawlins, G., ed.: Foundations of Genetic Algorithms, Indiana Univ., Morgan Kaufmann (1991) 316–337

182   Rodolfo Torres-Velázquez and Vladimir Estivill-Castro

5. Merz, P., Freisleben, B.: A genetic local search approach to the quadratic assignment problem. Bäck, T., ed.: Proc. 7th Int. Conf. Genetic Algorithms, Michigan State Univ., East Lansing, Morgan Kaufmann (1997) 465–472
6. Merz, P., Freisleben, B.: Fitness landscape analysis and memetic algorithms for the quadratic assignment problem. IEEE T. Evolutionary Computation 4 (2000) 337–352
7. Tsai, H.K., Yang, J.M., Kao, C.Y.: A genetic algorithm for traveling salesman problems. Spector, L., et al. eds.: GECCO-2001. Proc. Genetic and Evolutionary Conference, San Francisco, CA. Morgan Kaufmann (2001) 687–693
8. Rocha, M., Mendes, R., Cortez, P., Neves, J.: Sitting guests at a wedding party: Experiments on genetic and evolutionary constrained optimization. Congress on Evolutionary Computation CEC2001, Seoul, Korea, IEEE Press (2001) 671–678
9. Estivill-Castro, V., Torres-Velázquez, R.: Classical sorting embedded in genetic algorithms for improved permutation search. Congress on Evolutionary Computation CEC2001, Seoul, Korea, IEEE Press (2001) 941–948
10. Estivill-Castro, V., Torres-Velázquez, R.: How should feasibility be handled by genetic algorithms on constraint combinatorial optimization problems? the case of the *valued n-queens problem*. 2nd Workshop on Memetic Algorithms. WOMA II. GECCO-2001. (2001) 146–151
11. Wang, Y.K., Fan, K.C., Horng, J.T.: Genetic-based search for error-correcting graph isomorphism. IEEE T. Systems, Man and Cybernetics, Part B: Cybernetics 27 (1997) 588–597
12. Tsai, W.H., Fu, K.S.: Error-correcting isomorphisms of attributed relational graphs for pattern analysis. IEEE T. Systems, Man and Cybernetics 9 (1979) 757–768
13. Messmer, B., Bunke, H.: A decision tree approach to graph and subgraph isomorphism detection. Pattern Recognition (1999) 1979–1998
14. Aarts, E., Lenstra, J.: Introduction. Aarts, E., Lenstra, J., eds.: Local Search in Combinatorial Optimization, Wiley (1997) 1–17
15. Knuth, D.: Sorting and Searching. Volume 3 of The Art of Computer Programming. Addison-Wesley (1973)
16. Sedgewick, R.: Algorithms in C++. Addison-Wesley (1992)
17. Estivill-Castro, V., Wood, D.: Randomized adaptive sorting. Random Structures and Algorithms 4 (1993) 26–51
18. Croes, G.: A method for solving traveling-salesman problems. Operations Research 5 (1958) 791–812
19. Goldberg, D., Lingle, R.J.: Alleles, loci, and the traveling salesman problem. Grefenstette, J., ed.: Proc. Int. Conf. Genetic Algorithms and their Applications, Carnegie Mellon Univ., Lawrence Erlbaum (1985) 154–159
20. Baker, J.: Adaptive selection methods for genetic algorithms. Grefenstette, J., ed.: Proc. Int. Conf. on Genetic Algorithms and their Applications, Carnegie Mellon Univ., Lawrence Erlbaum (1985)
21. Wolpert, D.H., MacReady, W.: No free lunch theorems for optimization. IEEE T. on Evolutionary Computation 1 (1997) 67–82
22. Reingold, E., Nievergelt, J., Deo, N.: Combinatorial Algorithms, Theory and Practice. Prentice-Hall, Englewood Cliffs, NJ (1977)
23. Li, M., Vitanyi, P.: A theory of learning simple concepts under simple distributions and average case complexity for the universal distribution. Proc. 30th IEEE Symp. on Foundations of Computer Science, Research Triangle Park, NC. (1989) 34–39

Evolutionary Techniques
for Minimizing Test Signals Application Time

Fulvio Corno, Matteo Sonza Reorda, and Giovanni Squillero

Politecnico di Torino
Dipartimento di Automatica e Informatica
Corso Duca degli Abruzzi 24 I-10129, Torino, Italy
http://www.cad.polito.it/

Abstract. Reducing production-test application time is a key problem for modern industries. Several different hardware solutions have been proposed in the literature to ease such process. However, each hardware architecture must be coupled with an effective test signals generation algorithm. This paper propose an evolutionary approach for minimizing the application time of a test set by opportunely extending it and exploiting a new hardware architecture, named interleaved scan. The peculiarities of the problem suggest the use of a slightly modified genetic algorithm with concurrent populations. Experimental results show the effectiveness of the approach against the traditional ones.

1 Introduction

Before selling a device, the integrated-circuit producer needs to check the correctness of its manufacturing process, testing all possible defects. In recent years deep sub-micron manufacturing technology is enabling designers to put millions of transistors on each integrated circuit. Complex automatically synthesized finite-state machines (FSMs) acting as control units can be found in current designs. As a result of this skyrocketing complexity, *testing* devices, taking into account all possible defects, is an increasingly difficult task. Currently, the test process accounts for a relevant percentage of the total production cost, and, noticeably, the cost of test strictly depends on test-session length.

In the test process, the device is stimulated with a set of input signals and its response is compared with the expected one. However, since these devices are very complex, devising a set of test signals is a very hard task. The problem, in the general case, has been proven NP-Complete and, practically, devising the optimal test set is frequently infeasible.

The computer-aided design (CAD) community devoted many efforts to speed-up and simplify production tests. Some attempts were based on small modifications of the original device to ease the subsequent test set generation. Different hardware architectures have been proposed, each one characterized by advantages and drawbacks.

This paper presents an approach for generating a minimal set of signals for testing an electronic device exploiting the *interleaved scan* architecture. The approach exploits an evolutionary algorithm for driving the search process within the gigantic

S. Cagnoni et al. (Eds.): EvoWorkshops 2002, LNCS 2279, pp. 183–189, 2002.

space of all possible input signals. Recently, several successful results in the field of CAD have been reported using evolutionary algorithms [1], [2]. Evolutionary heuristics begin to appear as a reasonable alternative to traditional techniques.

Section 2 introduces the problem, while section 3 illustrates its evolutionary core. Section 4 reports some preliminary experimental results and section 5 concludes the paper.

2 Background

The most common test architecture is the one called *full scan*. When adopting this approach, the set of test signals is usually generated exploiting an *automatic test pattern generator* (ATPG). The main advantage of this technique is that, being well-known and consolidated, effective ATPGs able to generate minimal test sets are commercially available. On the other hand, even with optimal test sets, full-scan architecture may lead to unacceptably-long test sessions. In more details, Bushnell and Agrawal [3] suggested a formula for calculating the exact length of a test session for applying a given test set:

$$CC = (n_{ATPG} + 3) \cdot n_{FF} + 4 . \tag{1}$$

Where CC is the overall number of clock cycles required for running the test session (the test-session length), n_{ATPG} is the number of vectors in the test set and n_{FF} is the number of memory elements of the circuit. It can be easily noted that, when the number of memory elements is significant, even a small test set requires a long test session to be applied.

The *interleaved scan* is a new architecture recently proposed in [4]. Compared to the full-scan one, it permits to reduce the number of some specific operations. These operations, called *scan-in* and *scan-out*, are remarkably longer than any other operations performed during the application of the test set, thus, reducing *scan* operations leads to a reduction in overall test application time. Unfortunately, no commercial tools are able to systematically generate interleaved-scan test set, yet.

The problem of generating signals composing a test set is rather a peculiar problem. Each single solution encodes a whole set of signals, thus the dimension of the search space is gigantic. For the smallest circuit reported in this paper, it contains about 2^{30} points, while in normal cases it may easily exceed 2^{1000} points. Moreover, the function that evaluates each single solution requires high computational efforts. On a SPARC Workstation, it may be necessary several minutes to assess the effectiveness of a single set of signals. The fitness-landscape shape is unknown, but probably slightly deceptive. Finally, several different solutions are probably equivalent, although this fact can not be established by any *a-priori* analysis.

Despite these problems, literature reports several successful results using evolutionary algorithms for test-set generation, like [5]. Several approaches also exploited additional knowledge on the problem, as in [6]. Authors are actively examining this possibility, but, in the approach presented here, only one problem-specific heuristic is used. In contrast, the approach proposed here exploits a set of standard full-scan test signals generated by a commercial ATPG. The proposed evolutionary engine modifies them to fit the interleaved-scan approach, minimizing the required test application time.

3 Test Set Minimization

Starting from a full-scan test set, the proposed interleaved-scan test-set generation consists in two steps: select a subset of full-scan vectors and extend them by possibly adding new vectors. The two steps are performed iteratively: first, a vector is heuristically chosen. Then it is expanded. The process is reiterated until required. It is important to note that the final number of interleaved-scan vectors is likely to be larger than the number of full-scan ones. However, since the operations required to apply the two test sets are different, the overall number of clock cycles required to apply the interleaved-scan test may be lower.

In the first step, the most *promising* set of input signals is selected, i. e., the set of input signals able to detect most production errors and to excite most device behaviors. This is the only problem-specific heuristic adopted during minimization. The second step exploits a modified genetic algorithm (GA) [7], mixing hill-climbing with the concept of population.

At the beginning of the evolution process, a population of individuals is randomly built. Then, it is evolved following a generational replacement scheme, i.e., newly created individuals are never compared against old ones, but simply replace them. To enforce convergence, an elitist operator copies the fittest individual into each new population.

In the proposed framework, an individual in the population is a test set, i.e., a sequence of signals to be applied to the circuit. The *fitness* of an individual measures the effectiveness of the test set in detecting production errors. A penalty term slightly penalizes longer test set. New populations are generated from the current one through crossover operators, where the new sequence is generated from two parents, and mutations, where the new sequence is generated by modifying an existing one. Four crossover operators are chosen with equal probability:

- **Horizontal 1-cut crossover:** the new sequence is composed of some vectors coming from either parent, according to the position of one cut point randomly generated in the first individual, and another one randomly generated in the second. Horizontal 1-cut crossover operates in the domain of time, combining subsequences of consecutive stimula.

- **Horizontal 2-cut crossover:** the new sequence is composed of some vectors coming from either parent, according to the position of two cut points randomly generated in the first individual. The length of the new sequence is the longest between the two parent ones. Horizontal 2-cut crossover operates in the domain of space, as the horizontal 1-cut crossover.

- **Horizontal uniform crossover:** each vector in the new sequence is taken randomly from the first or from the second parent. The length of the new sequence is the longest between the two parent ones.

- **Vertical uniform crossover:** each vector in the new sequence inherits some bit from the first parent and some from the second. The length of the new sequence is the longest between the two parent ones: inputs taken from the shortest parent are completed with random values where needed. Vertical uniform crossover operates in the domain of space, joining together stimula applied to different inputs.

Three crossover operators were chosen for their ability to preserve parents' useful characteristic, both in the domain of space and time. The purpose of horizontal uniform crossover, on the other hand, is mainly to introduce variability. Three mutation operators are also implemented and are selected with equal probability:

- **Change mutation:** a vector in the sequence is replaced with a new randomly generated one.

- **Add mutation:** a random vector is added in a random position, shifting forward the subsequent vectors.

- **Delete mutation:** a randomly selected vector is removed from the sequence, shifting backward the subsequent vectors.

All these operators have already been successfully exploited for test-set generation, as in [8]. Individuals are probabilistically selected with a roulette wheel technique whereby fitter sequences are more likely to be selected.

Unlike standard generational GA, in each generation n_p new populations are built concurrently. Then, in each population the best individual, termed *champion*, is identified. Finally, all champions are compared and the fittest one is selected; its population is picked for survival, while all other populations are destroyed.

This change in the evolutionary process had significant effects, dramatically increasing the quality of the devised test sets. However, it is not easy to understand the motivations. Choosing the best population and discarding all others is a local-search, hill-climbing step. Under this aspect, the approach looks complementary to Sebag and Schoenauer's societies of hill-climbers [9]. Instead of evolving populations of hill climbers, the algorithm exploits a hill climbing amongst populations. However, the solution space is so different that no additional comparison may be made.

Another possible way to examine the proposed scheme is to consider each whole population as a single individual, or *super individual*. Only the chromosome of the champion is significant in the super individual, while other chromosomes are not considered in the fitness evaluation. Thus, in super individual most of genetic material can be classified as *non-coding*. Several studies investigated the effects of non-coding genes on GA performance. These unused sequences were expected to improve performances by providing a buffer against the disruptive effects of crossover; however, comparisons revealed little difference in overall GA performance [10], [11]. A later study on the details of reproduction events found that non-coding regions reduced the disruptive effects of crossover in building blocks [12]. In fact, non-coding regions reduced crossover total activity within building block regions, including crossover ability to construct new building blocks. However, it is quite hard to adapt the concept of building blocks in the proposed application.

Experimental evidence gathered using highly deceptive Kaufman's NK-landscape [13] suggests that the adoption of this paradigm leads to a faster converge than regular GAs, although toward a lower maximum. This may partially explain while it is effective in the interleaved-scan framework, where the solution space is so huge than fast convergence is an absolute requirement.

4 Experimental Results

A prototype implementing the proposed approach was built in C++. The prototype was tested on ISCAS89 circuits [14], a standard set of benchmarks in the electronic test field. The goal of the experiments was to assess the capability of the evolutionary engine to minimize test application time, by extending a full-scan test set to an interleaved-scan one.

Table 1: Experimental Parameters

Parameter	Value	Meaning
n_p	10	Number of concurrent populations
S_p	50	Population size
M	10^{-3}	Mutation probability

Table 1 shows the parameter values exploited in the experimental evaluation. The prototype required about 24 hours of CPU time, table 2 details the obtained results. The first two columns report the name of the circuit [Circuit] and the number of memory elements [nFF]. Next, test application times for the full scan [FULL] and interleaved scan [INTER] approaches are shown. Last column summarizes the gain in percentage attained by the proposed approach, i.e., the reduction in term of test application time. Results are also summarized in figure 1.

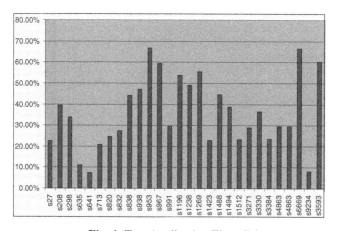

Fig. 1. Test Application Time Gain

Table 2. Experimental Results

Circuit	nFF	FULL	INTER	GAIN
s27	3	31	24	22.58%
s208	8	292	177	39.38%
s298	14	452	299	33.85%
s635	32	1988	1771	10.92%
s641	19	669	619	7.47%
s713	19	669	529	20.93%
s820	5	629	474	24.64%
s832	5	654	475	27.37%
s838	32	5540	3098	44.08%
s938	32	5636	2989	46.97%
s953	29	2933	974	66.79%
s967	29	3020	1225	59.44%
s991	19	593	418	29.51%
s1196	18	3010	1390	53.82%
s1238	18	3244	1649	49.17%
s1269	37	1484	660	55.53%
s1423	74	3482	2683	22.95%
s1488	6	850	470	44.71%
s1494	6	832	510	38.70%
s1512	57	4906	3765	23.26%
s3271	116	7428	5273	29.01%
s3330	132	29968	18998	36.61%
s3384	183	15193	11630	23.45%
s4863	104	6556	4610	29.68%
s4863	104	6556	4610	29.68%
s6669	239	10998	3682	66.52%
s9234	228	54952	50569	7.98%
s35932	1728	74308	29436	60.39%

Experimental data show a significant reduction in test application time for most of the circuits, an average of 36% with two peak gains of 66%. It can also be noted that, as the number of memory elements increase, the improvement for removing scan-in/scan-out set operations step up. Thus, the interleaved scan is expected to scale well with design complexity.

Few benchmarks, like s635 and s641, are so *hard-to-test* that it was almost impossible to extend the full-scan test set, thus, gain is strongly limited. On the other hand, benchmark s9234 deserves special consideration and is currently under study.

5 Conclusions

Interleaved scan is a promising test architecture for reducing the number of clock cycles required for a test session, a key problem for modern circuit producer, however, it requires an effective algorithm to generate appropriate test signals.

The paper presented an evolutionary approach tackling this problem. The proposed algorithm is an extension of a generational genetic algorithm, exploiting concurrent populations. Experimental results show the effectiveness of the approach against the traditional ones.

Authors are currently experimenting with the evolutionary algorithm on a larger set of circuits and on well-known test problems to better assess its characteristics.

References

1. F. Corno, M. Sonza Reorda, G. Squillero, "RT-Level ITC 99 Benchmarks and First ATPG Results," *IEEE Design & Test, Special issue on Benchmarking for Design and Test*, July-August 2000, pp. 44-53
2. R. Drechsler, "Evolutionary Algorithms for Computer Aided Design of Integrated Circuits", *International Symposium on IC Technologies, Systems and Applications*, pp. 302-311, 1997
3. M. L. Bushnell, V. D. Agrawall, *Essentials of Electronic Testing for Digital, Memory & Mixed Signals VLSI Circuits*, Kluwer Academic Publishing, 2000
4. F. Corno, M. Sonza Reorda, G. Squillero, "An Evolutionary Algorithm for Reducing Integrated-Circuit Test Application Time", to appear in *SAC2002: Symposium of Applied Computing*, Madrid 2002
5. F. Corno, G. Cumani, M. Sonza Reorda, G. Squillero, "ARPIA: a High-Level Evolutionary Test Signal Generator" *EvoIASP 2001: 3rd European Workshop on Evolutionary Computation applications to Image Analysis and Signal Processing*, Como (Italy), April 20, 2001
6. F. Corno, P. Prinetto, M. Rebaudengo, M. Sonza Reorda, R. Mosca, "Advanced Techniques for GA-based sequential ATPGs", *IEEE Design & Test Conference*, Paris (F), March 1996
7. E. Goldberg, *Genetic Algorithms in Search, Optimization, and Machine Learning*, Addison-Wesley, 1989
8. F. Corno, M. Sonza Reorda, G. Squillero, "RT-Level ITC 99 Benchmarks and First ATPG Results", *IEEE Design & Test — Special issue on Benchmarking for Design and Test*, 2000, pp. 44-53
9. M. Sebag, M., Schoenauer, "A Society of Hill Climbers", *ICEC'97: 4th IEEE International Conference on Evolutionary Computation*, Indiana, 1997
10. S. Forrest, M. Mitchell, "Relative building-block fitness and the building-block hypothesis", Foundations of Genetic Algorithms 2, 1992, pp. 109–126
11. A. Wu, R. K. Lindsay, "Empirical studies of the genetic algorithm with non-coding segments", *Evolutionary Computation*, 3(2), 1995
12. A. S. Wu, R. K. Lindsay, R. L. Riolo, "Empirical observations on the roles of crossover and mutation", *Proceedings of the 7th International Conference on Genetic Algorithms*, 1997, pp. 362–269.
13. S. A. Kauffman, "Adaptation on rugged fitness landscapes", *Lectures in the Sciences of Complexity*, vol. 1, pages 527-618, Addison-Wesley, 1989
14. F. Brglez, D. Bryant, K. Kozminski, "Combinational profiles of sequential benchmark circuits," *Proceedings International Symposium on Circuits and Systems*, 1989, pp. 1929-1934

Prediction and Modelling of the Flow
of a Typical Urban Basin through Genetic Programming

Julian Dorado[1], Juan R. Rabuñal[1], Jerónimo Puertas[2], Antonino Santos[1],
and Daniel Rivero[3]

[1] Univ. da Coruña, Fac. Informática, Campus Elviña, 15071 A Coruña, Spain
{julian, juanra, nino}@udc.es
[2] Univ. da Coruña, E.T.S.I.C.C.P., Campus Elviña, 15071 A Coruña, Spain
{puertas}@iccp.udc.es
[3] Univ. da Coruña, Fac. Informática, Campus Elviña, 15071 A Coruña, Spain
{infdrc00}@ucv.udc.es

Abstract. Genetic Programming (GP) is an evolutionary method that creates
computer programs that represent approximate or exact solutions to a problem.
This paper proposes an application of GP in hydrology, namely for modelling
the effect of rain on the run-off flow in a typical urban basin. The ultimate goal
of this research is to design a real time alarm system to warn of floods or
subsidence in various types of urban basin. Results look promising and appear
to offer some improvement over stochastic methods for analysing river basin
systems such as unitary radiographs.

1 Introduction

Hydrology is the science of the properties of the movement of the earth's water in
relation to land. A river basin is an area drained by rivers and tributaries. Run-off is
an amount of rainfall that is carried away from this area by streams and rivers. In the
study of an urban basin the streams and rivers are replaced by a sewage system. An
urban basin is influenced by the water consumption patterns of the inhabitants of a
city.

The modelling of run-off flow in a typical urban basin is that part of hydrology
which aims to model sewage networks. It aims to predict the risk of rain conditions
for the basin and to sound an alarm to protect from flooding or from subsidence.

The proposed approach to modelling this problem represents a building block
towards a model for a more ambitious task, that is the completely autonomous and
self-adaptive system which makes predictions of, in real time, sounding an alarm to
alert the authorities to the risk of flooding of the town.

This paper proposes a method to model runoff by means of GP. Modelling is
divided into a number of stages as discussed in section 2.

GP [1] is a search technique which allows the solution of problems by means of the
automatic generation of algorithms and expressions. These expressions are codified or
represented as a tree structure with its terminals (leaves) and nodes (functions).

The run-off flow in the urban basin is modelled both by GP with and without type
specification. Once the terminal and non-terminal operators are specified, it is

S. Cagnoni et al. (Eds.): EvoWorkshops 2002, LNCS 2279, pp. 190–201, 2002.

possible to establish types: each node will have a type, and the construction of child expressions with crossover or mutation operations needs to abide by the rules of the nodal type [2], i.e. respect those grammatical rules specified by the user or investigator. Moreover, both specified operator sets must fulfil two requirements: closure and sufficiency, that is, it must be possible to build right trees with the specified operators, and the solution to the problem (the expression we are looking for) must be expressed by means of those operators.

The automatic program generation is carried out by means of a process based on Darwin's evolution theory [3], in which, after subsequent generations, new trees (individuals) are produced from old ones by means of crossover, copy and mutation [4][5], based on natural selection: the best trees will have more chances of being chosen to become part of the next generation. Thus, a stochastic process is established in which, after successive generations, obtains a well adapted tree.

In the field of Hydraulic Engineering GP has automatically induced equations as accurate as, or more accurate than, the natural sedimentary particle settling velocity equations developed by human ingenuity [6][7]. GP research has so far only produced cumbersome expressions of accurate rainfall-runoff predictions at specific locations [8][9], and has clearly failed to shed light concerning the undelying hydrologic processes.

Our investigation now considers the rain-run-off transformation or clear case in which one of the variables (rain) is transformed into another one (flowing volume through a sewer). The transference function has many different conditions (degree of permeability of the urban truss, street slopes, irregular pavement surfaces, roof types...). The system is of such variability that it becomes impossible to define an equation capable of modelling the course of a drop of water from the moment it falls to the instant in which it enters the drain network. The course of water through the network is quite complex and, although there are equations for modelling it, they are subject to the use of adjustment parameters which make them dependent on a calibration process.

There are methods for calculating the rain runoff process [10] based on the use of transfer functions, usually called "unit hydrographs", sanctioned by experience and whose parameters are fixed by the morphologic characteristics of the study area (kinematical wave). Commercial packs for calculating sewage networks usually provide both modules, "unit hydrographs" and "kinematical wave".

The use of calculation methods based on physics equations, such as artificial neural networks and GP, is becoming widespread in various civil and hydraulic engineering fields. The process we are dealing with is particularly apt for this kind of calculation.

2 Description of the Problem

The goal is predicting and modelling the flow of a typical urban basin from the rain collected by a sensor located in the city sewer. The city of Vitoria (Spain) has been taken as a model for the various trials and data registers. Two signals are obtained for modelling: the first comes from a pluviometer which measures the quantity of rain, and the second one measures the flow level in the sewer. Both the signal corresponding to the flow and the one which corresponds to the pluviometer have been sampled at five minute intervals.

The flow level signal has two distinct causes and hence it can be argued that it can be decomposed into two signals, one representing the daily pattern of water use by inhabitants of the city, and the other which is due to rainfall. Therefore, the system can be divided into two parts: the modelling of the daily flow rate and the prediction of the flow level caused by the rain.

Since Vitoria city sewage system represents a distinct area, the oscillatory behaviour of the periodic flow level signal depends on the quantity of water consumed by its citizens. For instance, the volume is lower at night, while there is an increase during the day, with certain peak hours. The runoff caused by sporadic rainfall is added. Normally, both flows are studied separately, since their origins differ, and this is the methodology used in the present study.

3 Modelling the Daily Flow

The modelling of the daily flow is performed by the induction of a formula representing the desired signal, which is obtained from the whole flow signal, taking only those samples which were not affected by the rain signal, and finding the average value of the samples corresponding to each moment in a day.

Given that the samples were obtained every 5 minutes, we obtain a total of 288 samples per day, therefore the signal being modelled will have that length, as can be seen in figure 1, where the central plain corresponds to the night hours. Peaks corresponding to the morning shower (about 10:00) , lunch (15:00) and dinner (about 22:00) can also be observed. The first point corresponds to 15.00 (lunchtime). The measurement unit of the rain collected by the pluviometer is millimetres (mm) or litres per square metre (l/m^2).

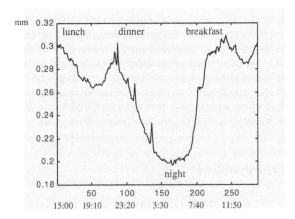

Fig. 1. Average daily flow

We must keep in mind that the use of these indexes and signal for learning will generate a formula whose results will be unforeseeable for values outside the range [0, 287]. This supposition is acceptable, given that no values outside that range are expected (the range is a complete day). However, there is a problem with limit values in the first and last samples, since it is possible that their modelled values are very distant, i.e., that there is discontinuity on first and last values when several days are simultaneously reproduced.

In order to avoid this discontinuity, part of the signal has been replicated at the beginning and the end for GP to learn to fit in firsts and last values when different days are reproduced. Actually, the whole signal has been repeated at the beginning and the end, so that the whole set of patterns is integrated by: (a) input vector values from −288 to 575; and (b) output vector, the previous signal repeated three times in the length corresponding to three days.

Having established the input and output set, a detailed GP encoding can now be established. A number of experiments arrived at a favourable choice of GP functions and terminals. For example:

$$\text{Function set:} \quad \{ \sin, \cos, +, *, \%, - \}$$

$$\text{Terminal set:} \quad \{ X, \pi, [-1,1] \}$$

Traditional arithmetic operators have been added as non-terminal elements and, given the periodical nature of the signal, the sine and cosine functions. Variable X is the sample whose output is needed; random numbers were generated in the interval [-1,1]; and π as a constant. Note that operator % represents the traditional protected division operator.

We have tested other configurations where we did not include the functions sine and cosine, as well as constant π, but it produces worse results, as is discussed next.

3.1 Results

GP experiments which lacked both: (a) the sine and cosine functions and (b) pi (π) as terminal, consistently resulted in a mean error level of 0.04. However, by introducing sine and cosine functions the same runs obtained a mean error level of 0.022; and by further introducing the pi (π) terminal the number of generations required to arrive at the 0.022 level was reduced on average for the same experiments. It is submitted that inclusion of these functions and of the pi constant is helpful because of to the periodic nature of the signals modelled in this study. With this kind of evaluation, we get an error of 0.04 in the set that did not include neither funcions sine and cosine nor constant π.

Once we included funciones sine and cosine as operators, thinking they would be useful due to the periodical nature of the signal, the system improved to get a mean error of 0.022. After that, we decided to try again, but including the constant π as terminal. We found that the system improved in the number of generations needed for achieving the solution, but not in error, showing that the constant was useful in the expression but not necessary, i.e., it is possible to obtain it with random values and arithmetic operators.

With this seemingly advantageous choice of function and terminal set, GP performance using several crossover and mutation rates, and different population sizes, was empirically investigated. The parameters which produce better results are the following:

- Population size: 2000 individuals.
- Crossover rate: 90 %
- Mutation rate: 3 %
- Replication rate: 7%
- Selection mode: Tournament.

The following expression corresponds to the best individual obtained simplified, out of 40 parallel independent runs:

$$\frac{1}{\pi}\cos\left(\cos\left(\frac{\sin(-0.0225\cdot(x+2\pi))+0.9992}{3.02673244}+0.97747\right)\right) \quad (1)$$
$$\cos(0.02996\cdot(\cos(0.0247\cdot(x+\pi))+\cos(0.0247\cdot x)+x-0.89227))$$

Without considering the effect of rain, the daily flow can be modelled from expression (1) taking input values between 0 and 287 (a whole day), and in comparison to the average daily flow, as observed in figure 2.

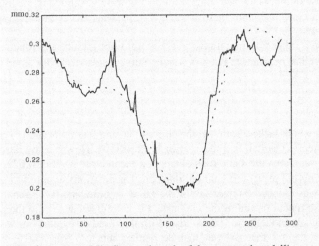

Fig. 2. Average daily flow and result of the tempered modelling

It must be noted that there exist no analytical expressions to predict the shape of the daily curve, since it depends on a variety of factors linked to the usage of water, i.e. the numerous factors unmeasured which can drive the behaviour of this signal: industrial or commercial activity in the area, permanent or weekend residence, hygiene or alimentary habits, etc. Generally speaking, there is a curve formed by a high level (day) and a low one (night). This way, the approach obtained fulfils the usual calculation standards. It is better to temper the adjustment of the average flow in order to take an average point, due to possible fluctuations in the flow variation depending on external circumstances, i.e. to average out the influence of the many unmeasured drivers.

4   Modelling the Rain-Flow Relationship

Once the daily flow rate effect has been modelled, a pre-processing of the original signal takes place in order to eliminate this daily component by using the modelled signal. This is done by means of placing the obtained signal upon the real one and calculating the difference between them. Once this is done, the resulting signal will only be due to the rain level which has fallen.

Now the goal is to predict the future flow rate resulting from the rain. Usually, these signals have a relationship which is shown in another signal which turns the pluviometer signal into that of the flow. This signal has the shape of decreasing exponential functions, acting as a response to the stimulus of a system which relates both signals.

Now there is a different approach: we are not trying to obtain an expression which relates the rain level to the flow rate, now the goal is to find a signal which models the desired system according to the relationship between both signals, showing the following form, with n as the index time and C_i and K_i as constants which characterize the system's response to the stimulus:

$$\sum_i c_i \cdot e^{-k_i \cdot n} \tag{2}$$

This expression models a filter among signals, that is, a transformation which generates an output signal (the flow rate, in this case) from an input one (the rain level). This transformation is characterized by an $h(n)$ signal which has that form. That is, now the goal is to find the system in fig. 3.

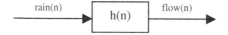
rain(n) → h(n) → flow(n)

Fig. 3. System to be solved

This will be a discret system, therefore the relationship between the signals is shown in an equation (3):

$$\sum_{i=0}^{M} b_i \cdot y(n-i) = \sum_{i=0}^{N} a_i \cdot x(n-i) \tag{3}$$

Where $x(n)$ is the input signal and b_i are the constants which relate the displaced input and output samples. M stands for the number of past samples of the generated output which are used, while N stands for the number of past samples of the input used to generate the output.

Discrete systems also have a representation in the Z domain. This representation is defined by a set of zeros (c_k, complex values), poles (d_k, complex values) and a constant K, according with the following expression:

$$H(z) = K * \frac{\displaystyle\prod_{K=0}^{M}(z - c_K)}{\displaystyle\prod_{K=0}^{N}(z - d_K)} \tag{4}$$

Zeros and Poles are complex values, and are usually represented in a 2-D diagram, called Zero-Pole diagram. As complex values, they have real and imaginary part, but pole´s absolute value never exceeds 1, because if any pole does, the system becomes unstable.

Note that expressions (2), (3) and (4) represent the same system: expression (2) as the system response, equation (3) in time domain, and equation (4) in Z (frequency) domain, so it is possible to obtain any of the expressions from another.

After many attempts, we found that the best way to find the goal system is to get the expression in the Z domain and then convert it to time domain. In order to get this equation, we need to locate the poles and zeros, as well as the system's constant. To make this, the following GP function-terminal configuration is used to get it:

Function set: {Filter, Pair, Pole, Zero, +, *, -, %}
Terminal set: {[-1, 1]}

Filter takes two children: the first one of the "FLOAT" kind which defines the system's constant and will be generated with arithmetic operators and real constants. The second one is of the "ZERO-POLE" type, and defines the system's set of poles and zeros. This second child can be one of the following: *Zero*, which defines a zero, *Pole*, which defines a pole, or *Pair*, which allows two other children of the same type (new zeros, poles or pairs). Given the form of the type system created, this node will be the root of each tree.

Pole takes two children of the "FLOAT" type. They characterize the location of a pole as its real (first child) and imaginary part (second child). These two children will be generated with arthmetic operators and real constants. Their evaluation causes the storing of the designated pole in an internal variable together with the conjugated one, in case of being complex.

Zero takes two children of the "FLOAT" type. They characterize a the location of a zero as its real (first child) and imaginary part (second child). These two children will be generated with arthmetic operators and real constants. Their evaluation causes the storing of the designated zero in an internal variable together with the conjugated one, in case of being complex.

The real interval between −1 and 1 is taken as the terminal element for generating random numbers within that range. This interval, with the arthmetic operators, allow the creation of the constants for defining the real and imaginary parts of zeros and poles, as well as for defining the system's constant.

This set has been used because it locates poles and zeros directly, while the crossover and mutation operations, which cause the poles and zeros to move effectively across the diagram, are more profitable. Besides, this representation offers an advantage: we know whether the system is going to be unstable before carrying out the simulation. This happens when a pole's absolute value is equal or bigger than one.

Thus, the trees generated will have a variable number of zeros and poles. The following steps shall be taken in order to evaluate each tree:

- Execution of the tree: This execution does not return any value, it only stores the system's poles, zeros and constant.
- If any pole has an absolute value which is equal or bigger than one, the tree is discarded, otherwise the system would become unstable. This is done by returning a very high value (infinite) as fitness.
- The equation is rebuilt with the differences defined by the system in poles and zeros.
- This equation emulates the system, returning the average error as fitness.

4.1 Results

The result which best represents the flow prediction is the one represented in fig. 4, where the poles are located at 0.5875, 0.08 + 0.552j and 0.08 + 0.5536j, and the zeros at –0.5333 + 0.792j, -0.5333 – 0.792j, -0.0229 + 0.0946j and 0.0229 –0.0946j.

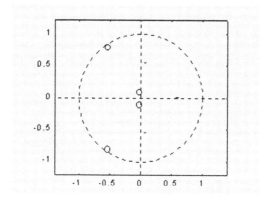

Fig. 4. Zeros and poles diagram of the resulting system

This system has transformed into Z:

$$H(z) = \frac{0.0347 + 0.0386 \cdot z^{-1} + 0.0337 \cdot z^{-2} + 0.0018 \cdot z^{-3} + 0.0005 \cdot z^{-4}}{1 - 1.3507 \cdot z^{-1} + 0.8578 \cdot z^{-2} - 0.4293 \cdot z^{-3} + 0.1109 \cdot z^{-4}} \qquad (5)$$

And its response to the unitary impulse can be observed in fig. 5. In hydrology, the response to the unitary impulse is called "unitary hydrograph". The unitary hydrographs which are commonly used belong to various families. The most frequent one is the exponential family, which produces Gamma and Nash unitary hydrographs. They are defined by exponential functions. In this case, the best adjusted feedback function belongs to the unitary hydrographs produced by stochastic hydrology [11], which use tools such as Autoregressive Moving Average (ARMA) analysis, which are closely linked to the analysis carried out in the present study.

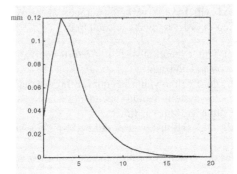

Fig. 5. Response to the unitary impulse of the found system

5 System Description

The system will admit two inputs: the time of day and the measured pluviometer level. The time of day will be a number between 0 and 287, so that it will coincide with each of the measurements taken. This value will be contrasted with the formula obtained in the prediction of the average flow dislodged at that time. The pluviometer input is contrasted with the system which has been developed, thus obtaining a prediction of the flow resulting form rain. Finally, the signal will be the addition of both (fig. 6).

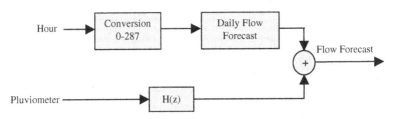

Fig. 6. Diagram of the system's global functioning

5.1 Example

The pluviometer signal is taken from fig. 7. This signal has been obtained at the time of 2.25, lasting 20 hours and 45 minutes, with a total number of 249 samples.

First, the necessary moment in time is taken. Given that the samples are consecutive from the initial sampling time (2.25) to the end, the system's inputs are taken form 137 to 97 (249 samples, returning to zero when we surpass 287). The signal of the flow dislodged daily at those times is modelled with these inputs. Second, the pluviometer signal is contrasted with the system, which renders a dislodged flow signal as a consequence of rain.

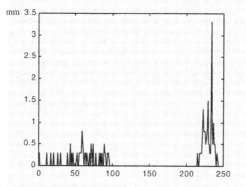

Fig. 7. Example of pluviometer signal

Finally, the resulting signal will be the addition of both, as shown in fig. 8, in comparison to the real signal obtained during those days and the signal modelled with a unitary SCS hydrograph. This kind of hydrograph is widely used in hydrology in order to forecast the flow.

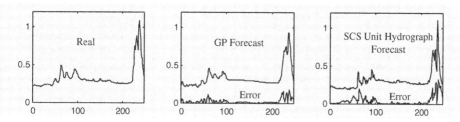

Fig. 8. Comparison between real signal, one obtained by GP and one modelled by means of a unitary hydrograph

Table 1 draws a comparison among the resulting signals, appearing to show that the evolutionary method yields better results than the traditional one.

Table. 1. Comparison between the evolutionary method and the unitary hydrograph

	GP Forecast	SCS Unit Hydrograph
Mean Error	2.736 E-2	4.343 E-2
Mean Square Error	1.857 E-3	5.707 E-3

6 Conclusions

In view of the results, the signal obtained approaches the real one, improving the method commonly used in hydrology and automating the process.

An expression which relates the time of day with the corresponding flow has been obtained for modelling the daily flow, by means of a feedback between the input and output values. In this case, the expression obtained is not easily analyzed, since no restrictions have been made on the way of building trees (no typed GP has been used).

A filter has been used for modelling the rain-flow relationship, which will produce an expression relating both of them. There are several filter-design techniques for obtaining this expression [12]. However, after many tests, it has been shown that the method with best results has been the one in which the GP typing power has been used for searching the system's poles and zeros.

The importance of GP is manifest, not only when trying to find relations among parameters, but also in order to obtain expressions with restrictions for the application of other techniques, such as the search for parameters in filter design.

7 Future Works

Once the modelling of a typical basin is obtained, the present technique will be applied to the modelling of other urban basins which do not follow a typical pattern, being affected by various physical phenomena which disturb the surface permeability, the distribution of water through the subsoil, pollution, etc.

Besides, the technique will be applied to rural basins, where there is no typical daily behaviour, given that there is no residual human activity. This will favour modelling on the one hand, but, on the other hand, the characteristics of the surface cause greater disturb in predictions than in urban areas, being very dependent on isolated natural phenomena (storms, droughts,…).

The goal is to implement an autonomous system which makes real-time predictions, being self-adaptive and correcting its parameters dynamically as the rain measurements take place. Thus, alarm systems which are able to predict the risk of rain conditions for the basin can be obtained.

Acknowledgements

The authors would like to thank Dr. Daniel howard (Head of Research and Development QinetiQ Software Evolution Centre) and the reviewers for their comments and suggestions.

References

1. Koza J., "Genetic Programming. On the Programming of Computers by means of Natural Selection". The Mit Press, Cambridge, Massachusetts, 1992.
2. Montana, D.J., "Strongly Typed Genetic Programming", Evolutionary Computation, 3(2):199-200, Cambridge, MA: The MIT Press, 1995.
3. Darwin C, "On the origin of species by means of natural selection or the preservation of favoured races in the struggle for life". Cambridge University Press, Cambridge, UK, sixth edition, 1864, originally published in 1859.

4. Fuchs, M., "Crossover Versus Mutation: An Empirical and Theoretical Case Study", 3rd Annual Conference on Genetic Programming, Morgan-Kauffman, 1998.
5. Luke, S., Spector, L., "A Revised Comparison of Crossover and Mutation in Genetic Programming", 3rd Annual Conference on Genetic Programming, Morgan-Kauffman, 1998.
6. Babovic, V., Keijzer, M., Aguilera, D.R., Harrington, J., "An Evolutionary Approach to Knowledge Induction: Genetic Programming in Hydraulic Engineering", Proceedings of the World Water & Environmental Resources Congress, May 21-24, 2001, Orlando, USA, 2001.
7. Babovic, V., Keijzer, M., Aguilera, D. R., Harrington, J., "Automatic Discovery of Settling Velocity Ecuations", D2K Technical Report, D2K-0201-1, 2001.
8. Babovic, V., Keijzer, M., "Declarative and Preferential Bias in GP-based Scientific Discovery", Genetic Programming and Evolvable Machines, Vol 3 (1) 2002.
9. Drecourt, J.P., "Application of Neural Networks and Genetic Programming to Rainfall-Runoff Modelling", D2K Technical Report 0699-1-1, Danish Hydraulic Institute, Denmark, 1999.
10. Viessmann, W., Lewis, G.L., Knapp, J.W., "Introduction to Hydrology", Harper Collins, 1989.
11. Loucks, D.P., "Stochastic Methods for Analyzing River Basin Systems", Tech. Rep. N.16, Cornell U., Ithaca, NY, 1969.
12. Liberali, V., Rossi, R., "Artificially Evolved Design of Digital Filters", Workshop on Advanced Mixed-Signal Tools, Munich, 2001.

Using EAs for Error Prediction
in Near Infrared Spectroscopy

Cyril Fonlupt[1], Sébastien Cahon[2], Denis Robilliard[1],
El-Ghazali Talbi[2], and Ludovic Duponchel[3]

[1] Laboratoire d'Informatique du Littoral
BP 719
62228 Calais Cedex, France
{fonlupt,robillia}@lil.univ-littoral.fr
[2] Laboratoire d'Informatique Fondamentale de Lille
Université de Lille 1
59655 Villeneuve d'Ascq Cedex, France
{talbi,cahon}@lifl.fr
[3] LASIR
Université de Lille 1
59655 Villeneuve d'Ascq Cedex, France
Ludovic.Duponchel@univ-lille1.fr

Abstract. This paper presents an evolutionary approach to estimate
the sugar concentration inside compound bodies based on spectroscopy
measurements. New European regulation will shortly forbid the use of
established chemical methods based on mercury to estimate the sugar
concentration in sugar beet. Spectroscopy with a powerful regression
technique called *PLS* (Partial Least Squares) may be used instead. We
show that an evolutionary approach for selecting relevant wavelengths
before applying *PLS* can lower the error and decrease the computation
time. It is submitted that the results support the argument for replacing
the *PLS* scheme with a GP technique.

1 Introduction

Spectroscopy is an interesting tool to get relevant information about the structure of complex compound bodies in a non destructive way. Among all spectroscopy techniques, the NIRS (Near Infrared Spectroscopy) is widely used as an efficient and low cost method and is especially suitable for analyzing compound bodies and giving meaningful information about the simple bodies. This paper proposes how to evaluate the quality control of sugar in sugar beet with evolutionary algorithms. Until very recently, established chemical methods estimated the concentration of sugar, but such methods were reliant on mercury as reagent. However, use of mercury as reagent will shortly be banned owing to new European regulation. Thus NIRS appears as a useful alternative to established methods for estimating sugar concentration.

Spectroscopy has been applied to estimate the sugar concentration inside sugar beet using the partial least squared (PLS) method [1,2]. We show in this

S. Cagnoni et al. (Eds.): EvoWorkshops 2002, LNCS 2279, pp. 202–209, 2002.
© Springer-Verlag Berlin Heidelberg 2002

article that Evolutionary Algorithms can enhance this technique by eliminating irrelevant wavelengths. After selecting a relevant terminal set, we present an on-going work on the application of GP for spectroscopy.

This paper is presented in the following way. Section 2 introduces the Near Infrared Spectroscopy methodology and the application of the method for estimating the sugar concentration. As the *PLS* is a very time-consuming method, we propose to decrease the set of wavelengths used to accelerate the computation. Section 3 deals with the use of an evolutionary approach as a feature selection algorithms. Section 4 presents our experimental results and preliminary results on our GP scheme can be found in section 5

2   The Near Infrared Spectroscopy

The first part of this section focuses on the NIRS (Near Infrared Spectroscopy) basic concepts and we deal in the second part with the *PLS* (Partial Least Squares) technique which is currently used for estimating the concentration of compound bodies from the information gathered by the spectroscopy method.

2.1   Principles of the NIRS Method

Near infrared Spectroscopy (NIRS) is a method which uses a beam of light in the near infrared range (700-1000nm) which passed through samples, and measurements of the absorption and scattering of the photons are made. As a matter of facts, some molecules and especially organic molecules absorb infrared radiation and start to oscillate. An illustration of near infrared spectroscopy is presented in Fig. 1.

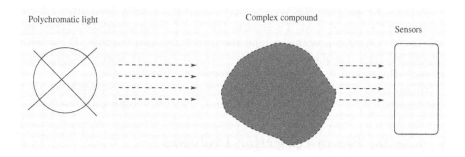

Fig. 1. Illustration of near infrared spectroscopy

The absorption spectrum as measured by a sensor is usually highly complex because more than several hundred of wavelengths are usually monitored. (1020 in our case)

More formally, let L the signal emitted by a polychromatic source, L_i the energy levels monitored on the sensors, C the sugar concentration. We face here an inverse problem that is finding out a function f such that:

$$f(L_{0001}, L_{0002}, \ldots, L_{1020}) = C$$

This symbolic regression problem is usually tackled using the *PLS* method introduced in the next section.

2.2   The PLS Method

It is not our aim to give a full detailed presentation of the partial least squares (*PLS*) method but only to provide the basic ideas of this technique.

PLS regression is an extension of the multiple linear regression model. In its simplest form, a linear model describes the linear relationship between a variable Y and a set of predictors X_i. Multiple regression tries to predict the relationship between several independent variables. *PLS* extends multiple linear regression without imposing any restrictions which are found in principal components regression or canonical correlation. To sum up, *PLS* can be seen as the least restrictive of the various multivariate extensions of the multiple linear regression model and as one of the most powerful regression method. It has now become a standard tool for modeling linear relations between multivariate measurements.

When the *PLS* scheme is applied to NIRS for estimating the sugar concentration, the RMS error is used as a measurement of the quality of the prediction.

$$\text{relative RMS} = \sqrt{\frac{1}{n} \sum_{i=1}^{n} (C_{\text{computed}} - C_{\text{expected}})^2}$$

In order to use the *PLS* scheme, data from the LASIR laboratory of the university of Lille are used. These data consist of about 2000 cases (one case meaning the sugar concentration and the associated wavelengths). These data are split into two sets of about 1000 cases each. The first set is used as a learning set while the second set is used as a validation set. (see Fig. 2).

We get a low 0.174% relative RMS prediction error on the validation set after applying the *PLS* method on the learning set. These are quite good results.

3   EAs for Feature Selection Problems

EAs in the data-mining field has been a hot research topic lately [3]. It is widely known that optimal feature selection is NP-hard [4]. Because of its practical importance, feature selection has been a very active field and GAs have received special attention as they are well suited for this problem (see [5,6]). As in the "sugar" problem the fitness function is a time consuming step (PLS), the time requirements for the algorithm grow dramatically with the number of features.

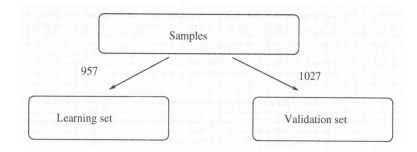

Fig. 2. Learning and validation sets

In our case, the feature selection could be used to decrease the complexity of the problem by eliminating the irrelevant features. A possible drawback might be the loss of some discriminating power. However as argued by [7], in view of the Occam's Razor, bias often causes many algorithms to prefer a small number of relatively predictive features to a large number of features.

Many feature selection oriented GAs are available like the wrapper methods or filter methods. We briefly present the representation, the operators and the fitness function used in our GA.

Representation of the chromosomes In our case, we use a very obvious representation. As more than 1000 features (1020 exactly) are available for selection, a string of bits is simply used to indicate corresponding features inclusion or exclusion.

A solution is simply represented by a string S of 1020 bits, $S = \{b_i\}, i = 1, 2, \ldots, 1020$ where $b_i = 1$ denotes the feature selection while $b_i = 0$ indicates that the feature is not selected. The length of the chromosomes is the number of features, as every gene in the chromosome represents the value of the feature occupying that position.

Size of the population Regarding the size of the population, it is not desirable to work with a large population because of the time-consuming fitness function required. Hence, we set up the population size to 100, which corresponds to one full day of computation on a 1.5 Ghz PC.

Genetic operators Standard genetic operators are used (one-point crossover and bit-flipping mutation). Their probabilities are fixed to 0.8 for crossover and 0.1 for mutation. Elitism was used throughout all these experiments. Uniform crossover has also been considered and has shown to perform a little better than standard one-point crossover.

Fitness function The fitness function is the *PLS* method with the relevant features selected by the chromosome representation. In order to speed up the computation, the fitness function has been parallelized using a PVM slave/master approach on 16 computation nodes. The spectroscopy database for these experiments consists of about 2000 test cases. Half of these partake in the learning phase and the other half in the validation set.

Tab. 3 sums up the main GA parameters.

Table 1. main GA parameters

Population size	100
Fitness function	*PLS* method
	over 1000 learning cases
Crossover	1-point crossover
	uniform crossover
Mutation	bit-flipping mutation
Crossover probability	0.8
Mutation probability	0.1 (per individual)
Max generations	1500

4 Experiments

As introduced in section 2, the best error using the *PLS* method on all wavelengths is 0.174%. Even if this error might seem already low, it is economically very interesting to improve precision because there are huge amounts of sugar beets produced in the world.

Tab. 4 presents the results we obtained using the GA approach for feature selection and the classical approach using only the *PLS* method. The GA selection brought an improvement from 0.17% to 0.124% relative RMS error.

Table 2. Results of the GA used as a feature selection method vs classical method (relative RMS error)

	PLS method (all wavelengths)	GA (features selection) + *PLS* 1-point crossover	GA (features selection) + *PLS* uniform crossover
Results	0.174%	0.14%	0.124%

Moreover, the best run only needed 272 wavelengths compared to more than 1000 for the standard *PLS* approach. On average, only about one third of all wavelengths are needed (*i.e.* it means that about two thirds of all wavelengths are not needed or useless). When we compare the best solutions to each others, it appears that usually the same wavelengths are selected (see Fig. 3 for an illustration).

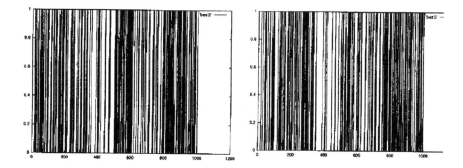

Fig. 3. Comparison of the two best chromosomes (a black line indicates a selected feature). Note that both chromosomes are very close

In one word, it appears that using GA as a feature selection method in a spectroscopy application has improved the performance and has shown that about two thirds of the wavelengths can be ignored.

5   Genetic Programming and Spectroscopy

In this section we investigate whether the GP scheme may be a more suitable choice than the *PLS*, because it can be argued that GP can theoreticaly model a non-linear regression task with higher precision than *PLS*.

Lots of work has been performed on the use of GP for knowledge discovery [8,9,10] and it has been shown that GP can rival older models like rule induction or decision trees. In that case, each individual can be seen as a "rule". The terminal set consists of the predicting attributes while the function set consists of mathematical comparison and logical operators.

The main problem is to select an appropriate terminal set. Even if the use of EA allows to only work with about 1/3 of all wavelengths, it seems it is not relevant to work with such a large terminal set. By using a multi-objective approach (not presented in this paper) based on several criteria (minimize prediction error, minimize interdependence between selected fatures), we were able to select a set composed of 4 wavelengths (394, 564, 653, and 914). The parameters used for the GP experiments are presented in Tab. 3.

Tab. 4 sums up the overall results (*PLS* alone, *PLS* with EAs, GP)

The GP results are not as good as the *PLS*+EA scheme nor the *PLS* alone. However, GP already comes close to *PLS*, based on only 4 wavelengths. A more elaborate terminal and/or function set might give better results.

Table 3. GP parameters

Population size	1000
Max. Gen.	200
Max. tree depth	15
Selection	Tournament
Function set	$+, -, \times, /, \exp, \log$
Terminal set	$L_{394}, L_{564}, L_{653}, L_{914}$

Table 4. Overall results

	PLS method (all wavelengths)	GA (features selection) + PLS	GP
Results	0.174	0.14	0.27

6   Conclusions

We show in this paper that using the robust GA heuristic with the very efficient *PLS* method enables to lower the estimation error on sugar concentration.
Moreover, this evolutionary approach also decreases the computation time as
only about 1/3 of the wavelengths are really useful. We propose to replace the
PLS estimation method by a model obtained by Genetic Programming. Although
preliminary GP results are not as good as the *PLS* scheme, this still could be a
promising way as only 4 wavelengths out of 1020 were used.

References

1. H. Swierenga, W.G. Haanstra, AP de Weijer, and LMC Buydens. Comparison of
   two different approaches toward model transferability in nir spectrocopy. *Applied
   Spectroscopy*, 52:7–16, 1998.
2. H. Swierenga, PJ de Groot, AP de Weijer, MWJ Derksen, and LMC Buydens. Improvement of PLS method transferability by robust wavelength selection. *Chemom.
   Intell. Lab. Syst.*, 14:237–248, 1998.
3. Maria Martin-Bautista and Maria-Amparo Vila. A survey of genetic feature selection in mining issues. In *1999 Congress on Evolutionary Computation*, pages
   1314–1321, 1999.
4. P.M. Narendra and K. Fukunaga. A branch and bound algorithm for feature subset
   selection. *IEEE Trans. on Computer*, 26(9):917–922, sep 1977.
5. H. Vafaie and K. De Jong. Robust feature selection algorithms. pages 356–363.
   IEEE Computer Society Press, 1993.
6. Michael L. Raymer, William F. Punch, Erik D. Goodman, Paul C. Sanschagrin,
   and Leslie A. Kuhn. Simultaneous feature scaling and selection using a genetic
   algorithm. In *[11]*, pages 561–567, 1997.
7. Kan Chen and Huan Liu. Towards an evolutionary algorithm: A comparison of
   two feature selection algorithms. In *1999 Congress on Evolutionary Computation*,
   pages 1309–1313, 1999.
8. J. Eggermond, A.E. Eiben, and J.I. van Hemert. Adapting the fitness function in
   GP for data-mining. In *[12]*, pages 193–202, 1999.

9. Alex Freitas. A genetic programming framework for two data mining tasks: Classification and generalized rule induction. In *[13]*, pages 96–101, 1997.
10. Celia Bojarczuk. Discovering comprehensible classification rules using genetic programming: a case study in a medical domain. In *[14]*, pages 953–958, 1999.
11. *Proceedings of the 7th International Conference on Genetic Algorithms*, East Lansing, Michigan, USA, July 1997. Morgan Kaufmann.
12. R. Poli, P. Nordin, W.B. Langdon, and T.C. Fogarty, editors. *Genetic Programming, proceedings of EuroGP'99*, volume 1598 of *LNCS*, Goteborg, Sweden, may 1999.
13. John R. Koza, Kalyanmoy Deb, Marco Dorigo, David B. Fogel, Max Garzon, Hitoshi Iba, and Rick L. Riolo, editors. *Proceedings of the Second Annual Conference on Genetic Programming*, Stanford University, CA, USA, Jul 1997. Morgan Kaufmann.
14. Wolfgang Banzhaf, Jason Daida, Agoston Eiben, Max Garzon, Vasant Honavar, Mark Jakiela, and Robert Smith, editors. *Proceedings of the Genetic and Evolutionary Computation Conference*, Orlando, Florida, USA, july 1999. Morgan-Kaufmann.

The Prediction of Journey Times on Motorways Using Genetic Programming

Daniel Howard and Simon C. Roberts

QinetiQ Software Evolution Centre, Building U50, QinetiQ,
Malvern WR14 3PS,UK tel: +44 1684 894480 dhoward@qinetiq.com

Abstract. Considered is the problem of reliably predicting motorway journey times for the purpose of providing accurate information to drivers. This proof of concept experiment investigates: (a) the practicalities of using a Genetic Programming (GP) method to model/forecast motorway journey times; and (b) different ways of obtaining a journey time predictor. Predictions are compared with known times and are also judged against a collection of naive prediction formulae. A journey time formula discovered by GP is analysed to determine its structure, demonstrating that GP can indeed discover compact formulae for different traffic situations and associated insights. GP's felxibility allows it to self-determine the required level of modelling complexity.

1 Introduction

The MIDAS system on the M25 motorway which encircles London produces minute averaged readings of traffic speed (this quantity will be referred to as velocity), traffic flow, occupancy, headway and vehicle counts by vehicle length. MIDAS loop sensors are incorporated into the tarmac for each lane of the motorway at roughly 500 metre intervals; quantities are measured directly rather than as quantities derived from other measurements.

Motorway users would like to know how long it takes to cover a journey from one junction of the M25 to another one, e.g. between points A & B just prior to the driver's journey, and to satisfy this want, it is necessary to predict the state of the motorway for the time of that future journey. This investigation aims to discover mathematical and logical relationships between predicted journey times and past and actual motorway journey data using minute averaged measurements of traffic measurements obtained from loop sensors on the road.

For the purpose of this investigation the journey time, J, is defined as the average time taken by a driver already on the motorway[1] to cover the distance between two junctions, e.g. junction A merging and junction B diverging.

This paper investigates two alternative classes of strategy for predicting journey times: (a) direct prediction of journey time, and (b) prediction of velocities for the stretch of motorway in the journey and computation of the journey time

---

[1] For this reason only three lanes, i.e. the offside and its two adjacent lanes, participate in the journey time calculation.

S. Cagnoni et al. (Eds.): EvoWorkshops 2002, LNCS 2279, pp. 210–221, 2002.

based on such velocity predictions. While prediction is immediate with the first strategy and is also applicable to journey times of any duration (including those in heavy traffic), the second strategy requires that journey times be computed from analysis of the motion of a virtual vehicle through the lane averaged velocity field, and therefore a time limit exists with this type of forecasting. A period of 15 minutes is chosen, to be of any practical use to a driver - and in heavy traffic either a naive velocity prediction can be assumed for the remainder of the journey, or a 30 minute velocity predictor can be developed for longer journeys.

Both alternatives take as input present and past traffic quantities for current and *downstream* stations. It is important to use downstream information because motorway queues and slowdowns (usually originating at junctions) travel upstream as waves at approximately 20 km/h and cross the path of the virtual vehicle. More work is involved in the second alternative because velocity predictions must be carried out at each loop sensor position of the virtual journey. In contrast, the first alternative simply makes one forecast for the entire journey.

2   Transforming MIDAS Data into a GP Terminal Set

MIDAS records minute-averaged traffic quantities ϕ_t^{sl} for minute t by a loop sensor s on lane l:

$$\phi_t^{sl} = (V_t^{sl}, O_t^{sl}, F_t^{sl}, C_t^{sl})$$

where V_t is velocity in km/h, O_t is the percentage lane occupancy, F_t is the flow rate in cars/minute, and C_t is the number of vehicles counted for a certain length class and so here the l superscript denotes the length class. Lanes are numbered from the offside fast lane[2]. Minute averaged velocities at a loop sensor are averaged again across offside and two adjacent lanes, and weighted by the number of cars in each lane, i.e. flow averaged,

$$V_t^s = \frac{F_t^{s1} V_t^{s1} + F_t^{s2} V_t^{s2} + F_t^{s3} V_t^{s3}}{F_t^{s1} + F_t^{s2} + F_t^{s3}}$$

Other participants in prediction are the total flow rate across the three lanes and the mean of minute averaged lane occupancy values:

$$F_t^s = \sum_{l=1}^{3} F_t^{sl} \qquad\qquad O_t^s = \frac{1}{3} \sum_{l=1}^{3} O_t^{sl}$$

Data is modified when $F_t^{sl} = 0$, V_t^{sl} is absent but a velocity is never-the-less required or when sensors fail momentarily and do not report their minute

---

[2] Between junctions 15 and 11 on the counter-clockwise carriageway of the M25, loop sensor locations between diverging and merging lanes of the junction have three lanes while all other loop junctions have four lanes. The fourth lane is used either to exit or to join the motorway.

averaged readings. Values at the previous minute or values from immediately adjoining sensors can recover these missing quantities. This study avoided times of day when a group of more than two adjacent sensors reported no readings for more than two consecutive minutes.

Two years of historical MIDAS data were inspected to find days when the sensors were fully operational and the traffic was of interest on the counter-clockwise carriageway of the M25 between junctions 15 and 11. September 1999 was a good month for data at busy periods between 13:00 hrs and 22:00 hrs.

Figure 3 illustrates the training data set (distance vs time, origin top left). Journey times between Jct. 15 and Jct. 11 took anywhere between 6 minutes to over 30 minutes. For the experiments reported in this proof of concept study GP trained on the period 13:00hrs to 22:00hrs on 2nd September.

Where should the predictor obtain present and past quantities? A rectangle can be set of laged time dimension: minute t to $t - T$, and distance from downstream: sensor s to $s + S$. Input traffic quantities from this space-time rectangle can be block averaged in time and in space to reduce the number of terminals for GP. For velocity predictors $S = 7$ and $T = 20$ while for journey time predictor all sensors in the journey were used, i.e. $S = 27$. Use of downstream sensors tries to capture the traffic queues upstream[3].

Figure 1 illustrates these rectangular areas. For some experiments, raw values were used whereas for other experiments averages for the quantity over uniform rectangular regions, i.e. box-averaged data over a rectangle with dimensions $b_s \times b_t$, were used. Figure 2 illustrates a special case where the predictor was based not only on the known data in the rectangle but on previous velocity predictions $(\hat{V}_t^s, \hat{V}_{t+1}^s...\hat{V}_{t+14}^s)$. Initially, this vector is set to V_t^s and becomes updated as the predictions arrive[4].

The first alternative strategy (see section 1) also uses as data input past journey times implemented as follows: iteratively look back to see which journey would have finished at the current minute and use this JT input. Repeat this to obtain a previous journey time, and so on.

2.1   Calculation of the Journey Time from the Velocity Field

This calculation moves the virtual vehicle one second at a time. While the vehicle travels between two loop sensors, distance is computed with the harmonic mean $V_t^{s,s+1}$ of the velocities from both sensors. Distance traveled d by the car at each second uses the conversion factor from km/h to m/s. Time t is updated to $t + 1$ every 60 seconds, and loop sensor index s is updated to $s + 1$ once d has reached the next sensor. Iteration finishes when the journey is at an end.

$$\frac{1}{V_t^{s,s+1}} = \frac{1}{2}\left(\frac{1}{V_t^s} + \frac{1}{V_t^{s+1}}\right) \text{ ; and } d = d + \frac{1000}{3600}V_t^{s,s+1}$$

---

[3] The number of downstream sensors included, e.g. 7, was insufficient to detect the shock wave fifteen minutes in advance. Current research is including more stations.

[4] this is a form of autoregressive prediction.

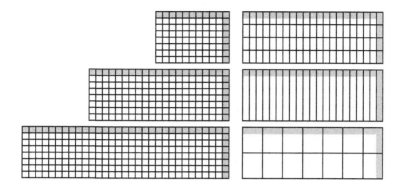

Fig. 1. Examples of raw and boxed-averaged inputs. Horizontal axis is the time axis and the vertical axis is the loop sensor axis. Each square either represents a raw value or a box-averaged quantity over a rectangle with dimensions $b_s \times b_t$. Grey cells are either at the sensor in question or at the current time. The velocity prediction is for the quantity 15 mins. into the future, the white cell on the extreme right of figure 2.

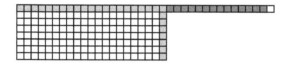

Fig. 2. Velocity prediction $\hat{V}_{t+15}^s$ which uses former predictions of velocity. Fourteen former predictions are available and indicated as dark grey cells.

3   Comparative Experimental Study of JT Prediction

GP runs explored evolution options: a velocity predictor; a velocity gradient predictor; a journey time predictor. Each driven by a fitness measure which used either velocity errors or journey time errors to indicate prediction accuracy in various combinations. They used steady-state GP method with a typical population size of 2000 individuals, cross-over dominance with small amounts of mutation, a breed tournament size of 4 and a kill tournament selecting two individuals at random and eliminating the weakest; functions: $+$, $-$, $*$, protected division, *min* and *max*, and at least 15 parallel independent runs were carried out for each experiment.

Both L2 norm $\| \cdot \|$ and max error $\max(\cdot)$[5] were applied to measure absolute and percent accuracy. However, the experiments investigated various fusions of $\| \cdot \|$ and $\max(\cdot)$ errors, where ϕ is the quantity whose error needs to be minimised[6], $\| \cdot \|$ is the root mean square error. If f is the Darwinian fitness, e_r

---

[5] the uniform or infinity norm $\| \cdot \|_\infty$
[6] V_s^t or ΔV_s^t or J^t for example.

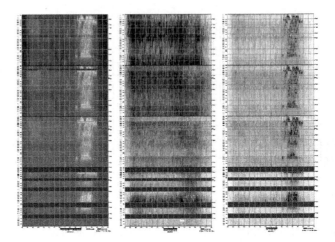

Fig. 3. Training data 06:00-22:00 hrs on Thursday 2nd September 1999, M25 counter-clockwise carriageway. Journey times are from loop 4940b to 4811b (junction 15 merging to junction 11 diverging). From left to right: velocity, flow and occupancy. From top to bottom: offside (fast) lane, offside-1 lane, offside-2 lane, offside-3 lane.

is the relative error and e_n is error relative to a naive solution, then for ϕ,

$$e_a = |\hat{\phi} - \phi| \qquad e_r = \frac{e_a}{\phi} \qquad e_n = |\hat{\phi} - \phi_N| \qquad ; \text{e.g. } \|e_a\| = \sqrt{\frac{1}{n}\sum e_a^2}$$

where ϕ_N is the naive approx., and $\hat{\phi}$ is the evolved value driven by one of $f =:$

$$- \left(\|e_a\|\|e_r\| + \max(e_a)\max(e_r) \right) \text{ ; or by } - \left(\|e_a e_r\| + \max(e_a e_r) \right) ;$$

$$\text{or defining } e_c \text{ as: if } \left(\frac{e_a}{e_n} \right) > 1 \text{ then } e_c = \left(\frac{e_a}{e_n} \right)^2 \text{ else } e_c = \left(\frac{e_a}{e_n} \right)$$

$$\text{by } - \|e_c\| \text{ ; or by } - \left(\|e_c\| + \frac{\max(e_c)}{100} \right)$$

The primary input data types were traffic velocity, V, occupancy, O, and flow, F, and each of these were averaged across the three offside lanes. Secondary data types were based on flow-velocity plot descriptors. Furthermore, intermediate predictions could be fed back in as input data. The experiments investigated which data types gave the most predictive power from:

$$\phi = \left(V, O, F, M = (F^2 + V^2), G = \frac{60F}{V}, \hat{V} \right)$$

JT prediction errors were obtained for the 10 best predictors from each parallel independent run, regardless of whether the evolution was driven by JT or by velocity errors. The following four error calculations were used for validation, again regardless of which error calculations drove evolution: $\max(e_a)$, $\max(e_r)$, $\|e_a\|$ and $\|e_r\|$. This posed a multi-objective optimisation problem in four dimensions and a *Pareto* ranking scheme was used to compare the different predictors. Pareto ranking was first performed on the $\max(e_r)$ and $\|e_r\|$ dimensions. A second independent Pareto ranking process was then performed on the $\max(e_a)$ and $\|e_a\|$ dimensions. Predictors were then ordered according to e_p rank and then e_a rank. The best predictor for the experiment was then subjectively designated from the top ranking predictors, and the associated JT errors were tabulated.

It is true of this time series that improving upon naive predictions is challenging. Thus, results of five naive formulae were compared to the evolved predictions. The quantities involved (ϕ_t) were either:

- velocity at the current sensor, or
- JT of the most recent journey *completed* by time t. For example, let the virtual journey which commenced 9 minutes ago take J_{t-9} minutes to complete, and that which commenced 8 minutes ago take J_{t-8} minutes to complete. If $J_{t-9} = 8.5$ and $J_{t-8} = 8.1$ then ϕ_t was set to J_{t-9}. However, if instead $J_{t-8} = 7.9$ then ϕ_t was set to J_{t-8}.

$$\text{Type 1} \quad \hat{\phi}_{t+15} = \phi_t \quad ; \quad \text{Type 2} \quad \hat{\phi}_{t+15} = 2\phi_t - \phi_{t-15}$$

$$\text{Type 3} \quad \hat{\phi}_{t+15} = (\phi_t + \phi_{t-15})/2 ; \quad \text{Type 5} \quad \hat{\phi}_{t+15} = \frac{1}{16} \sum_{i=0}^{i \leq 15} \phi_{t-i}$$

$$\text{Type 4} \quad \hat{\phi}_{t+15} = \phi_t/2 + (\phi_{t-5} + \phi_{t-10} + \phi_{t-15})/6$$

3.1   Results: (a) Velocity Predictor and (b) Direct JT Predictor

The results in Table 1 pertain to runs which evolved velocity predictors. Most runs used JT errors to drive the evolution. Other runs used velocity errors as the driving quantity, but the resulting predictors were still validated by calculating JT prediction errors. Experiments used raw input data - no window averaging. However, data reduction was investigated by fusing quantities, namely the flow-velocity gradient G or the flow-velocity magnitude M. Some experiments also investigated feeding intermediate predictions, $\hat{V}$, back in as input data. Note that the fitness measure is expressed as $-f$.

Table 2 gives the results from averaging the raw inputs over boxes of b_t minutes by b_s sensors. All runs used JT errors to drive the evolution. The table also includes results from using multidimensional boxes where b_t increased according to the Fibonacci sequence. Each box was aligned to terminate at the current time with larger boxes capturing more past data. The first box used the sensor data at time t, the second box used data from $t - 1$ to t, and the last box used data from T to t where T equaled the last Fibonacci number minus one. Hence, the input data was effectively biased towards the more recent data. The average and standard deviation over each box was input to GP. Table 2 also gives results of evolving JT predictors directly, and Table 3 for the naive predictions.

Table 1. Results

Predicting **VELOCITY from raw inputs**				J Errors for Best Run			
ID	input	T	S	fitness using J errors	$\max(e_a)$	$\max(e_r)$	$\|e_a\|$	$\|e_r\|$
1	VOF	20	7	$\|e_a\|\|e_r\| + \max(e_a)\max(e_r)$	141.0	0.196	44.9	0.065
2	VF	20	7	$\|e_a\|\|e_r\| + \max(e_a)\max(e_r)$	135.3	0.175	50.5	0.071
3	VOF	10	7	$\|e_a\|\|e_r\| + \max(e_a)\max(e_r)$	136.1	0.183	47.0	0.069
4	VOF	30	7	$\|e_a\|\|e_r\| + \max(e_a)\max(e_r)$	130.3	0.175	45.6	0.065
5	VOF	20	7	$\|e_a e_r\| + \max(e_a e_r)$	145.0	0.172	43.3	0.065
6	VOFM	20	7	$\|e_a\|\|e_r\| + \max(e_a)\max(e_r)$	144.2	0.194	48.6	0.068
7	OM	20	7	$\|e_a\|\|e_r\| + \max(e_a)\max(e_r)$	158.7	0.213	48.6	0.072
8	OG	20	7	$\|e_a\|\|e_r\| + \max(e_a)\max(e_r)$	156.5	0.221	60.3	0.084
9	VOG	20	7	$\|e_a\|\|e_r\| + \max(e_a)\max(e_r)$	127.2	0.169	45.9	0.066
10	OGM	20	7	$\|e_a\|\|e_r\| + \max(e_a)\max(e_r)$	163.5	0.197	50.8	0.073
11	VOFG	20	7	$\|e_a\|\|e_r\| + \max(e_a)\max(e_r)$	137.5	0.185	42.7	0.063
15	VOF$\hat{V}$	20	7	$\|e_a\|\|e_r\| + \max(e_a)\max(e_r)$	142.2	0.191	47.3	0.068
16	OGM$\hat{V}$	20	7	$\|e_a\|\|e_r\| + \max(e_a)\max(e_r)$	159.9	0.204	50.1	0.073
17	OM$\hat{V}$	20	7	$\|e_a\|\|e_r\| + \max(e_a)\max(e_r)$	137.8	0.213	53.6	0.079
ID	input	T	S	fitness using V errors	$\max(e_a)$	$\max(e_r)$	$\|e_a\|$	$\|e_r\|$
20	VOF	20	7	$\|e_a\|\|e_r\| + \max(e_a)\max(e_r)$	1476.5	1.646	808.6	1.104
21	VOF	20	7	$\|e_a e_r\| + \max(e_a e_r)$	561.1	0.766	249.4	0.350
22	VOF	20	7	$\|e_c\|$	219.0	0.228	60.2	0.080
23	VOF	20	7	$\|e_c\| + \max(e_c)/100$	226.0	0.216	61.11	0.081
24	VOFG	20	7	$\|e_c\|$	220.6	0.218	61.83	0.082
Predicting **VELOCITY GRADIENT**				J Errors for Best Run			
ID	input	T	S	fitness using V errors	$\max(e_a)$	$\max(e_r)$	$\|e_a\|$	$\|e_r\|$
25	VOF	20	7	$\|e_a\|\|e_r\| + \max(e_a)\max(e_r)$	224.4	0.270	65.6	0.088

Pred. **VELOCITY f. boxed inputs**, $T = 20$					J Errors for Best Run			
ID	input	b_t	b_s	S	fitness using J errors	$\max(e_a)$	$\max(e_r)$	$\|e_a\|$	$\|e_r\|$
27	VOF	1	2	7	$\|e_a\|\|e_r\| + \max(e_a)\max(e_r)$	154.9	0.208	45.9	0.066
28	VOF	1	3	7	$\|e_a\|\|e_r\| + \max(e_a)\max(e_r)$	131.3	0.169	48.7	0.070
29	VOF	1	8	7	$\|e_a\|\|e_r\| + \max(e_a)\max(e_r)$	111.9	0.160	40.0	0.059
30	VOF	2	2	7	$\|e_a\|\|e_r\| + \max(e_a)\max(e_r)$	137.1	0.197	44.5	0.065
31	VOF	2	3	7	$\|e_a\|\|e_r\| + \max(e_a)\max(e_r)$	134.9	0.189	45.8	0.067
32	VOF	2	4	7	$\|e_a\|\|e_r\| + \max(e_a)\max(e_r)$	135.6	0.176	48.4	0.068
33	VOF	3	2	7	$\|e_a\|\|e_r\| + \max(e_a)\max(e_r)$	124.5	0.164	42.6	0.064
34	VOF	3	3	7	$\|e_a\|\|e_r\| + \max(e_a)\max(e_r)$	132.1	0.180	47.4	0.070
35	VOF	3	4	7	$\|e_a\|\|e_r\| + \max(e_a)\max(e_r)$	147.2	0.186	49.9	0.069
36	VOF	3	2	15	$\|e_a\|\|e_r\| + \max(e_a)\max(e_r)$	123.0	0.164	44.0	0.064
37	VOF	3	4	15	$\|e_a\|\|e_r\| + \max(e_a)\max(e_r)$	108.8	0.154	44.5	0.068
38	VOF	3	8	15	$\|e_a\|\|e_r\| + \max(e_a)\max(e_r)$	121.1	0.142	45.1	0.064
40	VOF	1	4	7	$\|e_a e_r\| + \max(e_a e_r)/10$	153.3	0.164	44.5	0.061
41	VOF	2	4	7	$\|e_a e_r\| + \max(e_a e_r)/10$	155.2	0.178	44.1	0.063
42	VOF	3	4	7	$\|e_a e_r\| + \max(e_a e_r)/10$	137.1	0.181	38.5	0.055
43	VOF	4	4	7	$\|e_a e_r\| + \max(e_a e_r)/10$	138.8	0.157	43.6	0.062

Table 2. Results - continued...

Predicting VELOCITY, b_t based on the Fibonacci sequence, $S = 7$

ID	input	b_s	T	fitness using J errors	$\max(e_a)$	$\max(e_r)$	$\|e_a\|$	$\|e_r\|$
50	VOF	8	20	$\|e_a\|\|e_r\| + \max(e_a)\max(e_r)$	129.2	0.175	46.4	0.070
51	VOF	8	33	$\|e_a\|\|e_r\| + \max(e_a)\max(e_r)$	117.9	0.156	41.0	0.063
52	VOF	8	54	$\|e_a\|\|e_r\| + \max(e_a)\max(e_r)$	148.0	0.195	41.3	0.058
53	VOF	8	88	$\|e_a\|\|e_r\| + \max(e_a)\max(e_r)$	140.0	0.148	43.6	0.061
54	VOF	8	88	$\|e_a\|\|e_r\| + (\max(e_a)\max(e_r))/10$	138.7	0.165	41.1	0.059
55	VOF	8	88	$\|e_a e_r\| + \max(e_a e_r)/10$	134.5	0.164	40.0	0.057
56	VOF	4	88	$\|e_a\|\|e_r\| + \max(e_a)\max(e_r)$	130.7	0.174	41.7	0.059
57	VOF	4	88	$\|e_a\|\|e_r\| + (\max(e_a)\max(e_r))/10$	137.7	0.159	39.5	0.056
58	VOF	4	88	$\|e_a e_r\| + \max(e_a e_r)/10$	126.4	0.170	36.0	0.055

Predicting JOURNEY TIME					J Errors for Best Run			
ID	input	b_t	b_s	T	fitness using J errors	$\max(e_a)$	$\max(e_r)$	$\|e_a\|$	$\|e_r\|$
90	VOF	1	27	15	$\|e_a\|\|e_r\| + \max(e_a)\max(e_r)$	157.6	0.221	57.1	0.081
91	VOF	2	3	15	$\|e_a\|\|e_r\| + \max(e_a)\max(e_r)$	173.5	0.275	64.2	0.094
92	VOF	2	9	15	$\|e_a\|\|e_r\| + \max(e_a)\max(e_r)$	149.3	0.265	53.3	0.077
93	VOF	2	3	9	$\|e_a\|\|e_r\| + \max(e_a)\max(e_r)$	183.9	0.256	67.2	0.093
94	VOF	2	9	9	$\|e_a\|\|e_r\| + \max(e_a)\max(e_r)$	156.6	0.202	53.9	0.076
95	VOF	2	3	5	$\|e_a\|\|e_r\| + \max(e_a)\max(e_r)$	184.1	0.284	58.4	0.084
96	VOF	2	9	5	$\|e_a\|\|e_r\| + \max(e_a)\max(e_r)$	156.6	0.257	59.6	0.086
97	VOF	2	3	15	$\|e_a e_r\| + \max(e_a e_r)/10$	151.7	0.206	54.0	0.077
98	VOF	2	9	15	$\|e_a e_r\| + \max(e_a e_r/10)$	142.4	0.241	47.1	0.070

Table 3. Top: JT errors for naive predictors of velocity and JT errors for naive predictors of JT. Naive type 2 predicts negative velocities during congested periods to the extent that virtual journeys cannot be completed. Bottom: comparison and percentage improvements for the best GP evolved predictor against the best naive predictor.

Velocity Predictor					JT Predictor				
Naive type	$\max(e_a)$	$\max(e_r)$	$\|e_a\|$	$\|e_r\|$	Naive type	$\max(e_a)$	$\max(e_r)$	$\|e_a\|$	$\|e_r\|$
1	242.5	0.316	72.0	0.095	1	383.4	0.582	120.6	0.164
2	—	—	—	—	2	483.0	0.655	112.8	0.148
3	270.7	0.281	86.0	0.111	3	383.1	0.600	143.0	0.198
4	266.6	0.236	79.1	0.100	4	369.2	0.579	135.6	0.187
5	287.0	0.244	87.2	0.110	5	376.3	0.590	143.5	0.199

Predictor type	Velocity	Journey Time
Best experiment	58	98
Best naive type	1	2
$\max(e_a)$ (%)	52	30
$\max(e_r)$ (%)	54	37
$\|e_a\|$ (%)	50	42
$\|e_r\|$ (%)	58	47
typical $\max(\cdot)$ (%)	55	40
typical $\|\cdot\|$ (%)	65	50

3.2   Comparison of Both Prediction Strategies

Results could be analysed to compare the accuracy of a velocity predictor against the accuracy of a direct JT predictor. Such a comparison would reveal that the best JT predictor (Exp98) gave similar $\|\cdot\|$ errors to the worst velocity predictors (Exp28 and Exp34) but that it gave worse $\max(\cdot)$ errors.

However, the validity of this comparison is questionable and instead comparison between the two evolution schemes can judge each scheme against its associated set of naive predictors as follows:

– The evolved velocity predictors (driven by JT errors) were consistently better than all the naive predictors. The evolved predictors typically gave $\max(\cdot)$ errors which were 55% of the naive $\max(\cdot)$ errors and $\|\cdot\|$ errors which were 65% of the naive $\|\cdot\|$ errors.
– The evolved JT predictors were consistently better than all the naive predictors. The evolved predictors typically gave $\max(\cdot)$ errors which were 40% of the naive $\max(\cdot)$ errors and $\|\cdot\|$ errors which were 50% of the naive $\|\cdot\|$ errors.

The percentage improvements in each error measure for the best evolved velocity predictor (Exp58) against the best naive velocity predictor, and the best evolved JT predictor (Exp98) against the best naive JT predictor.

The above comparisons suggest that evolving JT predictors directly leads to more accurate predictions than by evolving velocity predictors. However, the JT predictor implementation may not generalise across journeys traversing a different number of sensors. While this is not a problem for velocity predictors since they only process data local to a given sensor.

3.3   Comparison of Results Using Different Error Types

Table 1 shows that driving the evolution with JT errors (Exp1 and Exp5) was better than driving with velocity errors (Exp20 to Exp25). Velocity error schemes tended to be dominated by very large $\max(e_a)$ and $\max(e_r)$ errors at the transition from a free-flowing to a congested period. The evolution consequently tended to reduce this by generally underestimating velocity and thus giving poor $\|\cdot\|$ errors. This explains the extremely large JT errors for Exp20. This problem was eased by using a case-based fitness measure (Exp21) and by driving the evolution to out-perform naive velocity predictions (Exp22 and 23). Exp25 shows that the evolution of a velocity gradient predictor instead of a velocity predictor (Exp20) also greatly improved the results, but not to the accuracy achieved by driving with JT errors.

Driving the evolution with JT errors was probably better because individual velocity predictions were less critical in this scheme. Journey time calculation requires the participation of multiple velocity predictions and so GP had the freedom to manipulate these implicit low-level predictions to merit overall journey time prediction.

3.4   Comparison Results from Error Fusion Schemes

The default fitness measure (e.g. used in Exp1) typically resulted in:

$$\max(e_a)\max(e_r) > 10\|e_a\|\|e_r\| \tag{1}$$

The $\max(\cdot)$ component was subsequently weighted by 0.1 to counteract this bias (e.g. used in Exp54). This weighting improved the $\|\cdot\|$ errors slightly but produced marginally greater $\max(\cdot)$ errors (compare Exp53 to Exp54 and Exp56 to Exp57). The default fitness measure used the absolute errors and relative errors independently, by combining errors across all training cases. Conversely, the case-based fitness measure (e.g. used in Exp5) fused the absolute and relative errors per training case. The case-based measure thus forced the predictors to be more accurate when the journey time was short, i.e. in free-flowing traffic. The results show that the case-based fitness measure gave lower $\|\cdot\|$ errors and that it was never dominated on $\max(\cdot)$ errors. This can be seen by comparing the following pairs of experiments: (1, 5), (32, 41), (35, 42), (54, 55), (57, 58), (91, 97) and (92, 98).

3.5   Comparison of Input Data Types

This section discusses the quantities shown in Table 4 pertaining to the selection of input data types. The table gives the IDs of the experiments which allow a comparative analysis of each quantity.

Table 4. Experiment IDs for a comparative analysis of input data types.

Quantity	Comparative experiments
occupancy, O	(1, 2)
flow-velocity magnitude, M	(1, 6), (1, 7), (8, 10)
flow-velocity gradient, G	(1, 8), (1,11), (7, 10), (16, 17), (22, 24)
time window, T	(1, 3, 4), (50, 51, 52, 53)
sensor window, S	(33, 36), (35, 37)
prediction feedback, $\hat{V}$	(1, 15), (7, 17), (10, 16)
time box size, b_t	(27, 30, 33), (28, 31, 34), (32, 35), (40-43)
sensor box size, b_s (V predictor)	(27-29), (30-32), (33-35), (36-38), (53, 56), (54, 57), (55, 58)
(J predictor)	(91, 92), (93, 94), (95, 96), (97, 98)

The simplest data reduction removed some of the primary input data types but proved to be a poor method because V and F gave the most predictive power, and even the omission of O significantly increased the $\|\cdot\|$ errors. Decreasing T and S proved to be better for reducing the amount of input data, but this also tended to an accuracy trade-off. Merging V and F into a single quantity by using a flow-velocity plot descriptor not only reduced the size of the search space but it also tied V and F together for each time and sensor, thus saving GP from having to discover the underlying correlations between the associated V and F values. The magnitude M proved to be better than the gradient G for this task, but both quantities gave worse prediction accuracy than when using V and F independently. Furthermore, the plot descriptors failed to significantly change the prediction accuracy when they augmented an input data set.

The boxed input proved to be the best data reduction. Furthermore, comparing Exp29 with Exp50 shows that the box-averaged method was better than the Fibonacci method when all other variables were equal. This suggests that the predictions did not benefit from using box standard deviations in addition to box averages. However, the Fibonacci method produced the best overall velocity predictor (Exp58) because the amount of input data increased less than linearly with T, and so the method could be used with larger T values. For example, increasing T from 54 to 88 just added a single time step.

The results showed no consistent variation in prediction accuracy with b_t or b_s. For example, the comparative experiments using the Fibonacci method (Exp53 to Exp58) showed that the $\| \cdot \|$ errors decreased when b_s was decreased but that the $\max(\cdot)$ errors could increase. Conversely, the journey time predictor experiments (Exp91 to Exp98) showed that the $\| \cdot \|$ errors tended to decrease when b_s was increased and variation in prediction accuracy with b_t or b_s was marginal. Significant data reduction could be obtained by using relatively large box sizes, e.g. setting $b_s = 8$ gave averaging across all sensors when $S = 7$. The Fibonacci method had the advantage that only b_s needed to be manually set.

Feeding intermediate predictions, $\hat{V}$, back into the input tended to reduce journey time prediction accuracy. The intermediate predictions, this "autoregression", probably complicated the search space by increasing the size of the input data set and by allowing premature predictions to be an encumbrance.

3.6   Interpretation of Predictors

The best predictor in Exp29 (pred. velocity; $(b_t \times b_s) = (1 \times 8)$ thin strips) was analysed. The evolved tree was 75 nodes in size but it simplified to 33 nodes or:

$$g_0 = \min\left(O^s_{t-8}, \frac{V^s_{t-5}}{F^s_{t-20}}\right) \qquad g_1 = \max\left(O^s_{t-7}, g_0\right) \qquad g_2 = \min\left(O^s_{t-1}, g_1\right)$$

$$g_3 = O^s_{t-3} + \min\left(O^s_{t-12}, O^s_{t-1}\right) \qquad g_4 = \min\left(0.735, \frac{V^s_{t-19}}{V^s_{t-2}}\right)$$

$$g_5 = -0.2704 - \frac{F^s_{t-18} - O^s_t + V^s_t + g_2}{g_4} \qquad V_{t+15} = \max\left(F^s_{t-16}, g_3, g_5\right)$$

GP functions $min(\cdot, \cdot)$ and $max(\cdot, \cdot)$ fuse sub-predictors. Discovering which paths activated over the data revealed 8 active sub-predictors as in Table 5. Computing the path frequency in partitioned regions of the flow-velocity regime revealed that paths B1 and B2 dominate most cases but that B4 and B5 become highly active for peculiar flow-velocity regimes.

Table 5. Four out of the eight paths/branches that are active in the data.

velocity predictor equation family	four active paths/branches	
$-0.2704 + 1.36054 \left(-F^s_{t-18} - O^s_t + X + V^s_t \right.$	B1:$X = O^s_{t-1}$	B2:$X = O^s_{t-7}$
$\left. Z + O^s_{t-3} \right)$	B4:$Z = O^s_{t-12}$	B5:$Z = O^s_{t-1}$

4 Conclusions

Conclusions are possible from comparative analysis across many experiments:

- Predictor type: GP evolved velocity, velocity gradient and JT predictors. The lowest prediction errors were with velocity predictors. The best velocity predictor (Exp58) gave absolute errors of 36.0s ($\| \cdot \|$) and 126.4s ($\max(\cdot)$) and relative errors of 5.5% ($\| \cdot \|$) and 17.0% ($\max(\cdot)$). Both $\max(\cdot)$ errors occurred at the onset of the evening congestion at 1641h. The actual JTs ranged from 454.9s (7.6 minutes) to 1174.3s (19.6 minutes).
- Error type: Prediction errors drove the evolution to better solutions via the *fitness measure*. JT errors proved to be better drivers than velocity errors.
- Error fusion: Various formulae for fusing prediction errors into an overall fitness measure were explored. The formulae combined absolute and relative errors, and involved $\| \cdot \|$ and $\max(\cdot)$ error calculations. Marginally better predictions resulted from fusing the errors per training case as opposed to combining errors across all training cases.
- Input data types: Combinations of input data types were investigated. Omitting any one of the primary data types (velocity, flow or occupancy) resulted in less accurate predictions. Various data reductions were compared to allow a sufficient amount of input data to be presented without swamping the search space. Box averaging proved to be the most beneficial.

The experiments demonstrate the flexibility of GP where many different solution options can be explored by varying:

- fitness measure to evolve different predictor types or to steer the evolution to tackle particular problem aspects.
- terminal set to investigate different combinations of input data.
- function set to investigate different low-level manipulations of the input data.

References

1. Moorthy C K and Radcliffe B G (1988): Short term traffic forecasting using time series methods, Tansportation Planning & Technology, Vol. 12, pg. 45-46.
2. Mahalel D and Hakkert A S (1985): Time Series Model for Vehicle Speeds. Transportation Research Vol. 19B, no 3, pg. 217-225.

The Boru Data Crawler
for Object Detection Tasks in Machine Vision

Daniel Howard[1], Simon C. Roberts[1], and Conor Ryan[2]

[1] QinetiQ Software Evolution Centre, Building U50, QinetiQ PLC,
Malvern WR14 3PS,UK, tel: +44 1684 894480
dhoward@qinetiq.com
[2] Department of Computer Science, University of Limerick, Ireland

Abstract. A 'data crawler' is allowed to meander around an image deciding what it considers to be interesting and laying down flags in areas where its interest has been aroused. These flags can be analysed statistically as if the image was being viewed from afar to achieve object recognition. The guidance program for the crawler, the program which excites it to deposit a flag and how the flags are combined statistically, are driven by an evolutionary process which has as objective the minimisation of misses and false alarms. The crawler is represented by a tree-based Genetic Programming (GP) method with fixed architecture Automatically Defined Functions (ADFs). The crawler was used as a post-processor to the object detection obtained by a Staged GP method, and it managed to appreciably reduce the number of false alarms on a real-world application of vehicle detection in infrared imagery.

1 Introduction

The perennial problem with machine vision is to distil features which characterise an object from a huge amount of available information, i.e. the specification of an intelligent data reduction mechanism. This data reduction must be discovered because it cannot be determined a priori. To deal with this surfeit of information, a way of selecting a sub-set of data, which is representative of an object, needs to be derived which allows the object to be identified.

The current algorithm was inspired by the story of Brian Boru [1], an Irish chieftain who had a serious predicament: the Vikings, a more numerous and powerful foe, did as they pleased in Ireland. How to expel them? Brian trained his men in the weapons of the Vikings, the battle axe and the long ship. By this strategy, and at the Battle of Clontarf in 1014, his army once and for all defeated the Vikings and expelled them from Ireland.

The story motivated the first author to wonder whether this lesson in history could be applied to object detection. Perhaps the image could be considered as an enemy who frustrates efforts to recognise the objects which lie within it. In this gedanken experiment, the image employs the following weapon to defeat any analysis:

S. Cagnoni et al. (Eds.): EvoWorkshops 2002, LNCS 2279, pp. 222–232, 2002.

- If an object is viewed from too close, there isn't enough contextual information upon which to make an identification.
- If an object is too far away, the reverse is true. There is now too much information and identification is equally difficult.

This becomes apparent in figure 1. It is clearly impossible to identify the vehicle object when the image is viewed from too close.

Fig. 1. The 'weapon' of the image is discovered when considering two different views of a target: up close and from afar.

Animal vision most probably "defeats" the image when the eye somehow identifies a number of detailed image clues which the brain then uses to "imagine" or to construct a model of the object from afar. The brain projects this model onto the world and this can be said to be: vision. Optical illusions are obvious examples. However, even when two people recognise an unambiguous object, they may be see slightly differently versions of it and project different imaginations.

2   Boru Data Crawler

The Boru Data Crawler is based on this idea of using both views: near and far, and of using a small number of image clues to connect these views to detect the object. The central problem then becomes how to identify the required image clues. However. evolution with GP resolves this difficult issue. The image is crawled, inspected, from very close to discover a few useful clues, and then these clues are used to examine the image from afar.

The evolutionary process is driven by the global objective of reducing misses and false alarms. It is implemented using GP with Automatically Defined Functions (ADFs) [2]. Evolution provides answers to questions such as: how should this image crawler be guided in its search for the clues?, how far should it travel and turn?; what should excite it sufficiently to cause it to deposit a flag which marks the existence of an image clue?

3   Objectives

The Boru data crawler was implemented to detect vehicles in airborne recoinassance infrared imagery as discussed in references [3,4]. In this application GP trains in a supervised way to detect objects in infrared imagery taken by a low flying aircraft. The objects are GP determined vehicle parts. Vehicles appear in many sizes and orientations, in many perspectives and thermal states, next to various types of objects such as buildings and roofs which cast thermal shadows on the vehicles, and in many environments and weather conditions. Moreover, the infrared line scan sensor produces a jittered image and one which needs aspect ratio correction (achieved crudely by repeating pixel lines).

The motivation behind the Boru Data Crawler was to further reduce the amount of data presented to any human whose task it is to find objects in images. Although a high false alarm rate is acceptable, it could be argued that a further reduction in false alarm rates would be desirable by analysisng the surroundings or context of each suspicious pixel. For example, in [3,4] cars had been detected on roof tops, or as parts of the central reservation of a road and it was hoped that such obvious mistakes could be repaired with some method of context analysis.

Therefore, the Boru data crawler was devised as a post processor to the pixels that are raised by the Staged GP method [3,4], and with the objective of reducing the false alarm rate of that method while maintaining the same level of false negatives, i.e. same number of vehicle detections/indications.

Thus, a suspicious pixel, or starting point for the Boru Data Crawler, is a pixel that the Staged-GP pre-processor vehicle detector has identified to possibly be inside or on a vehicle.

4   Structure and Representation

Each individual data crawler in the population, is a GP tree structure equipped with two result producing branches and two ADF branches:

- First Result Branch (FRB)
- Turn Decision Branch (TDB)
- Mark Decision Branch (MDB)
- Second Result Branch (SRB)

FRB is allowed to call TDB and MDB many times but otherwise the branches are unrelated. SRB works on the 2D memory devised by FRB as will be explained shortly. Each branch determines a specific property of the crawler, and each has its own set of terminals and functions. Moreover, some have access to task specific working memories, see table 1.

FRB together with the ADFs TDB and MDB guides the data crawler to explore the image in the near and deposit a number of "flags" or image clues. Once this process is completed, the discriminat SRB looks from afar at this binary 2D map of "flags" and "no flags" to decide whether the original starting point was correctly indicated as a true positive or whether it was a false alarm.

Data crawler decisions, e.g. whether or not to mark the image with a flag, require a memory of past events which allows the data crawler to consult and integrate previous information prior to a decision. Each individual maintains the memories as listed in table 1 to save information about its past experience.

Table 1. Result producing branches, ADFs, and memories.

First Result Branch (FRB)   WORKING memory
Turn Decision Branch (TDB)   FLAG memory
Mark Decision Branch (MDB)   MOVE memory
Second Result Branch (SRB)

branch	functions	terminals
FRB	$GL2$, $GL3$, MDB	M, TDB
TDB	+, -, *, / $(x/0 = 1)$, $\min(A, B)$, $\max(A, B)$, if $(A < B)$ then C else D	μ_{11}^C, σ_{11}^C, μ_{11}^N, σ_{11}^N, μ_{11}^E, σ_{11}^E, μ_{11}^S, σ_{11}^S, μ_{11}^W, σ_{11}^W, READFLAG, READMOVE, altitude in feet (250:650)
MDB	+, -, *, / $(x/0 = 1)$, $\min(A, B)$, $\max(A, B)$, if $(A < B)$ then C else D, WRITEMEM, READMEM.	μ_{disk}, σ_{disk}, altitude in feet (250:650), READFLAG, READMOVE
SRB	+, -, *, / $(x/0 = 1)$, $\min(A, B)$, $\max(A, B)$, if $(A < B)$ then C else D	10 flag based terminals (see text), altitude in feet (250:650), 10 area statistics (textural stats) taken over 5 car length square area

Two parent Boru Data Crawlers exchange genetic material with a crossover operator that respects branch typing.

4.1   First Result Branch (FRB)

The Boru Data Crawler has some similarity with the Santa Fe trail ant that was used by Koza [5]. It shares in common with the crawler the glue functions $GL2$ and $GL3$ and the M or 'move terminal' to move the crawler one pixel in the direction of its travel. However, the L (left) and R (right) terminals used to turn the crawler are replaced by TDB which first turns and then moves the crawler.

The FRB is executed until the data crawler has moved a predetermined number of times, i.e. $9 * w$ times, where w is the width of a vehicle (this width is automatically scaled by aircraft altitude so that w is smaller at $600ft$ and larger at $300ft$). FRB is iteratively evaluated until the crawler executes the predetermined number of moves. In future research the number of moves will not be predetermined but will also be allowed to evolve.

4.2   Turn Decision Branch (TDB)

TDB first decides on a direction of travel and then moves the crawler in this direction. Terminals μ_{11}^C, σ_{11}^C are averages and standard deviations obtained from the pixel values in an eleven pixel square window centred on the crawler. Labels C, N, E, S, W stand for centre, north, east, south and west locations as shown in Fig. 2. The window size gets smaller for images taken at aircraft altitudes higher than 300 ft.

TDB is always evaluated four times, once for each permutation of the order of the terminals at directions N, E, S, and W, and this produces four outputs. In the current implementation, the chosen data crawler direction is given by the permutation returning the largest output and follows certain rules.

After every Boru Data Crawler moves, i.e. when the FRB invokes M or TDB, the MOVE memory is updated. The MOVE memory is a trail memory which consists of:

1. last 10 directions, e.g. F,R,B,B,F,F,B,L,R,L is stored as [1,2,3,3,1,1,3,4,2,4];
2. last 10 μ_{11}^C;
3. last 10 σ_{11}^C;

Recorded are the local turns of the crawler: F, R, B, L (forward, right, back, left) rather than the absolute directions N, E, S, and W. At the start of travel, the direction memory locations are initialised to 0 and all ten values in each statistic memory location are set to the initial μ_{disk}^C and σ_{disk}^C.

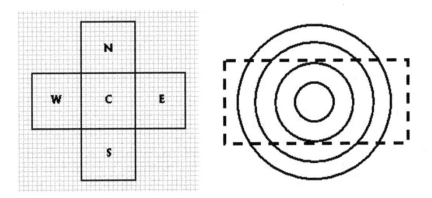

Fig. 2. Left: TDB terminals: mean and standard deviation computed over square areas in front, behind, to the right, to the left, and centred on the pixel. These area are the 'sensors' of the Boru Data Crawler. Right: Four concentric pixel rings centred on a vehicle with diameters: 0.5, 1.0, 1.5 and 2.0 vehicle widths.

4.3 Mark Decision Branch (MDB)

MDB determines whether a flag should be left at the current pixel position in the trail to indicate presence of an "image clue". If the MDB returns a positive number it deposits a flag. This decision is based on pixel data and trail data.

MDB acts as an "IF" statement in the FRB. Placing a flag executes a THEN subtree, otherwise an ELSE subtree is executed. The executed portion is returned to FRB.

Terminals μ_{disk}, σ_{disk} in Table 1 are averages and standard deviations that are calculated over a small disk centred on the pixel, the disk diameter is half the width of a car. Its diameter corresponds to the smallest disk in Fig. 2 and is scaled by altitude so that at higher altitudes this disk gets smaller. The disk is also distorted inot an ellipse when the image is at perspective (aircraft roll).

The FLAG memory is updated following the call to MDB. This memory consists of:

1. last 10 MDB results, e.g. [-1,-1,1,1,1,1,-1,1,-1,-1];
2. last 10 μ_{disk};
3. last 10 σ_{disk};

Here -1 stands for "no flag", and 1 stands for "flag set". Initially, all values in the vector of results are set to -1 and in the other two memory vectors are set to the initial μ_{disk} and σ_{disk} respectively.

READMEM and WRITEMEM write to and read from the WORKING memory. This consists of three locations or slots which can be written to or read from. All three memory locations are initialised to 0.0 before the crawl.

Note that successive calls to MDB from FRB will not move the data crawler. The result of the repeated calls, however, can differ from one another because the FLAG memory changes between calls. Consecutively calls to MDB may allow FRB to reset the FLAG memory.

4.4   Second Result Branch (SRB)

SRB returns a real numbered value that can be either positive (target is present) or negative (no target is present). The SRB is only executed when the FRB finishes execution. Note that the Boru Data Crawler may revisit a cell, thus the trail may become a loop. In such cases and those where the trail crosses itself the crawler can overwrite a cell with a flag or remove a flag if it decides that there should be no flag.

The FRB marks a 2D map of square dimensions. On this map, the presence of a flag is indicated by a 1 at (i,j) - the map is preset to 0 values. If the map reveals fewer than T_F or threshold number of flags, e.g. 4 flags, then the suspicious pixel (the starting pixel at the centre of this map) is labeled "negative" (no target present) and the fitness evaluation moves on to classify the next suspicious pixel. In such a case the SRB never gets invoked.

If FRB marks the map with greater than T_F flags, then the SRB works with this map. It defines statistical measures based on the distribution of the flags.

Textural statistics over the square pixelated area surrounding each starting pixel may help GP to cluster false alarms and true positives, e.g. such that the thermal return for a sheep in heavy rain not be compared to that for a vehicle in a city (both things will not occur together). In SRB the information gathered from the crawl is being utilised to discriminate false alarms from targets. For this reason it was deemed appropriate to include textural statistics as terminals in SRB, see Table 1. These textural statistics are taken over an area 5 car lengths square.

The ten flag based terminals in Table 1 require computation of the centre of mass of the flags. They are based on statistical measures over the vector of distances between the centre of mass and each flag. All ten are positive in value:

- **GP terminal 0**: the number of marked points;
- **GP terminal 1**: the distance from this centre of mass to the marked point that is furthest;
- **GP terminal 2**: the longest distance between any two marked points;
- **GP terminal 3**: the average distance between any two marked points;
- **GP terminal 4**: the standard deviation in distance between any two marked points;
- **GP terminal 5**: arithmetic mean μ;
- **GP terminal 6**: the standard deviation σ;
- **GP terminal 7**: the degree of asymmetry of a distribution around its mean or measure of the skweness;
- **GP terminal 8**: the relative peakedness or flatness of a distribution compared with the normal distribution or kurtosis;
- **GP terminal 9**: the vector is sorted in ascending order. If the number of entries is odd it gets rid of the highest value and divides the set into two halves, obtains a μ for each half respectively and compute a co-variance.

5   GP Implementation

The GP run parameters are given in table 2. Branch typing crossover restricted exchange of material between principal subtrees of the same type, e.g. a FRB only with a FRB. Each branch was assigned a probability to participate in crossover. The FRB had double probability (40 %) relative to any of the other three branches. Truncation mutation takes any node (and associated principal subtree) and replaces it by a terminal from the relevant terminal set.

Table 2. GP run parameters

parameter	setting
kill tournament	size 2 for steady-state GP
breed tournament	size 4 for steady-state GP
regeneration	90% x-over, 0% clone, 10% truncation mutation
max generations	20
max tree size	1000 (any branch)

Suspicious points (fitness cases) were processed in turn, and the fitness measure computed once all fitness cases had been processed:

$$\frac{\alpha TP}{\beta FP + cars} \tag{1}$$

Computation of a "true positive" (TP) in the fitness formula was associated with anything "hitting" the vehicle. Different values of α and of β could be set and the objective was to reduce the FP without losing any TP, i.e. the optimisation of α and β. The level of success was according to how many FP remained. Section 4.1 prescribed a maximum number of moves, or length of travel, equivalent to nine car widths, but there must also exist a practical limit for the number of flags or "image clues" that are allowed. This was set to be 50. Should MDB threaten to exceed this threshold, the FRB literally aborted and immediately returned the map produced so far.

6   Experimental Results

The first experiments were conducted on a rural image pre-processed by the Staged GP method. This had returned 11 TP but 690 FP 'clustered' into about 50 groups. A parameter search for α and β was conducted to determine if it was possible to get a handle on the sensitivity of results to initial conditions.

6.1 $\alpha = 1, \beta = 1$

The initial experiments all used a population of 500 individuals running for 20 generations. The experiments were terminated at 20 generations regardless of the current fitness of the best individual in the population. The best individual produced scored 11-340[1]. Many individuals appeared that managed a much better score for false positives, but always at a cost of TPs. A large variety of other inviduals appeared with this setting for α and β, including the following:

− 9-2; 5-1; 9-0; 10-140

The first three show that it is possible to drastically reduce the FP rate, but at an apparent cost to the TP rate. The fourth individual above only incorrectly classified one TP, but incorrectly classified almost 21% of the FPs. These results suggest, not surprisingly, that the values for α and β required tuning, but also that individuals can trade off one part of their score for another. Consider the top ten performing individuals from one experiment:

− 5-2; 5-3; 4-2; 5-6; 5-10; 5-10; 5-12; 5-14; 3-8; 4-17.

The variety of scores suggests that GP had difficulty in balancing them. A good example of this is the difference between individuals #3 & #4. The GP

---

[1] This is intended to be read as 11 True Positives and 340 False Positives

classifies individual #3 as being more fit, even though it has fewer TPs correctly classified. Clearly, α needs to be higher than β, but how much? If it is set to too high a value, one would expect to see individuals with a score of 11 TPs, but uselessly high FPs. The next set of experiments examined the effect of varying the value of α.

6.2   Varying α

This set of experiments involved running 32 different parameter sets, each run an average of approximately 16 times, for a total of 534 separate experiments. The remit for this experiment was to determine which parameter settings could yield a perfect score on the TP measure (i.e. 11) while minimising the FP measure. Seven different parameter setting produced acceptable results. An acceptable result was to reduce the FP measure to less than 5% of the initial rate of 690 as given to this detector, while maintaining 100% of the TP rate. All these experiments were run with the same parameters as the initial ones. Table 3 illustrates the successful runs. The value of β was left at 1 for all of these experiments.

Table 3. Acceptable individuals from the second set of experiments.

α	Individual	% of FPs	% of Total FPs
1.2	11 - 9	1.3	0.225
2.6	11 - 16	2.3	0.4
2.8	11 - 13	1.9	0.325
3.5	11 - 18	2.6	0.45
3.6	11 - 22	3.2	0.55
4.5	11 - 11	1.6	0.275
5	11 - 22	3.2	0.55

In all cases where an individual towards the end of a run attained a score of 11 in the TP measure, they also scored less that 5% in the FP measure.

Figure 3 illustrates the behaviour of a Boru Data Crawler produced with an α of 2.6, a β of 1 and a population of 500 running for 50 generations. Black pixels on the trail indicate that the crawler deposited a flag at that location. It illustrates the behaviour of the same crawler on different images. Although these are different vehicles, the crawler takes a similar, circular path around the suspicious point. The power of the Boru Data Crawler is illustrated in figure 4 where it became possible to work out that what appeared to be a vehicle was really part of a roof. The vehicle detection task in these operational images is extremely challenging.

Sometimes when the crawler is applied to the *same* image, but with different starting positions, in each case, it takes a different route, but still identifies the target as a vehicle.

At least four flags had to be set for the SRB or second result branch to be executed. Experiments suggested that the number of individuals that *don't* place

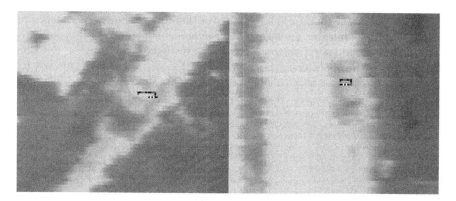

Fig. 3. Similar routes taken by the same data crawler on vehicles in different images.

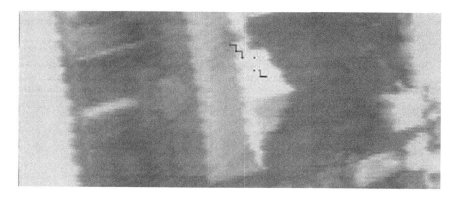

Fig. 4. Case of vehicle part detected on a roof is faithfully rejected.

enough flags varies from about 10% to 20% of the total population. The longer the experiment was run, the more likely individuals were to place the threshold number, i.e four flags. For *best-so-far* individuals, the number of individuals that didn't place enough flags varied between 15% and 20%. The vast majority of individuals that didn't set enough flags and still managed to be reported as a *best-so-far* individual appeared in the first, randomly created population. As the population evolves, individuals are more likely to employ the second result branch.

Virtually all of the *best-of-run* individuals observed performed some retracing of their steps. Interestingly, little if any looping behaviour was exhibited. Possibly this behaviour was as a result of the individuals considering the contents of their memory before making a decision on the next step.

The next step was to investigate the performance of the system when applied to several images at a time. Three images took part: edge of suburbia; rural; industrial. These experiments were conducted using an identical set up to the earlier ones, but took considerably longer than the previous experiments due to the additional images. This and similar experiments reduced FP typically by 30% and have motivated a more extensive investigation to optimise the Boru Data Crawler. Its speed could be improved with convolutions in hardware, e.g. use of the MMX. Use of smaller data windows for simpler statistics may also dramtically improve speed. More than one crawler could participate or an iteration performed between SRB and FRB to reinforce a tenuous decision with additional crawls to gather more clues.

References

1. Llywelyn, Morgan: Brian Boru Emperor of the Irish, O'Brien Press, Dublin (1999).
2. Koza J. R.: Genetic Programming II: Automatic Discovery of Reusable Programs (1994) MIT Press.
3. Howard D. and Roberts S. C.: A staged genetic programming strategy for image analysis. Proceedings of the Genetic and Evolutionary Computation Conference, Orlando, Florida (1999) 1047–1052 Morgan Kaufmann
4. Roberts S. C. and Howard D.: Evolution of vehicle detectors for infrared line scan imagery. Joint Proceedings of the European Workshop on Evolutionary Image Analysis, Signal Processing and Telecommunications, Göteborg, Sweden (1999) 111–125 Springer LNCS.
5. Koza J. R.: Genetic Programming: On the Programming of Computers by Means of Natural Selection (1992) MIT Press.

Surface Profile Reconstruction from Scattered Intensity Data Using Evolutionary Strategies

Demetrio Macías, Gustavo Olague, and Eugenio R. Méndez

División de Física Aplicada,
Centro de Investigación Científica y de Educación Superior de Ensenada,
Apdo. Postal 2732, 22800 Ensenada, B. C., México.
(dmacias, olague, emendez)@cicese.mx

Abstract. We present a study of rough surface inverse scattering problems using evolutionary strategies. The input data consists of far-field angle-resolved scattered intensity data, and the objective is to reconstruct the surface profile function that produced the data. To simplify the problem, the random surface is assumed to be one-dimensional and perfectly conducting. The optimum of the fitness function is searched using the evolutionary strategies (μ, λ) and $(\mu + \lambda)$. On the assumption that some knowledge about the statistical properties of the unknown surface profile is given or can be obtained, the search space is restricted to surfaces that belong to that particular class. In our case, as the original surface, the trial surfaces constitute realizations of a stationary zero-mean Gaussian random process with a Gaussian correlation function. We find that, for the conditions and parameters employed, the surface profile can be retrieved with high degree of confidence. Some aspects of the convergence and the lack of uniqueness of the solution are also discussed.

1   Introduction

The interaction of electromagnetic waves with randomly rough surfaces is a subject of importance in many scientific disciplines. It finds applications, for example, in the remote sensing of the ocean surface and the Earth's terrain, and in studies of astronomical objects with radio waves. In optics, it has applications in surface metrology, in the evaluation of optical components and systems, and in the development of paints and coatings, to mention but a few. The direct scattering problem, that consists of finding the scattered field or the scattered intensity from knowledge of the surface profile, its optical properties, and the conditions of illumination, has been widely studied in the past [1,2]. Approximate methods of solution, such as perturbation theory [2,3] and the use of the Kirchhoff boundary conditions [1,2], have lead to a better understanding of different aspects of the phenomenon, but more rigorous methods [4,5] are required to account for the multiple scattering phenomena that give rise to coherent effects such as that of enhanced backscattering [6,7].

The inverse scattering problem, on the other hand, that consists of the reconstruction of the profile of a surface from scattering data has also received some

S. Cagnoni et al. (Eds.): EvoWorkshops 2002, LNCS 2279, pp. 233–244, 2002.

attention, but due to the difficulties involved little progress has been made. Many of the works on inverse scattering have focused on the retrieval of some statistical properties of the surface under study [8,9,10,11,12]. The reconstruction of surface profiles from far-field amplitude data has been considered by Wombel and DeSanto [13,14], Quartel and Sheppard [15,16], and Macías et al. [17]. From a practical stand point, the need to have amplitude, rather than intensity data, constitutes an important drawback that all these methods share. Optical detectors are not phase sensitive; they respond only to intensity and special schemes need to be used to determine the phase of the optical field. The other important limitation is that these methods are based on simple approximate models for the interaction between the incident light and the surface, and fail when multiple scattering is important.

In the present work, the inverse scattering problem is viewed as a problem of constrained optimization. We implement, and study the performance of two inversion algorithms based on evolutionary strategies taking, as the input, far-field intensity data. Unlike previous works on the subject, the method does not rely on approximate expressions for the field-surface interaction. Furthermore, although at present we can not demonstrate this, our experience with the method indicates that the convergence to the optimum improves when multiple scattering occurs. Since, in general, the use of intensity data implies that the solution to the problem is not unique, the improvement in the convergence is possibly due to a reduction in the number of solutions in the presence of multiple scattering.

The organization of this paper is as follows. Section 2 deals with the direct scattering problem. We introduce the notation and discuss the complexity of the relation between the surface profile and the far-field intensity. In Sect. 3, we approach the inverse scattering problem as an optimization problem, describing in some detail the procedure to generate and mutate the surfaces. Some problems occurring when intermediate recombination is employed are also pointed out. Section 4 is devoted to the description of the algorithms studied, and typical results are presented and discussed in Sect. 5. Finally, in Sect. 6, we present our main conclusions.

2   The Direct Scattering Problem

We consider the scattering of light from a one-dimensional, perfectly conducting, randomly rough surface defined by the equation $x_3 = \zeta(x_1)$. The region $x_3 > \zeta(x_1)$ is vacuum, the region $\zeta(x_1) > x_3$ is a perfect conductor, and the plane of incidence is the x_1x_3-plane. With reference to Fig. 1, the surface is illuminated from the vacuum side by an s-polarized plane wave making an angle θ_0 with the x_3 axis.

The scattering amplitude $R_s(q|k)$ can be written in the form [5]

$$R_s(q|k) = \frac{-i}{2\alpha_0(k)} \int\limits_{-\infty}^{\infty} dx_1 \left[\exp\{-i\alpha_0(q)\zeta(x_1)\}F(x_1|\omega)\right] \exp\{-iqx_1\}, \quad (1)$$

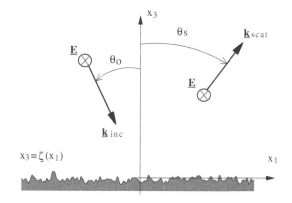

Fig. 1. Geometry of the scattering problem considered.

where

$$F(x_1|\omega) = \left(-\zeta'(x_1)\frac{\partial}{\partial x_1} + \frac{\partial}{\partial x_3}\right) E_2(x_1, x_3)\Bigg|_{x_3=\zeta(x_1)} \tag{2}$$

represents the source function, $E_2(x_1, x_3)$ is the only nonzero component of the electric field, $\alpha_0(k) = \sqrt{\omega^2/c^2 - k^2}$, and $\alpha_0(q) = \sqrt{\omega^2/c^2 - q^2}$. The angles of incidence θ_0 and scattering θ_s are related to the components of the incident and scattered wavevectors that are parallel to the mean surface through the expressions

$$k = \frac{\omega}{c}\sin\theta_0, \qquad\qquad q = \frac{\omega}{c}\sin\theta_s, \tag{3}$$

where c is the speed of light and ω is the frequency of the optical field. The far-field intensity $I_s(q|k)$ is defined as the squared modulus of the scattering amplitude $R_s(q|k)$.

At this point, it is important to mention that the source function $F(x_1|\omega)$ is unknown, and this constitutes the main difficulty for solving direct scattering problems. The classical analytical approaches to the problem are based on approximate expressions for this unknown boundary condition. Rigorous approaches, on the other hand, involve the solution of an integral equation for the determination of $F(x_1|\omega)$, and this has to be done numerically. So, the relation between the surface profile function, the conditions of illumination, and the source function is not straightforward.

Consider for a moment a special situation for which the relation between the surface profile function $\zeta(x_1)$ and $F(x_1|\omega)$ is known. Let us further assume that the variation of $\alpha_0(q)$ in Eq. (1) can be neglected. Then, the scattered intensity would be given by the squared modulus of the Fourier transform of the quantity within square brackets. Thus, the phase information of the scattering amplitude has been lost, complicating the inversion of the data. For one-dimensional objects, such as the ones we are considering, it has been demonstrated [18,19] that the reconstruction of a function from the modulus of its Fourier transform does not yield a unique answer. For our problem, this means that it might be possible to find two or more surfaces that produce the same far-field intensity pattern. Some of these solutions may be trivially related, like an overall vertical shift of the profile, but others may have little correlation with the profile that generated the data.

3 Inverse Scattering as an Optimization Problem

In this section, we formulate the problem of inverse scattering as an optimization problem. It is assumed that we have access to far-field angle-resolved scattered intensity data corresponding to several angles of incidence. The goal is to retrieve the unknown surface profile function from these data. Some constraints on the kind of surface that we seek are introduced in order to reduce the search space.

3.1 The Fitness Function

As we have discussed, although the far-field scattered intensity depends on the surface profile function in a complicated way, the direct problem can be solved fairly rigorously by numerical methods [5]. The closeness of a proposed profile, $z_c(x_1)$, to the original one can be estimated through the difference between the measured angular distribution of intensity, $I^{(m)}(q|k)$, and the angular distribution of intensity, $I^{(c)}(q|k)$, obtained by solving the direct scattering problem with the trial profile $z_c(x_1)$. The goal then would be to find a surface for which the condition $I^{(c)}(q|k) = I^{(m)}(q|k)$ is satisfied. When this happens, and if the solution to the problem is unique, the original profile has been retrieved.

We, thus, define our fitness (objective) function as:

$$f(\zeta(x_1)) = \sum_{i=1}^{N_{\text{ang}}} \int \left| I_s^{(m)}(q|k_i) - I_s^{(c)}(q|k_i) \right| dq, \qquad (4)$$

where N_{ang} represents the number of angles of incidence considered, and the $k_i's$ are related to those angles through Eq. (3). Note that in our definition of the fitness function we require that the proposed surface reproduces the "measured" scattering data for several angles of incidence. The satisfaction of this requirement should reduce the number of possible solutions and, hopefully, produce a unique one. The inverse scattering problem can be viewed, now, as the problem of minimizing $f(\zeta(x_1))$.

3.2 Representation and Constraints

To deal with the scattering problem numerically, the surface must be sampled. From the preceding discussion it seems natural to choose the surface heights, evaluated at the sampling points, as the object variables that will change in each iteration of the algorithm. However, changing these numbers independently of each other would lead to surfaces with abrupt height changes, which does not correspond to the physical situation of interest. One way to avoid this problem is to restrict the search space to randomly rough surfaces that belong to a certain class. We are, thus, faced with a problem of constrained optimization. In our case, we have chosen the target surface as a realization of a stationary, zero-mean Gaussian random process with a Gaussian correlation function.

Surfaces belonging to this class can be generated numerically with the spectral method described in Refs. [4] and [5]. Correlated random numbers that represent the surface heights at the sampling points can be obtained through the expression [5]

$$\zeta_n = \frac{\delta}{\sqrt{L}} \sum_{j=-N/2}^{N/2-1} \frac{[M_j + iN_j]}{\sqrt{2}} \sqrt{g(|q_j|)} \exp\{iq_j\chi_n\}. \tag{5}$$

Here, N represents the total number of points on the surface, L represents its length, $\chi_n = -L/2 + (n - 0.5)\Delta x$ are the sampling points spaced by Δx along x_1, $q_j = -\pi/\Delta x + 2\pi(j - 0.5)/L$ are the sampling points in Fourier space, and $\zeta_n = \zeta(\chi_n)$. The random sets $\{M_j\}$ and $\{N_j\}$ contain statistically independent random Gaussian variables with zero mean and unit standard deviation, and $g(|q_j|)$ represents the power spectral density of surface heights. In order to produce a set of real random numbers $\{\zeta_n\}$, it is required that the complex array $\{M_j + iN_j\}$ be Hermitian. The first and second derivatives of the profile function, which are required for the direct scattering calculations, can be obtained by differentiation of the Eq. (5). The numerical generation of the surface through Eq. (5) is shown, schematically, in Fig. 2. The Fourier transform of the product between the complex array $\{M_j + iN_j\}/\sqrt{2}$ and the square root of the power spectrum of the surface $g(|q_j|)$, produces the real array $\{\zeta_n\}$ that represents the surface heights.

With this notation the fitness function takes the form

$$f(\zeta_n) = \Delta q \sum_{i=1}^{N_{\text{ang}}} \sum_{j=1}^{N_{\text{ff}}} \left| I_s^{(m)}(q_j|k_i) - I_s^{(c)}(q_j|k_i) \right|, \tag{6}$$

where $\Delta q = 2\pi/L$, N_{ff} is the number of far-field angles for which one has data, and the calculated intensity is determined through the expression

$$I_s^{(c)}(q_j|k_i) = \frac{1}{4\alpha_0^2(k_i)} \left| \Delta x \sum_{i=1}^{N} [F_n \exp\{-i\alpha_0(q_j)\zeta_n - iq_j\chi_n\}] \right|^2, \tag{7}$$

where $F_n = F(\chi_n|\omega)$.

The fitness function determines some general properties of the problem that may be useful in the selection of the optimization algorithm. The search space of the function $f(\zeta_n)$ consists of all the random rough surfaces that belong to the specified statistical class. We first note that $f(\zeta_n)$ is N-dimensional with continuous parameters ζ_n. Also, from the form of Eqs. (6) and (7), and the fact that the determination of F_n involves the solution of an integral equation, it is

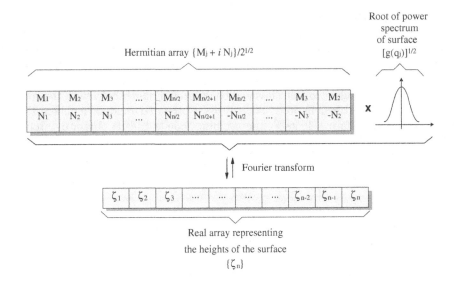

Fig. 2. Illustration of the numerical generation of the random surface.

clear that the derivatives of the fitness function cannot be determined analytically. Furthermore, the numerical evaluation of the derivatives of $f(\zeta_n)$ is not a practical proposition.

3.3   Recombination and Mutation

In the search of the optimum, new trial surfaces related to a previous set of surfaces (population), must be generated. Changes to the old population can be introduced by selectively combining some of these surfaces (recombination), by introducing random changes on them (mutation), or both.

In an intermediate recombination operation, the new surface is the result of a linear combination of Gaussian processes. The result, therefore, is also a Gaussian random processes. However, after the first iteration the surfaces are partially correlated, and the statistics of the resulting surface differ from those of the parent surfaces. In other words, the new generation of surfaces does not belong the statistical class specified for the search space, even though the surfaces of the initial population belong to this class. Thus, if possible, the use of intermediate recombination operations should be avoided.

Mutations, on the other hand, can be introduced by changing some of the elements of the Hermitian array $\{M_l + iN_l\}$ employed in the generation of a given surface [see Fig. 2 and Eq. (5)]. In this case, provided that the new numbers, M_j and N_j, are zero-mean Gaussian-distributed random numbers with unit standard deviation, and the Hermiticity of the array is conserved, the new surface will belong to the statistical class specified for the search space.

4   Description of the Algorithms

In this section, we describe the two evolutionary strategies employed in the search of the optimum. We also discuss, briefly, our first attempts to solve the problem using the downhill simplex method proposed by Nelder and Mead [20,21].

In the simplex method, the number of vertices of the simplex is equal to the dimensions of $f(\zeta_n)$ plus one. Each vertex has an associated value of the fitness function $f(\zeta_n)$, obtained through the evaluation of the $N+1$ random surfaces $\{\zeta_n\}$. In the simplex implementation we explored, a new surface is generated through a reflection operation [21], and the random surface associated with the worst value of $f(\zeta_n)$ is discarded.

We found that the downhill simplex algorithm did not converge to a solution, even after a large number of iterations. Since the reflection operations involved consist of linear combinations of the surface population, the lack of convergence is possibly due to the problem pointed out Sect. 3.3. For this reason, evolutionary strategies were preferred over the simplex method and genetic algorithms. Also, in our the evolutionary strategies, only mutation was used.

Since the time of computation required to find the optimum increases with the number of sampling points on the surface, in order to keep the problem to a manageable size, we chose a surface with $N = 128$ sampling points. To produce a mutation, we first choose randomly some (typically 20) members of the Hermitian array, and substituted them by newly generated numbers with the same statistics. Carrying out the operations illustrated in Fig. 2 [or Eq. (5)] with the new array, we obtain a mutated function profile $\{\zeta_n^{(2)}\}$ that belongs to the same statistical class as the original one. Other mutation schemes can be implemented, but we decided to use the present one due to its relative simplicity.

Using Schwefel's notation [22], the evolutionary strategies explored in this work are the $(\mu + \lambda)$ and the (μ, λ) strategies. In this notation, μ is the number of surfaces in the initial population and λ is the number of surfaces in the population generated through the mutation process. It is worth mentioning that we have previously used λ to denote the wavelength of the light, which is the usual notation in optical work. It is believed that due to the different context in which the two quantities are employed, use of the same symbol to denote both should not lead to much confusion.

In our implementation of the $(\mu + \lambda)$ strategy, we chose $\mu = \lambda$. The first step is the generation of an initial population consisting of μ random surfaces belonging to the specified statistical class. An intermediate population with λ elements is created through the mutation process. From the union of the initial and the intermediate populations we select a secondary population that consists of the μ surfaces with the lowest associated values of $f(\zeta_n)$. This secondary set of surfaces constitutes the starting point for the next iteration of the algorithm. The process continues until the termination criterion is fulfilled, which in our case was set by the maximum number of iterations g.

In the (μ, λ) strategy, we kept $\lambda = 10\mu$ over the entire process. We start with μ random surfaces, and λ new surfaces are generated through the mutation process. The outcome is a secondary population that is evaluated, and from which we select the μ best random surfaces for the next iteration of the algorithm. As with the $(\mu + \lambda)$ strategy, the process continues until the termination criterion is fulfilled,

5   Results and Discussion

In principle, the data that serves as input to the algorithm should be obtained experimentally. However, in order to study and optimize the algorithms, in these preliminary studies we use data obtained through a rigorous numerical solution of the direct scattering problem [5].

For the two strategies explored, each element of the initial population consisted of a realization of a zero-mean stationary Gaussian-correlated Gaussian random process with a $1/e$-value of the correlation function $a = 2\lambda$ and standard deviation of heights $\delta = 0.5\lambda$. For the $(\mu + \lambda)$ strategy we chose $\mu = \lambda = 100$, whereas for the (μ, λ) strategy we set $\mu = 10$ and $\lambda = 100$. In both cases, the maximum number of iterations was $g = 300$, which also provided the termination criterion. The time of computation is similar with the two algorithms, but the elitist strategy $(\mu + \lambda)$ uses twice as much memory as the non-elitist strategy (μ, λ).

In the numerical experiments we considered five different random profiles to be retrieved. In each case, we used the described algorithms to search for the solution, starting from 30 different initial states chosen randomly. Not in all of the 30 attempts to recover the profile the algorithms converged to the correct one. However, we found that a low value of $f(\zeta_n)$ corresponded, in most cases, to a profile that was close to the original one. So, the final value of $f(\zeta_n)$ was used as the criterion to decide whether the function profile had been reconstructed or not.

For brevity, we only present representative results corresponding to one of the five surfaces studied. The surface profile used to generate the original scattering data is shown in Fig. 3. The surface was sampled at intervals of $\lambda/10$ and contains $N = 128$ points.

The data from which the profile is to be recovered were obtained by illuminating the surface in Fig. 3 from four different directions, defined by the angles of incidence $\theta_0 = -60°$, $\theta_0 = -30°$, $\theta_0 = 0°$, and $\theta_0 = 40°$. In Fig. 4, we show the scattering pattern produced by the surface shown in Fig. 3 for the case of normal incidence.

In Fig. 5 we present results obtained with the two evolutionary strategies we are studying. To facilitate the visualization of the results the original profile is shown in both graphs with a solid line. The vertical displacement of the profile in Fig. 5(a) is quite understandable, as the far-field intensity is insensitive to such shifts. On the other hand, the displacement is unimportant for practical profilometric applications. In this case the two algorithms retrieve the profile quite well; there are only some subtle differences that are perhaps more noticeable near the ends of the surface.

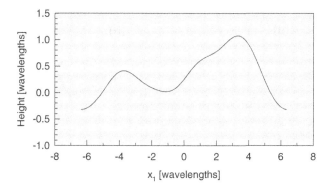

Fig. 3. The profile used in the generation of the scattering data.

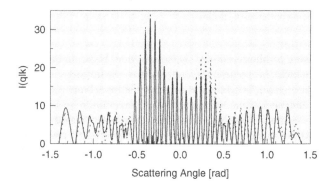

Fig. 4. Scattered intensity produced by the surface depicted in Fig. 3 for the case of normal incidence (solid curve). The dotted curve represents the scattering data produced by the surface shown with the circles in Fig. 6(a) under the same conditions of illumination.

An interesting result that also illustrates the lack of uniqueness of the solution when intensity data are used, is shown in Fig. 6. In Fig. 6(a) we present a curve with the original profile, and a curve with the recovered profile using the (μ, λ) strategy. The initial population differs from the one used in the examples of Fig. 5, with the rest of the parameters being the same. One can see that, in this case, the recovered profile does not resemble the original one.

The scattering patterns, obtained with the two profiles shown in Fig. 6(a) are shown in Fig. 4. The similarity between the two patterns is evident, illustrating the fact that two different profiles can generate scattering data that are practically undistinguishable.

A curious property of the scattering problem is also illustrated in Fig. 6. In Fig. 6(b), we present the profile recovered in Fig. 6(a), reflected with respect of both axis [that is, we replace $z_c(x_1)$ by $-z_c(-x_1)$]. Surprisingly, one observes

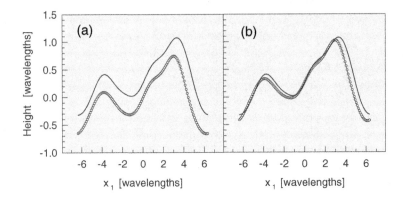

Fig. 5. Reconstruction of the surface profile using (a) the (μ, λ) strategy and, (b) the $(\mu + \lambda)$ strategy. The original profile is plotted with a solid line and the reconstructions are shown by the curves with circles.

that the resulting profile resembles the sought one. It can in fact be shown that in situations in which the polarization effects are not important and the Kirchhoff approximation is valid, the far-field intensity is invariant under this kind of operation [23]. Such situations, which are relatively simple for the direct scattering theories, lead to multiple solutions of the inverse scattering problem. Polarization and multiple scattering effects are then expected to reduce the number of possible solutions to the inverse problem.

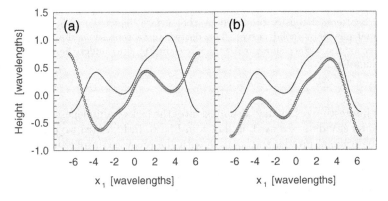

Fig. 6. (a) Reconstruction of the surface profile using the (μ, λ) strategy and a different initial population than in Fig. 5. The original profile is plotted with a solid line and the reconstruction is shown by the curve with circles. (b) Profile representing the profile shown in (a) reflected with respect to both axes (circles). The original profile is shown with a solid line.

6   Summary and Conclusions

We have presented a study of two evolutionary algorithms to solve inverse scattering problems. Starting from far-field intensity data and using both, the $(\mu+\lambda)$ and (μ,λ) strategies, we have successfully retrieved the surface profile that generated the scattering data. The time of computation is similar, but the elitist strategy $(\mu+\lambda)$ uses twice as much memory as the non-elitist strategy (μ,λ).

The solution of the inverse problem is not necessarily unique. We have found that the fitness function has many local minima, and that the initialization of the algorithm plays an important role in the search of the solution. However, in most cases there seems to be a unique global minimum and, in the rare cases in which we have found two solutions, it is possible to perform further tests to decide on the best one. Although the problem has many facets and can be rather complex, the results obtained so far are encouraging.

The success of an inversion scheme based on intensity information opens the possibility of implementing such a procedure experimentally. However, further work is needed, not just regarding the physical aspects of the problem, which have been simplified by our assumptions, but also regarding other aspects related to the performance of the evolutionary algorithms.

Acknowledgments

The authors are grateful to CONACyT (México) for financial support.

References

1. P. Beckmann and A. Spizzichino, *The Scattering of Electromagnetic Waves from Rough Surfaces*, (Pergamon Press, London, 1963), p. 29.
2. J. A. Ogilvy, *Theory of wave scattering from random rough surfaces*, (Institute of Physics Publishing, Bristol, 1991), p. 277.
3. Rice, S. O.: Reflection of electromagnetic waves from slightly rough surfaces, Commun. Pure Appl. Math. **4**, 351, 1951.
4. Thorsos, E. I.: The validity of the Kirchhoff approximation for rough surface scattering using a Gaussian roughness spectrum, J. Acoust. Soc. Amer. **83**, 78 (1988).
5. Maradudin, A. A., Michel, T., McGurn, A. R., Méndez, E. R.: Enhanced backscattering of light from a random grating, Ann. Phys. (N. Y.) **203**, 255 (1990).
6. McGurn, A. R., Maradudin, A. A., Celli, V.: Localization effects in the scattering of light from a randomly rough grating, Phys. Rev. B**31**, 4866, 1985.
7. Méndez, E. R., O'Donnell, K. A.: Observation of depolarization and backscattering enhancement in light scattering from Gaussian random surfaces, Opt. Commun. **61**, 91, 1987.
8. Chandley, P. J.: Determination of the autocorrelation function of height on a rough surface from coherent light scattering, Opt. Quant. Electron. **8**, 329, 1976.
9. Welford, W. T.: Optical estimation of statistics of surface roughness from light scattering measurements, Opt. Quant. Electron. **9**, 269, 1977.
10. Chandley, P. J.: Determination of the probability density function of height on a rough surface from far-field coherent light scattering, Opt. Quant. Electron. **11**, 413, 1979.

11. Elson, J. M., Bennett, J. M.: Relation between the angular dependence of scattering and the statistical properties of optical surfaces, J. Opt. Soc. Am. **69**, 31, 1979.
12. Stover, J. C., Serati, S. A.: Calculation of surface statistics from light scatter, Opt. Eng. **23**, 406, 1984.
13. Wombel, R. J., DeSanto, J. A.: Reconstruction of rough-surface profiles with the Kirchhoff approximation, J. Opt. Soc. Am. A **8**, 1892, (1991).
14. Wombel, R. J., DeSanto, J. A.: The reconstruction of shallow rough-surfaces profiles from scattered field data, Inverse Problems **7**, L7, (1991).
15. Quartel, J. C., Sheppard, C. J. R.: Surface reconstruction using an algorithm based on confocal imaging, J. Modern Optics **43**, 496, (1996).
16. Quartel, J. C., Sheppard, C. J. R.: A surface reconstruction algorithm based on confocal interferometric profiling, J. Modern Optics **43**, 591, (1996).
17. Macías, D., Méndez, E. R., Ruiz-Cortés, V.: Numerical study of an inverse scattering algorithm for perfectly conducting one-dimensional surfaces, in *Scattering and Surfaces Roughness*, A. A. Maradudin and Z. H. Gu, eds., Proc. SPIE **4100**, 57-64 (2000).
18. Walther, A.: The question of phase retrieval in optics, Opt. Acta **10**, 41,(1963).
19. O'Neill, E., Walther, A.: The question of phase in image formation, Opt. Acta **10**, 33,(1963).
20. Nelder, J., Mead, R.: A simplex method for function optimization *Computer Journal*, **7**, 308,(1965).
21. Press, W. H. (editor) *Numerical Recipes in Fortran*, (Cambridge University Press, Cambridge, 1992) p. 402.
22. H. P. Schwefel, *Evolution and Optimum Seeking*, (John Wiley & Sons Inc., NY, 1995), p. 444.
23. Macías, D., Méndez, E. R.: Unpublished work (2001).

Detection of Incidents on Motorways in Low Flow High Speed Conditions by Genetic Programming

Simon C. Roberts and Daniel Howard

QinetiQ Software Evolution Centre, Building U50, QinetiQ,
Malvern WR14 3PS,UK, tel: +44 1684 894480
dhoward@qinetiq.com

Abstract. Traditional algorithms which set a lower speed limit on a motorway to protect the traffic against collision with a queue are not successful at detecting isolated incidents late at night, in low flow high speed conditions. The Staged Genetic Programming method is used to detect an incident in this traffic regime. The evolutionary engine automatically decides the time duration for the onset of an incident. This method successfully combines traffic readings from the MIDAS system to predict a variety of late night incidents on the M25 motorway.

1   Introduction

The aim of this investigation is to detect motorway incidents during quiet periods such as late at night when the traffic flow is low and the traffic speed is high. In such periods the standard queue protection algorithms, which are decades old, such as California (USA) [1] and HIOCC (UK) [2] do not apply.

This proof of concept investigation considers using GP to evolve an incident detector for such late night periods on the M25 motorway which encircles London. Traffic on large parts of the M25 is managed with the MIDAS system. On each lane and approximately at 500 metre intervals, MIDAS loop sensors incorporated into the road surface compute minute averaged readings of traffic quantities, e.g. vehicle speed (this quantity will be referred to as velocity), flow rate, occupancy, and for these experiments also the headway; quantities are measured directly rather than as quantities derived from other measurements.

Figure 1 illustrates typical late night incidents and requires explanation. The X axis is discrete time, and the Y axis is discrete distance along the motorway. Each square represents a minute and a sensor. The figure is like a surface plot for one of the quantities in one of the motorway lanes. In this case velocity is shown, black is fast and white is slow. The origin of the coordinate system is always at the top left.

The sought incident detector should detect the incident as early as possible, i.e. smallest x value, to raise an alarm and provide a few minutes for the MIDAS operator to take action. For example, the operator could react by positioning a motorway camera to take a closer look, or he could alert the Police who need

S. Cagnoni et al. (Eds.): EvoWorkshops 2002, LNCS 2279, pp. 245–254, 2002.

time to travel to the scene of the incident. Incidents have many causes, e.g. from drivers swerving to avoid animals crossing the road to serious accidents.

GP is used in a supervised learning role. Hence, it requires a 'truth' from which GP may learn to detect an incident. Unfortunately, until now very few incidents have been reported. Even then their reasons are not always correctly recorded. This investigation manually scrutinised the data and marked up an incident 'truth'. Three years of motorway data were inspected for velocity in all lanes. Incidents were then graded by a subjective measure of their significance, and this level of significance is incorporated into the GP fitness measure to prevent the situation whereby GP may trade-off missing a significant incident to detect two insignificant ones [3].

Currently our team is successfully developing a comprehensive detector which is based on the supervised learning of three years of marked data covering more than one thousands incidents. This paper, however, presents the results for the far smaller proof of concept investigation which preceeded our current work.

GP discovers complex and non-linear mathematical relationships between the MIDAS data to determine whether an incident, e.g. an accident, has occurred. A specific technique has been chosen to facilitate this objective. The *Staged Genetic Programming Method* [4] developed by these authors for certain image anaylsis tasks, is applied to this problem because the precise onset of incidents is often poorly defined. The detection task is tackled by a two-stage evolution strategy. The first stage evolves for itself which traffic data best characterises an incident onset by distinguishing data proximate to incidents from a sample of non-incident data. The second stage is then required to minimise the false alarm rate whilst retaining at least a single detection point per incident. This means that the evolutionary process has the freedom to detect an incident as early as is feasibly possible.

As with our current work, the incidents were also *graded* subjectively in the proof of concept study - grade 1 is the most obvious incident type. This grading rewards detection of more obvious incidents above the detection of marginal or 'difficult to tell' incidents and is very important to guide the fitness measure in the second GP stage.

The first stage is a coarse detection required to identify all of the incidents in the data at the expense of producing many false alarms. The second stage provides a finer level of discrimination to reduce the false alarm rate. Hence, the classifier is a combination or fusion of the best detector from the first stage and the best detector from the second stage. The Genetic Programming runs use standard genetic operators, tournament selection, and a steady-state GP algorithm.

2   Traffic Data Input to GP

Traffic data was taken from 60 consecutive stations along the M25 between stations 4727 and 5010. Both carriageways were used for the 7 nights in August 1998 shown in Table 1. The nightly periods started on the dates and times shown and terminated the following day at the given stop time. Different days used slightly different times to avoid periods of extensive missing data. Any missing data in the input was corrected by averaging across adjacent minutes and stations. This correction procedure was applied iteratively to correct longer periods of missing data, e.g. for the offside lane.

Fig. 1. This figure is an amalgamation of other figures in order to illustrate six typical grade ≤ 3 late night or early morning incidents (the concept of subjectively grading incidents is explained in the text). Whiter is slower taffic, darker is faster traffic. The blue colour, which cannot be clearly seen in a monochrome image, denotes absent flow which occurs when there is no traffic. Note the incident at the bottom right; this is a lane closure; note the motion of the works vehicle which deposits the traffic cones clearly seen as a diagonal series of white boxes (top-left to bottom-right trajectory), i.e. at times the works vehicle is nearly stationary over the lane sensors.

Table 1. Night traffic data taken from August 1998.

date	carriageway	start time	stop time	no. of incidents
3	A	2130	0600	2
5	A	2130	0600	0
9	A	2130	0600	2
10	A	2300	0600	0
11	A	2130	0600	3
26	A	2130	0600	0
27	A	2130	0600	2
3	B	2130	0600	0
5	B	2130	0600	5
9	B	2130	0600	2
10	B	2300	0600	3
11	B	2145	0600	5
26	B	2130	0600	5
27	B	2130	0600	3

Table 1 also gives the number of incidents which were input to GP for each night and carriageway. The incidents were identified by manually scanning through the data. An incident was deemed to be represented by abnormal data on any of the lanes.

The incidents were classified according to five grades of severity. Grade 1 incidents were major incidents affecting upstream stations and other lanes whereas grade 5 incidents were minor appearing as single blips in one lane only. In addition, grade 10 identified spurious data and 11 were roadworks. All the identified incidents are given in Table 2. GP was exposed to 32 incidents of grade 1 to 4 but the other incidents were deemed to be periods of indifference where GP should not be rewarded or punished.

The traffic data comprised minute-averaged flow, occupancy, speed and headway on the 3 outer lanes and traffic counts for 4 grades of vehicle length. GP simultaneously processed this data for the current minute and T previous minutes and for the current station and one downstream station.

3   Fitness Measure

A GP chromosome was said to have detected an incident when it output a positive value. In other words, the traffic data being processed at the given time and station was deemed to represent an incident. If an actual incident had occurred then the output was called a *true positive*, otherwise it was called a *false positive* or a false alarm. After a chromosome had processed all the training cases, let TP denote the number of resulting true positives and FP denote the number of false positives.

The task was to distinguish incidents of grade 1 to 4 from non-incidents. Furthermore, lower numbered incident grades were more important than higher

Table 2. Manually identified incidents on carriageways A and B. The incident duration is shown in hours and was defaulted to 1 hour if not given.

carriageway A					carriageway B				
date	time	station	grade	duration	date	time	station	grade	duration
3	2320	4989a	1	-	3	2358	4836b	5	-
4	0015	4802a	11	3	4	0350	4752b	5	-
4	0046	5010a	5	-	5	2334	4955b	3	-
4	0319	4927a	4	-	5	2337	4940b	4	-
9	2100	4917a	10	2	6	0040	4757b	2	-
9	2253	4912a	3	-	6	0336	4955b	4	-
10	0203	4935a	3	-	6	0515	4955b	4	-
10	0332	4985a	5	-	9	2243	4945b	4	-
10	2320	4927a	11	6	10	0015	4935b	3	-
11	0405	4787a	10	0.5	10	0111	4883b	5	-
11	2200	4742a	1	-	10	0302	4998b	5	-
11	2232	4762a	2	-	10	2135	4935b	11	7
12	0540	4912a	3	-	10	2210	4935b	10	-
26	2120	4981a	11	8	11	0221	4757b	4	2
27	0435	4832a	5	-	11	0249	4742b	5	-
28	0035	4932a	3	-	11	0346	5002b	4	-
28	0215	4959a	4	-	11	0431	4949b	2	-
					11	2132	5002b	5	-
					11	2258	4949b	2	-
					11	2340	4949b	2	-
					12	0050	4945b	3	-
					12	0128	4752b	3	2
					12	0142	4963b	5	-
					12	0227	4949b	3	2
					12	0228	5002b	5	-
					12	0451	4949b	5	-
					12	0454	4907b	5	-
					26	2346	4888b	2	-
					26	2352	4797b	2	-
					27	0116	4945b	5	-
					27	0137	4935b	5	-
					27	0149	4863b	5	-
					27	0227	4949b	4	-
					27	0311	5002b	3	-
					27	0324	4949b	4	-
					27	2151	4832b	1	1.5
					28	0030	4955b	5	-
					28	0038	4963b	5	-
					28	0142	4959b	5	-
					28	0257	4737b	4	-
					28	0436	4737b	2	1

numbered grades. These issues were captured by the following fitness measure which drove the evolution runs,

$$\text{fitness} = \frac{TPS}{(TPS_{max} + \beta FP)} \tag{1}$$

where TPS was the score for the incidents detected and TPS_{max} was the score obtained when all incidents were detected. TPS was calculated as follows to weight the incidents according to grade.

$$TPS = \sum_{i=0}^{i<TP} 5 - \text{grade of incident}_i \tag{2}$$

The variable β was used to balance the importance of incident detection against the expense of detecting false alarms. For example, as β tended to zero the number of false alarms became irrelevant.

The *figure of merit* (FOM) is used in the results tables to compare different GP runs. The FOM is the same as the fitness when $\beta = 1$. Note that the minimum FOM is 0 for the case when no incidents are detected or when the number of false alarms approaches infinity, and the maximum FOM is 1 when all incidents are detected with no false alarms.

4   Evolving First Stage Detectors

Even though incidents happen at a single time and location, the incidents generally manifest themselves in the traffic data over multiple minutes and a number of stations (more upstream stations than downstream). Consequently GP was trained to detect the onset of each incident by sweeping from 5 minutes before to 5 minutes after the incident time, and by sweeping from 4 upstream stations to 4 downstream stations for each minute. However, an incident was said to be detected if a GP chromosome returned a positive output for any one of these times and locations. Therefore, TP represented the number of incidents detected and had a maximum value of 32. Non-incident training data was sampled at 10 minute steps using all possible stations at each minute, giving a total of 21,181 non-incident training cases.

Approximately 60 GP runs were conducted for the first stage each using $T = 3$. A first stage detector's task was to detect *all* incident onsets whilst producing a minimal false alarm rate. The fitness variable β was thus set to low values and the range 0.01 to 0.1 tended to give the best results. Higher settings caused incidents to be missed whilst lower settings tended to give an excessive false alarm rate. The population size was set to 1000.

5   Validating First Stage Detectors

The evolved first stage detectors were validated on the 32 incidents and a maximum score of 67 was achieved if all incidents were detected (i.e. $TPS_{max} = 67$).

The detectors processed all non-incident traffic data in validation which totalled to 200,010 cases. Figure 2 plots the incident score against false alarm rate for the best 10 detectors from each GP run, i.e. 10 fittest of each independent run.

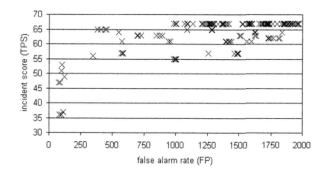

Fig. 2. Incident score against false alarm rate for the first stage detectors.

The lowest false alarm rate when all incidents were detected was 987 out of the 200,010 non-incident cases.

6 Evolving Second Stage Detectors

However, another first stage detector was judged to be the best because it produced 142 hits distributed across 30 incidents and it gave only 382 false alarms, and sacrificed only two grade 4 incidents in order to achieve this (i.e. $TP = 30$ and $TPS = 65$). As already explained, even by careful manual inspection it would difficult to determine whether these grade 4 incidents are true.

Second stage detectors were trained to reduce these false alarms whilst retaining at least a single hit per incident. Note that TPS_{max} reduced to 65 because of the two grade 4 incidents missed by the first stage detector.

The parameters β and T were optimised by conducting at least 50 GP runs for various settings. The detection performance was largely insensitive to T between values of 2 and 8. Table 3 shows the results for varying β with a population size of 1000. The table gives the average FOM of the best detector from each of the associated GP runs, the result for the detector which gave the maximum FOM from all of the runs, and the maximum TPS which was achieved for an FP limit of 0, 7 and 14. These false alarm limits were chosen because GP processed 7 nights worth of traffic data on both carriageways. Hence, the results in the last three columns give the best detection rates for the following false alarm rates: no false alarms, a single false alarm per night on average, and a single false alarm per night and carriageway on average.

Table 3. Variation of β with $T = 4$.

β	Average FOM	Best result: FOM	TPS	FP	max TPS when FP $\leq$: 0	7	14
0.1	0.592	0.821	64	13	-	-	64
0.3	0.723	0.908	59	0	59	63	64
0.5	0.743	0.912	62	3	54	62	64
0.7	0.755	0.859	61	6	51	61	62
1.0	0.751	0.879	58	1	57	60	60
2.0	0.706	0.785	51	0	51	53	53

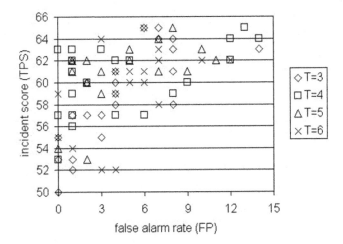

Fig. 3. Incident score against false alarm rate for the second stage detectors.

Table 3 shows that β must exceed 0.1 in order to achieve no false alarms. The last three columns show that fewer incidents were detected for a given FP limit as β was increased. A good balance between maximising detected incidents and minimising false alarms was achieved by setting β to 0.5, and this setting gave the maximum FOM in the *Best result* column.

More GP runs were conducted with $\beta = 0.5$ and a population size of 4000. Figure 3 plots the resulting incident scores against false alarm rate for the best detector from each GP run for various T values.

Table 4. Best second stage detectors.

FOM	TPS	TP	FP
0.969	63	28	0
0.941	64	29	3
0.915	65	30	6

Recall that two grade 4 incidents were missed at the first stage and so a second stage detector could identify a maximum of 30 incidents with a TPS of 65. This was achieved at the expense of giving 6 false alarms, i.e. less than a single false alarm per night on average. These false alarms are arguably grade 5 incidents and two of them actually refer to the same event which was detected at adjacent minutes and stations. The maximum FOM was achieved when two further grade 4 incidents were missed to bring TPS down to 63 but all false alarms were avoided.

The incidents detected by the best second stage detector (i.e. the one which gave the maximum FOM) are shown in Table 5. The *lag* column gives the lag in minutes between the incident onset and the time of detection. This was negative when the incident started to manifest itself in the traffic data before the onset time recorded for training, e.g. because of a gradual onset spread across lanes. The column shows that most incidents were detected within two minutes. The last column gives the number of hits per incident and shows that grade 1 incidents received multiple hits but higher grades tended to be hit only once.

Table 5. Incidents detected by the best second stage detector. The lag is in minutes and the number of detections per incident is given.

date	time	station	grade	lag	no. of hits	date	time	station	grade	lag	no. of hits
3	2320	4989a	1	-1	5	4	0319	4927a	4	0	1
9	2253	4912a	3	0	4	10	0203	4935a	3	0	1
11	2200	4742a	1	0	3	11	2232	4762a	2	0	1
12	0540	4912a	3	1	1	28	0035	4932a	3	0	1
5	2334	4955b	3	2	1	5	2337	4940b	4	-1	1
6	0040	4757b	2	1	2	6	0336	4955b	4	5	1
6	0515	4955b	4	2	2	9	2243	4945b	4	0	3
10	0015	4935b	3	0	2	11	0346	5002b	4	0	1
11	0431	4949b	2	5	1	11	2258	4949b	2	1	1
11	2340	4949b	2	1	1	12	0050	4945b	3	0	1
12	0128	4752b	3	3	1	12	0227	4949b	3	-4	4
26	2346	4888b	2	-5	1	26	2352	4797b	2	5	1
27	0227	4949b	4	0	1	27	0311	5002b	3	0	1
27	2151	4832b	1	0	6	28	0436	4737b	2	3	1

7   General Conclusions

The task was to detect incidents on the M25 at periods of low traffic occupancy. This was approached by training GP to detect the onset of incidents which occurred during the night (approximately between 2200h and 0600h) whilst producing a near zero false alarm rate.

The precise onset of incidents was often poorly defined and so the task was tackled by a two-stage evolution strategy. The first stage evolved for itself which traffic data best characterised an incident onset by distinguishing data proximate to incidents from a sample of non-incident data. The second stage was then required to minimise the false alarm rate whilst retaining at least a single detection per incident.

The best first stage detector missed two grade 4 incidents (the least obvious incidents) and gave 382 false alarms from 7 nights worth of data using both carriageways. The best second stage detector missed another two grade 4 incidents but abolished all the false alarms. Other second stage detectors retained all incident detections but gave 6 false alarms.

8   Current Work

The activity of manually marking more than 1000 incidents over three years of data as described in the introduction has shed light into the nature of the incidents, which in turn has increased the subjective threshold of what constitutes an incident. Thus, the scaleup to more data has resulted in better performance: fewer misses and fewer false alarms. Tests are being performed to assess generalisation. Other modelling parameters such as the effect of weather: rain and fog measurements, may be incorporated in future as inputs to these models.

Acknowledgment

The authors wish to thank the UK Highways Agency for facilitating the data for this study which is highly specific to MIDAS and to the M25 motorway.

References

1. Payne H. J., Goodwin D. N. and Teener M. D. (1975): Evaluation of existing incident detection algorithms. Technology Service Corporation, Santa Monica, California.
2. Collins J. F., Hopkins C. M. and Martin J. A. (1979): Automatic incident detection - TRRL algorithms HIOCC and PATREG. TRL Report SR 526, Transport Research Laboratory, Cowthorne, UK.
3. Howard D., Roberts S. C. and R. Brankin (1999): Evolution of Ship Detectors for Satellite SAR Imagery. Proceedings of EuroGP99, Götteborg, Sweden.
4. Howard D. and Roberts S. C. (1999): A staged genetic programming strategy for image analysis. Proceedings of the Genetic and Evolutionary Computation Conference, Orlando, Florida, 1047–1052 Morgan Kaufmann.

Image Filter Design with Evolvable Hardware

Lukáš Sekanina

Faculty of Information Technology
Brno University of Technology
Božetěchova 2, 612 66 Brno, Czech Republic
sekanina@fit.vutbr.cz

Abstract. The paper introduces a new approach to automatic design of image filters for a given type of noise. The approach employs evolvable hardware at simplified functional level and produces circuits that outperform conventional designs. If an image is available both with and without noise, the whole process of filter design can be done automatically, without influence of a designer.

1 Introduction

Image recognition is a problem that has to be solved successfully in various industrial applications, namely in automatic traffic sign recognition, car registration number recognition or in the automatic control of the producing line in a factory where correct and damaged products have to be detected.

The whole recognition system has to be extremely accurate. For example, only one wrong decision of the million is acceptable in case of recognition of the correct/damaged component on the production line. The quality of recognition algorithm strongly depends on quality of the images coming from a camera since these algorithms are commonly designed for idealized images.

Images usually acquired through modern cameras may be contaminated by a variety of noise sources (e.g. photon or on-chip electronic noise) and also by distortions such as shading or improper illumination. Therefore, a preprocessing unit (image filter) has to be incorporated before recognition to improve image quality.

This paper deals with filters for smoothing images. Industry calls for automatic design of such filters since (1) the system should adapt to changing environment autonomously (e.g. to the changes of illumination or after replacement of a damaged camera) and (2) it is expensive to pay designers when no standard solution can be easily adopted.

A new approach to automatic design of image filters for a given type of noise is introduced. The approach employs *evolvable hardware* at functional level and produces circuits that outperform conventional designs in terms of the resulting image quality and implementation cost in most cases. We do not know any work that is related to evolutionary design of image filters at hardware level at the moment. Available evolutionary designs reported in past years are oriented towards filters for one-dimensional signals or to specific image operators. Our

S. Cagnoni et al. (Eds.): EvoWorkshops 2002, LNCS 2279, pp. 255–266, 2002.
© Springer-Verlag Berlin Heidelberg 2002

solution is based on simple functions (binary operations or 8bit adders) that may be effectively implemented in low-cost, commercial off-the-shelf hardware devices like FPGA (Field Programmable Gate Array).

The next section briefly summarizes conventional image filters while the basic principles of evolvable hardware are described in Section 3. Some of the already published evolutionary approaches to filter and image operator design are mentioned in Section 4. Section 5 introduces experimental framework of our approach. Section 6 reports evolved designs that are discussed in Section 7. And finally, conclusions and problems for future work are given in Section 8.

2   Conventional Design of Image Filters

The conventional approach to image filter design is explained in many textbooks, e.g. in [1,2]. An image can be filtered either in the frequency or in the spatial domain. We are interested in the spatial domain where the input image x convolves with the filter function h. In discrete convolution, the kernel is shifted over the image and multiplies its values with the corresponding pixel values of the image. A kernel is a small matrix of numbers whose members define weights of accounted pixels. Let $y(i, j)$ denotes a pixel value of the resulting image at position (i, j). For a square kernel of size $M \times M$, we can calculate the output image with the following formula:

$$y(i,j) = \sum_{m=-\frac{M}{2}}^{\frac{M}{2}} \sum_{n=-\frac{M}{2}}^{\frac{M}{2}} h(m,n)x(i-m,j-n)$$

Various standard kernels exist for specific noise, where the size and the form of the kernel determine the characteristics of the operation. The filter can be applied on an already filtered image repeatedly. In contrast to the frequency domain, it is possible to implement non-linear filters in the spatial domain. In this case, the summations in the convolution function are replaced with some kind of non-linear operator (e.g. Median or Kuwahara filter). Another advanced filters like non-linear mean filter or averaging using a rotating mask are given in [1,2].

Let us briefly describe mean (denoted as $FA1$ in the paper), mean-2 ($FA2$), mean-4 ($FA4$) and median (FME) filters since reported results will be compared with them in the next sections. Consider M=3 for the paper. The idea of mean filtering is simply to replace each pixel value in an image with the mean (average) value of its neighbors, including itself. Mean-2 and mean-4 filters take some pixels in account several times and produce better results than mean filter for Gaussian noise since their coefficients are derived from the curve of Gaussian distribution. The kernels are defined as:

$$FA1 = \frac{1}{9}\begin{pmatrix} 1 & 1 & 1 \\ 1 & 1 & 1 \\ 1 & 1 & 1 \end{pmatrix} \qquad FA2 = \frac{1}{10}\begin{pmatrix} 1 & 1 & 1 \\ 1 & 2 & 1 \\ 1 & 1 & 1 \end{pmatrix} \qquad FA4 = \frac{1}{16}\begin{pmatrix} 1 & 2 & 1 \\ 2 & 4 & 2 \\ 1 & 2 & 1 \end{pmatrix}$$

In the case of median filter, a pixel value is replaced with the median of neighboring values. A median filter is much better at preserving sharp edges than the mean filter since it does not create the new (potentially unrealistic) pixel values.

3 Evolvable Hardware

Evolvable hardware (EHW) may be considered as a technology, which enables to establish an evolvable system with the ability of hardware on-line adaptation to dynamically changing environments [3]. A circuit connection of the fast reconfigurable circuit (whose configuration bits are encoded in a chromosome) is autonomously synthesized by an evolutionary algorithm. In the case of a single fitness function, the approach is usually called evolutionary circuit design. Evolution is free to explore many unconventional solutions beyond the scope of conventional engineering design and thus should introduce a new quality to solution. Real-world applications of EHW are summarized in [4].

Miller and Thomson have introduced *Cartesian Genetic Programming* (CGP) [5] that was recently applied by several researchers especially for evolutionary design of combinational circuits [6,14]. Reconfigurable circuit is modeled as an array of u (columns) $\times$ v (rows) programmable elements (gates). The number of circuit inputs and outputs is fixed. Feedback is not allowed. A gate input can be connected to the output of some gate in the previous columns or to some of circuit inputs. L-back parameter defines the level of connectivity and thus reduces/extends the search space. For example if $L=1$, only neighboring columns may be connected; if $L=u$, the full connectivity is enabled. For a given application, designer has to define: the number of inputs and outputs, L, u, v and a set of functions performed by programmable elements (typically binary operations over two or three inputs). In other words, these parameters define a configuration of the programmable circuit (see Figure 1).

The idea of EHW at *functional level*, where the programmable elements include functions like adders, multipliers, dividers, sine or cosine generators over floating point numbers, was initially introduced in [7].

4 Evolutionary Filter and Image Operator Design

In the case of the spatial domain, image filters and image operators are designed similarly. The designer usually determines M and the values of the kernel. This is a very time consuming job, especially when the noise type is unknown. Thus evolutionary design of either the kernel or the whole function (i.e. a circuit at hardware level) offers an alternative approach. The resulting structure evolves from primitives instead of calculating coefficients for a general-purpose model. Evolved solutions (circuits) should be more efficient than conventional design in terms of performance and implementation cost.

Authors in [8] evolved circuits for edge detection using elementary binary operations supported in FPGAs while another edge detectors (also evolved in

FPGA) were represented as 2D arrays of integers that defined the convolution kernel [9]. Evolutionary optimization of soft morphological filters for archive film restoration was extended to temporal domain in [10].

At least one paper at every conference on EHW was devoted to filter design in history: Evolvable System: From biology to hardware conference ICES96 (1 paper), ICES98 (1), ICES00 (1), ICES01 (1); NASA/DoD Workshops on Evolvable hardware 1999 (3), EH00 (1), EH01 (3). Proposed approaches however deal with one-dimensional signals only. Miller used pure gate array and CGP to filter simple signals [11]. In [12] the authors implemented a simple filter as well as whole evolutionary algorithm in the FPGA. Genetic programming approach to analog filter design is explained in detail in [13]. In [14] the authors used CGP for the design of finite impulse response digital filters with reduced power consumption. Resulting design is automatically transformed to VHDL and synthesized. The filter evolves from primitives like adder, subtractor or shifters.

5   Image Filter Evolution: Experimental Framework

The goal is to evolve a digital circuit operating as an image filter for a given type of noise. Gray-scale (8bits/pixel) images of size $N \times N$ (N=256) pixels are considered in the paper. The pixel value is filtered using 3×3 neighborhood. The new pixel value is available on the 8bit output of the circuit. The circuit input consists of nine pixel values. Based on initial experiments, the following parameters were set up as default:

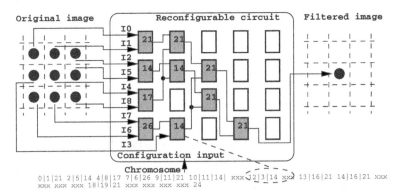

Fig. 1. An example of the reconfigurable circuit and its configuration for the image filter. Nine inputs (pixel values) are used to calculate a new (filtered) pixel value. Parameters: 9 inputs, 1 output, circuit topology 5×4, L-back=1. Only utilized programmable elements are marked.

5.1   Reconfigurable Circuit

Parameters of the reconfigurable circuit according to CGP are: 9 inputs (8bits), 1 output (8bits), $u = 10$ (columns), $v = 4$ (row), L-back = 2. A programmable element has two inputs and operates over 8 bits. Table 1 lists functions supported in the programmable element. Circuit inputs correspond with the pixels of the kernel according to the Figure 1. The proposed architecture operates rather like parallel gate level evolution than functional level evolution. In the case of adders (functions 14, 20, 21 and 30), only lower 8bits are considered as output. Except the adders, the elements have trivial hardware implementation.

Table 1. A list of functions implemented in a programmable element. The inputs a and b and the outputs operate over 8bits. Symbols used: $>>$ right shifter, $<<$ left shifter, $\wedge$ binary AND, $\vee$ binary OR, $\oplus$ binary exclusive-OR, $+$ 8bit adder, $\bar{a}$ is a binary negation of a. Constants are given in a hexadecimal system.

0	$a >> 1$	1	$a >> 2$	2	$a >> 4$
3	$\bar{a}$	4	$a << 1$	5	$a << 2$
6	$a << 4$	7	$(a << 4) \vee (a >> 4)$	8	0
10	FF	11	AA	12	55
13	33	14	$(a + b + 1) >> 1$	15	$\bar{a} \vee b$
16	$\overline{a \wedge b}$	17	$(a \wedge 0F) \vee (b \wedge F0)$	18	$(a \wedge CC) \vee (b \wedge 33)$
19	$(a \wedge AA) \vee (b \wedge 55)$	20	$a + b$	21	$(a + b) >> 1$
22	$a \vee b$	23	$a \wedge b$	24	$a \wedge \bar{b}$
25	$\bar{a} \wedge b$	26	$a \oplus b$	27	$\overline{a \vee b}$
28	$\overline{a \oplus b}$	29	$a \vee \bar{b}$	30	$((a + b) >> 1) + 1$

5.2   An Evolutionary Algorithm

Chromosome encoding: A chromosome is a fixed-size string of integers, containing $u \times v$ genes (corresponding to the programmable elements in the reconfigurable circuit) and one place devoted to the index of the element representing the circuit output (see a chromosome in Figure 1). A gene is described by three values: the position of the first input, the position of the second input and a number of the function applied on inputs. Thus genotype is of fixed length while phenotype is variable length since all the programmable elements need not be used.

Population: Population size is 16. Initial population is generated randomly, but only the function 21 was used in some runs (see Section 7). The evolution was typically stopped (1) when no improvement of the best fitness value occurs in the last 50000 generations, or (2) after 500000 generations.

Genetic operators: Mutation of two randomly selected gates is applied per circuit. A mutation always produces a correct circuit configuration. Crossover is not used. Four of the best individuals are utilized as parents and their mutated versions build up the new population (deterministic selection with elitism).

Fitness function: Various approaches exist to measure image visual quality. The *signal-to-noise ratio* or the pure *average difference per pixel (dpp)* are the commonest ones. We chose the second approach. Let *orig* denote an original image without any noise (e.g. Lena), *noise* denotes the original image corrupted with noise of type *Xxx* (e.g. LenaXxx), and *filtered* denotes an image filtered using some filter *Fyy* (e.g. lenaXxxFyy). The filter is trained using Lena256 and Lena256Xxx images for *Xxx* noise. Only the area of 254 x 254 pixels is filtered because the pixel values at the borders are ignored. To obtain the fitness value, the differences between pixels of the filtered and original image are added and the sum is subtracted from a maximum value (representing the worst possible difference: #grey_levels × #pixels):

$$FitnessValue = 255.(N-2)^2 - \sum_{i=1}^{N-2} \sum_{j=1}^{N-2} |orig(i,j) - filtered(i,j)|$$

6   Results

For our experiments, we consider three types of noise denoted as: $G16$ (Gaussian with a mean of zero and a standard deviation of 16), $R32$ (uniform random with parameter 32) and $N1$ (block uniform random). Images with $G16$ and $R32$ noise were generated from originals using Adobe Photoshop program. We designed the $N1$ noise for testing purposes to model random defects in the image. $N1$ noise is generated as $R32$ noise but applied only for randomly selected blocks of the image. The rest of the image preserves.

We have evolved more than one hundred image filters and 20 of them are presented. The test set contains the following images: Lena (popular for testing), Man (a man), Bld (a building), Cpt (a capacitor), and Rel (a relay). Cpt and Rel images were acquired through a camera and the system for automatic recognition of damaged/correct product on the production line. The best designs as well as results of traditional filters (typed as bold) are sorted according to their ability to remove general noise in Table 2. The ranks for a given type of noise are presented in the last three columns. As an example, the complete results for $G16$ noise are given in Table 3.

It was detected after analysis of the evolved designs that some filters do not employ all the functional elements effectively. For instance, the filter F20 uses two functional elements with the same inputs (marked elements in the Figure 3). These functional elements can be omitted (i.e. replaced by a direct connection) since they do not influence the output of the filter. The number of functions used in the evolved designs after manual optimization is listed in the column #O of the Table 2.

Table 2. A list of evolved and conventional filters sorted according to their ability to filter general noise. Columns have these purposes: TNF – trained for noise (or tested for noise for conventional filters); dpp – dpp for trained image; $u \times v$ – circuit topology; L – L-back parameter; $\#E$ – the number of functional elements used in the evolved design; $functions\ used$ in the evolved designs; $\#O$ – the number of functional elements used after manual optimization; $gener$ – the generation where the solution occured; $G16, R32, N1$ – the final rank for a given noise type and all the test images.

Filter	TFN	dpp	u × v	L	#E	functions used	#O	gener	G16	R32	N1
F24	G16	6.362	10x4	2	21	8, 17, 22, 21(12), 14(6)	14	185168	1	2	3
F20	G16	6.358	10x4	2	17	17, 21(14), 30(2)	15	79369	2	3	7
F26	G16	6.358	10x4	2	19	17, 18(2), 21(12), 14(4)	14	24853	4	4	8
F25	G16	6.356	10x4	2	24	21(16), 14(8)	22	13415	6	6	5
F21	G16	6.354	20x2	4	19	21(17), 30(2)	18	133224	7	1	12
F14	G16	6.401	20x2	4	14	17, 18, 21(9), 23, 30(2)	12	13678	3	16	4
F24A	G16	6.388	10x4	2	14	21(14)	14		5	12	6
F11	G16	6.354	10x4	2	26	21(19), 22, 30(6)	24	81532	8	5	10
FA4	G16	6.437							9	14	2
F23	N1	6.060	40x1	40	10	18, 21(9)	9	42772	13	18	1
F15	G16	6.363	40x1	40	18	21(18)	18	51356	10	11	13
F21A	G16	6.384	20x2	4	19	21(18)	18		11	8	17
F13	G16	6.360	20x2	4	22	21(16), 22, 30(5)	21	43749	14	7	15
F16	G16	6.367	40x1	40	16	21(16)	16	71518	12	9	16
F6	G16	6.400	50x1	50	17	17,21(10),22,24,29,30(3)	16	24693	15	17	11
F18	N1	6.077	40x1	40	9	21(9)	9	141744	16	19	9
F27	R32	6.926	10x4	2	25	18, 21(10), 14(14)	23	128025	18	10	19
F17	R32	6.977	40x1	40	14	2, 20, 21(12)	14	43549	17	13	18
F19	R32	6.981	40x1	40	16	10, 17(3), 21(12)	16	38007	19	15	20
F8	G16	6.640	50x1	50	12	7, 17, 21(9), 30	12	70304	20	21	14
FA2	G16	6.469							21	20	22
F22	N1	6.283	40x1	40	7	21(7)	7	35872	22	23	21
FA1	G16	6.655							23	22	24
FME	G16	7.157							24	24	23

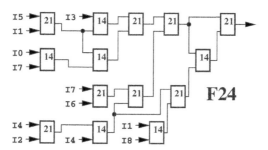

Fig. 2. The best filter evolved for the G16 noise. F24 (with topology 10×4, L-back=2) employs after optimization only functions 21 and 14.

Table 3. Columns 2-6 report *dpp* for a given filter and image with *G*16 noise. *Total − dpp* is an average of *dpp* for all the test images. The last column shows the standard deviation calculated using the best known *dpp* value (typed as bold) for a given image.

Filter	LenaG16	CptG16	RelG16	ManG16	BldG16	Total-dpp	Std. deviation
no filter	12.857	12.754	11.994	12.860	12.526	12.598	6.081
F24	6.362	6.046	4.930	8.243	7.763	6.669	0.082
F20	6.358	6.038	4.879	8.324	7.919	6.704	0.141
F14	6.401	6.109	4.998	8.256	7.762	6.705	0.115
F26	6.358	6.045	4.879	8.327	7.921	6.706	0.143
F24A	6.388	6.125	4.951	8.302	7.771	6.707	0.111
F25	6.356	6.065	4.913	8.310	7.893	6.707	0.135
F21	**6.354**	6.039	4.848	8.351	7.950	6.708	0.156
F11	6.354	6.058	4.877	8.356	7.914	6.712	0.146
FA4	6.437	6.094	4.981	**8.196**	7.875	6.717	0.140
F15	6.363	6.090	4.889	8.357	7.885	6.717	0.140
F21A	6.384	6.103	4.872	8.408	7.833	6.720	0.137
F16	6.367	6.084	4.876	8.372	7.915	6.723	0.152
F23	6.446	6.166	5.050	8.243	7.763	6.734	0.146
F13	6.360	6.073	4.879	8.376	7.985	6.735	0.179
F6	6.400	6.148	4.983	8.260	7.937	6.746	0.169
F18	6.438	6.171	5.023	8.277	7.958	6.773	0.192
F17	6.410	6.197	**4.804**	8.460	8.021	6.778	0.223
F27	6.411	**6.023**	4.882	8.475	8.143	6.787	0.261
F19	6.388	6.052	4.859	8.461	8.215	6.795	0.285
F8	6.640	6.328	5.158	8.511	**7.640**	6.855	0.283
FA2	6.469	6.105	4.856	8.516	8.414	6.872	0.381
F22	6.711	6.477	5.396	8.556	7.748	6.978	0.406
FA1	6.655	6.270	4.858	9.054	9.022	7.172	0.748
FME	7.157	6.837	5.820	9.160	8.042	7.403	0.828

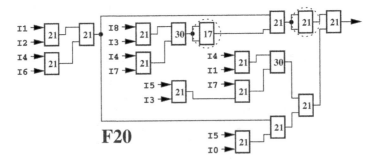

Fig. 3. Evolved filter F20 (with topology 10×4, *L*-back=2) employs only functions 17, 21 and 30. Marked functions may be omitted.

7   Discussion

7.1   Performance

Evolved filters are compared with conventional approaches in this section. However, we know the type of noise a priori and thus efficient conventional solutions may be prepared for comparison. It is not a case of real world situation, where we may not know anything about the noise and suitable conventional solution may not exist at all. Experiments performed with EHW at simplified functional level allowed us to predict that it is possible to evolve:

1. a filter that exhibits less *dpp* than conventional filters (like mean filters or median) for *each* given training image (without exceptions!), and
2. a filter that exhibits in average less *dpp* than conventional filters for a given noise and test images (e.g. $F24$ for $G16$ noise, $F21$ for $R32$ noise or $F23$ for $N1$ noise).

The result (1) is important especially for a production line where an image recognition system operates with very similar images corrupted by the same noise. Then the evolution leads to very efficient filters since they are not trained only for a given noise, but also for a given class of images (e.g. only capacitors). On the other hand, the result (2) proves generality of the evolved filters.

The filter $F24$ (Figure 2) ranked among the best known filters independently of the noise type in the image. It seems to be very general image filter and may probably be considered as a first solution pattern when the type of noise is unknown a priori. It consists only of 14 functions after manual optimization but evolution needed 21 functions to ensure the same behavior.

7.2   Hardware Requirements

Successful evolution requires adders, i.e. the function 21 at least. The approach does not work without adders as well. The function 21 is an 8bit adder equipped with a right shifter to simulate "average of two" operator. Another improvement is due the functions 30 ("average of two plus one") and 14 ("average of two with a carry"). We have manually replaced all the "average of two plus one" in the filter $F21$ and all the "average of two with a carry" functions in the $F24$ filter by "average of two" function to establish the filters $F21A$ and $F24A$ with uniform (and so cheaper) implementations than $F21$ and $F24$. The *dpp* of the filters $F21A$ and $F24A$ increased about 0.5-2% according to type of noise. But the resulting filter $F24A$ still exhibits better results than any of the conventional filters. If the initial population consists of the gates with the function 21 only, faster convergence occurs. The reason is evident: evolved filters are based on this function.

Evolved filters are compared with conventional $FA4$ filter that ranked as the best of conventional filters in the Table 2. The $FA4$ filter with tree architecture requires four 8bit adders, two 9bit adders, one 11b adder, one 12b adder and four shifters. A cost of hardware implementation of some evolved filters (e.g. $F23$,

$F14$, $F22$, $F18$) is evidently cheaper than for $FA4$ filter. As far the $F23$ filter produces the best dpp for the $N1$ noise and, furthermore, its implementation (nine 8bit adders and one logical operation) is cheaper than implementation of $FA4$ filter, the $F23$ outperforms conventional design totally. A cost of the cheapest circuit implementations of the filters for $G16$ and $R32$ noise are comparable with a cost of $FA4$ filter. Optimized $F14$ filter for G16 noise requires at least eleven 8bit adders and a simple logical function 17 while the $F24A$ filter for $R32$ noise consists of 14 adders. These preliminary considerations about hardware cost will be followed by detailed analysis in future work.

As claimed in some papers (e.g. [6]), a sufficient number of the gates is important for efficient evolution. We have applied 40 gates for different topologies (40x1, 20x2 and 10x4) and L-back parameters with similar results, but the dpp was significantly worse for 20x1 gates.

7.3 Experiments with the Simulation Model

The fitness calculation is very time consuming since a circuit simulator has to calculate $(N-2)^2$ pixel values. Two approaches were applied to speed up the fitness evaluation: (1) if deterministic selection is used, about 42% of fitness evaluations need not be finished because the fitness value reached after calculation of some number of pixels is worse than the worst already known solution needed for selection. (2) As far the functional element operates over 8 bits and *unsigned int* type occupies 32 bits, four independent pixels can be simulated in the circuit simulator concurrently.

Some experiments did not lead to efficient designs: (1) When we use four training images in the fitness function, only an average filter $F8$ has been evolved. (2) If programmable elements of the reconfigurable circuit operate on four or two bits, the dpp is more than 20% higher than for the filters operating on 8bits. (3) If a small training image (32 x 32 and 62 x 62 pixels were tested) is considered for fitness calculation, very efficient filter for a given image is evolved. However the filter fails for another images since it is not general. The optimal size of the training image is a question for future research.

It is also interesting that the best filters for $R32$ noise were evolved using training image with $G16$ noise while the best solutions for $G16$ (and $N1$) noise were evolved using training images with $G16$ ($N1$) noise.

7.4 The Evolvable Component for Image Pre-processing

It seems that the proposed approach can be useful also for other problems of image pre-processing like removal of other types of noise, edge detection, illumination enhancement or image restoration. That is also the reason why all the functions are still supported in the programmable element although many of them have not been used. These "useless" functions for image filters may be important e.g. for edge detection. The goal is to develop the *evolvable component* [15] for image pre-processing (with fixed architecture of the reconfigurable circuit and genetic operators) that should be reused in future designs only by redefinition of the fitness function.

8   Conclusions

In this paper, we experimentally proved that efficient circuit implementations of image filters can be evolved. It was shown for three different types of noise. Evolved filters outperform conventional designs in terms of average difference per pixel. Implementation cost is similar or better than in conventional approaches. It is important that evolved filters are built only from simple primitives like logical functions or 8bit adders that are suitable for hardware implementation. If an image is available both with and without a noise, the whole process of filter design can be done automatically and remotely, without influence of a designer. We plan these possible applications of evolved digital filters at the moment:

- The best-evolved filters will be translated to VHDL, synthesized and stored in a library to be reused instead of conventional filters for a given noise type.
- A new filter will be evolved for a given situation in an industrial application (e.g. for a given camera placed on a production line, which defines certain type of noise that should be removed). In a case of the changes of working conditions, the evolution will be restarted manually to find a better filter. Suitable reconfigurable system for an evolutionary design of image filters will be based on the Virtex Xilinx platform or implemented using the technique proposed in [16].

Acknowledgment

The research was performed with the Grant Agency of the Czech Republic under No. 102/01/1531 *Formal approach in digital circuit diagnostic – testable design verification.* The author would like to acknowledge Dr. Jim Tørresen for continuous support during his work with Department of Informatics, University of Oslo, funded by The Research Council of Norway. The author would like to acknowledge Dr. Otto Fučík for inspiration too.

References

1. Šonka, M., Hlaváč, V., Boyle R.: Image Processing, Analysis and Machine Vision. Chapman & Hall, University Press, Cambridge (1993)
2. Russ, J., C.: The Image Processing Handbook (third edition). CRC Press LLC (1999)
3. Sanchez, E., Tomassini, M. (Eds.): Towards Evolvable Hardware: The Evolutionary Engineering Approach. LNCS 1062, Springer-Verlag, Berlin (1996)
4. Tørresen, J.: Possibilities and Limitations of Applying Evolvable Hardware to Real-World Applications. In: Proc. of the Field Programmable Logic and Applications FPL2000, LNCS 1896, Springer-Verlag, Berlin (2000) 230–239

5. Miller, J., Thomson, P.: Cartesian Genetic Programming. In: Proc. of the Genetic Programming European Conference EuroGP 2000, LNCS 1802, Springer-Verlag, Berlin (2000) 121–132

6. Miller, J., Job, D., Vassilev, V.: Principles in the Evolutionary Design of Digital Circuits – Part I. In: Genetic Programming and Evolvable Machines, Vol. 1(1), Kluwer Academic Publisher (2000) 8–35

7. Murakawa, M. et al.: Evolvable Hardware at Function Level. In: Proc. of the Parallel Problem Solving from Nature PPSN IV, LNCS 1141, Springer-Verlag Berlin (1996) 62–72

8. Hollingworth, G., Tyrrell, A., Smith S.: Simulation of Evolvable Hardware to Solve Low Level Image Processing Tasks. In: Proc. of the Evolutionary Image Analysis, Signal Processing and Telecommunications Workshop EvoIASP'99, LNCS 1596 Springer-Verlag, Berlin (1999) 46–58

9. Dumoulin, J. et al.: Special Purpose Image Convolution with Evolvable Hardware. In: Proc. of the EvoIASP 2000 Workshop, Real-World Applications of Evolutionary Computing, LNCS 1803, Springer-Verlag, Berlin (2000) 1–11

10. Harvey. N, Marshall, S.: GA Optimization of Spatio-Temporal Gray-Scale Soft Morphological Filters with Applications in Archive Film Restoration. In: Proc. of the Evolutionary Image Analysis, Signal Processing and Telecommunications Workshop EvoIASP'99, LNCS 1596 Springer-Verlag, Berlin (1999) 31–45

11. Miller, J.: Evolution of Digital Filters Using a Gate Array Model. In: Proc. of the Evolutionary Image Analysis, Signal Processing and Telecommunications Workshop EvoIASP'99, LNCS 1596 Springer-Verlag, Berlin (1999) 17–30

12. Tufte, G., Haddow, P.: Evolving an Adaptive Digital Filter. In: Proc of the Second NASA/DoD Workshop on Evolvable Hardware, IEEE Computer Society, Los Alamitos (2000) 143–150

13. Koza, J. et al.: Genetic Programming III : Darwinian Invention and Problem Solving. Morgan Kaufmann Publishers (1999)

14. Erba, M. et al.: An Evolutionary Approach to Automatic Generation of VHDL Code for Low-Power Digital Filters. In: Proc. of the Genetic Programming European Conference EuroGP 2001, LNCS 2038, Springer-Verlag, Berlin (2001) 36–50

15. Sekanina, L., Sllame, A.: Toward Uniform Approach to Design of Evolvable Hardware Based Systems. In: Proc. of the Field Programmable Logic And Applications FPL 2000, LNCS 1896, Springer-Verlag, Berlin (2000) 814–817

16. Sekanina, L., Růžička, R.: Design of the Special Fast Reconfigurable Chip Using Common FPGA. In: Proc. of the Design and Diagnostic of Electronic Circuits and Systems IEEE DDECS'2000, Polygrafia SAF Bratislava, Slovakia (2000) 161–168

A Dynamic Fitness Function Applied to Improve the Generalisation when Evolving a Signal Processing Hardware Architecture

Jim Torresen

Department of Informatics, University of Oslo
P.O. Box 1080 Blindern, N-0316 Oslo, Norway
jimtoer@ifi.uio.no
http://www.ifi.uio.no/~jimtoer

Abstract. Evolvable Hardware (EHW) has been proposed as a new method for designing electronic circuits. In this paper it is applied for evolving a prosthetic hand controller. The novel controller architecture is based on digital logic gates. A set of new methods to incrementally evolve the system is described. This includes several different variants of the fitness function being used. By applying the proposed schemes, the generalisation of the system is improved.

1 Introduction

There are many roads into making embedded systems for signal processing. The traditional method is to be running software on a Digital Signal Processor (DSP). A new alternative method is to apply digital logic gates assembled using evolution – named Evolvable Hardware (EHW). There are many applications for signal processing systems. One - implied in this paper, is prosthetic hand control.

To enhance the lives of people who have lost a hand, prosthetic hands have existed for a long time. These are operated by the signals generated by contracting muscles – named electromyography (EMG) signals, in the remaining part of the arm [1]. Presently available systems normally provide only two motions: Open and close hand grip. The systems are based on the user adapting *himself* to a fixed controller. That is, he must train himself to issue muscular motions trigging the wanted action in the prosthetic hand. Long time is often required for rehabilitation.

By using EHW it is possible to make the *controller* itself adapt to each disabled person. The controller is constructed as a pattern classification hardware which maps input patterns to desired actions of the prosthetic hand. Adaptable controllers have been proposed based on neural networks [2]. These require a floating point CPU or a neural network chip. EHW based controllers, on the other hand, use a few layers of digital logic gates for the processing. Thus, a more compact implementation – by an ASIC (Application Specific Integrated Circuit), can be provided making it more feasible to be installed inside a prosthetic hand.

S. Cagnoni et al. (Eds.): EvoWorkshops 2002, LNCS 2279, pp. 267–279, 2002.

Experiments based the EHW approach have already been undertaken by Kajitani et al [3]. The research on adaptable controllers is based on designing a controller providing six different motions in three different degrees of freedom. Such a complex controller could probably only be designed by *adapting* the controller to each dedicated user. It consists of AND gates succeeded by OR gates (Programmable Logic Array). The latter gates are the outputs of the controller, and the controller is evolved as one complete circuit. The simulation indicates a similar performance as artificial neural network but since the EHW controller requires a much smaller hardware it is to be preferred. The EHW architecture applied in this paper is an extension of this controller.

One of the main problems in evolving hardware systems seems to be the limitation in the chromosome string length [4,5]. A long string is normally required for representing a complex system. However, a larger number of generations is required by genetic algorithms (GA) as the string increases. This often makes the search space becoming too large. Thus, work has been undertaken to try to diminish this limitation. Various experiments on speeding up the GA computation have been undertaken – see [6].

Incremental evolution for EHW was first introduced in [7] for a character recognition system. The approach is a divide-and-conquer on the evolution of the EHW system, and thus, named *increased complexity evolution*. It consists of a division of the *problem* domain together with incremental evolution of the hardware system. Evolution is first undertaken individually on a set of basic units. The evolved units are the building blocks used in further evolution of a larger and more complex system. The benefits of applying this scheme seem to be general and include a *simpler* as well as *smaller* search space (fitness landscape) compared to when conducting evolution in one single run. The goal is to develop a scheme that could evolve systems for complex real-world applications.

In this paper, it is applied to evolve a prosthetic hand controller circuit. A new EHW architecture as well as how incremental evolution is applied are described. Further, new dynamic fitness functions are introduced. These should improve the generalization performance of gate level EHW and make it a strong alternative to artificial neural networks.

The next two sections introduce the concepts of the evolvable hardware based prosthetic hand controller. Results are given in Section 4 with conclusions in Section 5.

2   Prosthetic Hand Control

The research on adaptable controllers presented in this paper provides control of six different motions in three different degrees of freedom: Open and Close hand, Extension and Flection of wrist, Pronation and Supination of wrist. The data set consists of the same motions as used in earlier work [3], and it has been collected by Dr. Kajitani at National Institute of Advanced Industrial Science and Technology (AIST) in Japan. The classification of the different motions could be undertaken by:

- **Frequency domain:** The EMG input is converted by Fast Fourier Transform (FFT) into a frequency spectrum.
- **Time domain:** The absolute value of the EMG signal is integrated for a certain time.

The latter scheme is used since the amount of computation and information are less than in the former.

The published results on adaptive controllers are usually based on data for non-disabled persons. Since you may observe the hand motions, a good training set can be generated. For the disabled person this is not possible since there is no hand observe. The person would have to by himself distinguish the different motions. Thus, it would be a harder task to get a high performance for such a training set but it will indicate the expected response to be obtainable by the prosthesis user. This kind of training set is applied in this paper. Some of the initial results using this data set can be found in [8].

2.1   Data Set

The collection of the EMG signals are undertaken using three sensors for each channel. The difference in signal amplitude between the two of them, together with using the third as a reference, gave the resulting EMG signal. The absolute value of the EMG signal is integrated for 1 s and the resulting value is coded by *four* bits. To improve the performance of the controller it is beneficial to be using several channels. In these experiments *four* channels were used in total, giving an input vector of 4 x 4 = 16 bits. A subset of the training set input – consisting of preprocessed EMG signals, is given in Fig. 1. For each motion, 10 samples are included.

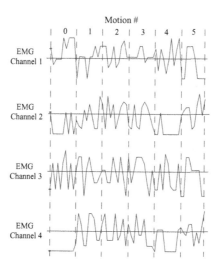

Fig. 1. EMG signals from the training set. 10 samples of data for each motion.

The *output* vector consists of one binary output for each hand motion, and therefore, the output vector is coded by *six* bits. For each vector only *one* bit is "1". Thus, the data set is collected from a disabled person by considering one motion at a time in the following way:

1. The person contracts muscles corresponding to one of the six motions. A personal computer (PC) samples the EMG signals.
2. The key corresponding to the motion is entered on the keyboard.

For each of the six possible motions, a total of 50 data vectors are collected, resulting in a total of: 6 x 50 = 300 vectors. Further, *two* such sets were made, one to be used for evolution (training) and the others to be used as a separate test set for evaluating the best circuit *after* evolution is finished.

3   An Architecture for Incremental Evolution

In this section, the proposed architecture for the controller is described. This includes the algorithm for undertaking the incremental evolution. This is all based on the principle of *increased complexity evolution.*

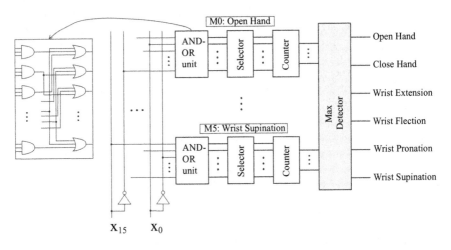

Fig. 2. The digital gate based architecture of the prosthetic hand controller.

The architecture is illustrated in Fig. 2. It consists of one subsystem for *each* of the six prosthetic motions. In each subsystem, the binary inputs $x_0 \ldots x_{15}$ are processed by a number of different units, starting by the AND-OR unit. This is a layer of AND gates followed by a layer of OR gates. Each gate has the same number of inputs, and the number can be selected to be two, three or four. The outputs of the OR gates are routed to the Selector. This unit selects *which* of these outputs that are to be counted by the succeeding counter. That is, for each new input, the Counter is counting the number of *selected* outputs being "1" from the corresponding AND-OR unit. Finally, the Max Detector outputs which counter – corresponding to *one* specific motion, is having the largest value. Each output from the Max Detector is connected to the corresponding motor in the prosthesis. If the Counter having the *largest* value corresponds to the correct hand motion, the input has been correctly classified. One of the motivations for introducing the selectors is to be able to adjust the *number* of outputs from each AND-OR unit in a flexible way. A scheme, based on using multi-input AND gates together with counters, has been proposed earlier [9]. However, the architecture used in this paper is distinguished by including OR-gates, together with the selector units involving incremental evolution. The incremental evolution of this system can be described by the following steps:

1. **Step 1 evolution.** Evolve the AND-OR unit for each subsystem *separately* one at a time. Apply *all* vectors in the training set for the evolution of each subsystem. There are no interaction among the subsystems at this step, and the fitness is measured on the output of the AND-OR units.
2. **Step 2 evolution.** Assemble the six AND-OR units into one system as seen in Fig. 2. The AND-OR units are now fixed and the *Selectors* are to be evolved in the assembled system – in one common run. The fitness is measured using the same training set as in step 1 but the evaluation is now on the output of the Max Detector.
3. The system is now ready to be applied in the prosthesis.

In the first step, subsystems are evolved separately, while in the second step these are evolved together. The motivation for evolving separate subsystems – instead of a single system in one operation, is that earlier work has shown that the evolution time can be substantially reduced by this approach [6,7].

The layers of AND and OR gates in one AND-OR unit consist of 32 gates each. This number has been selected to give a chromosome string of about 1000 bits which has been shown earlier to be appropriate for GA. A larger number would have been beneficial for expressing more complex Boolean functions. However, the search space for GA could easily become too large. For the step 1 evolution, each gate's *inputs* are determined by evolution. The encoding of each gate in the binary chromosome string is as follows:

Input 1 (5 bit)	Input 2 (5 bit)	(Input 3 (5 bit))	(Input 4 (5 bit))

As described in the previous section, the EMG signal input consists of 16 bits. Inverted versions of these are made available on the inputs as well, making up a total of 32 input lines to the gate array. The evolution is based on gate level building blocks. However, since several output bits are used to represent one motion, the signal resolution becomes increased from the two binary levels.

For the step 2 evolution, each line in each selector is represented by *one* bit in the chromosome. This makes a chromosome of 32 x 6 bits= 192 bits. If a bit is "0", the corresponding line should *not* be input to the counter, whereas if the bit "1", the line *should* be input.

3.1   Fitness Measure

In step 1 evolution, the fitness is measured on all the 32 outputs of each AND-OR unit. As an alternative experiment, we would like to measure the *fitness* on a limited number (16 is here used as an example) of the outputs. That is, each AND-OR unit still has 32 outputs but only 16 are included in the computation of the fitness function, see Fig. 3. The 16 outputs not used are included in the chromosome and have *random* values. That is, their values do not affect the fitness of the circuit. After evolution, all 32 outputs are applied for computing the performance. Since 16 OR gates are used for fitness computation, the "fitness measure" equals 16.

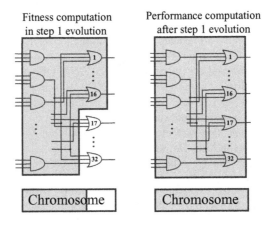

Fig. 3. A "fitness measure" equal to 16.

This could be an interesting approach to improve the generalisation of the circuit. Only the OR gates in the AND-OR unit are "floating" during the evolution since all AND gates may be inputs to the 16 OR gates used by the fitness function. The 16 "floating" OR-gates then provide additional combination of these *trained* AND gates.

3.2   Original Fitness Function

The fitness function is important for the performance of GA in evolving circuits. For the step 1 evolution, the fitness function – applied for each AND-OR unit separately, is as follows for the motion m ($m \in [0, 5]$) unit:

$$F_1(m) = \frac{1}{s} \sum_{j=0}^{50m-1} \sum_{i=1}^{O} x + \sum_{j=50m}^{50m+49} \sum_{i=1}^{O} x + \frac{1}{s} \sum_{j=50m+50}^{P-1} \sum_{i=1}^{O} x$$

$$\text{where } x = \begin{cases} 0 \text{ if } y_{i,j} \neq d_{m,j} \\ 1 \text{ if } y_{i,j} = d_{m,j} \end{cases}$$

where $y_{i,j}$ in the computed output of OR gate i and $d_{m,j}$ is the corresponding target value of the training vector j. P is the total number of vectors in the training set ($P = 300$). As mentioned earlier, each subsystem is trained for one motion (the middle expression of F_1). This includes outputting "0" for input vectors for other motions (the first and last expressions of F_1).

The s is a scaling factor to implicit emphasize on the vectors for the motion the given subsystem is assigned to detect. An appropriate value ($s = 4$) was found after some initial experiments. The O is the number of outputs included in the fitness function and is either 16 or 32 in the following experiments (referred to as "fitness measure" in the previous section).

The fitness function for the step 2 evolution is applied on the complete system and is given as follows:

$$F_2 = \sum_{j=0}^{P-1} x \quad \text{where } x = \begin{cases} 1 \text{ if } d_{m,j} = 1 \text{ and } m = i \text{ for which } \max_{i=0}^{5}(Counter_i) \\ 0 \text{ else} \end{cases}$$

This fitness function counts the number of training vectors for which the target output[1] being "1" equals the id of the counter having the maximum output.

3.3   Variable Fitness Function in Step 1 Evolution

Instead of measuring the fitness on a fixed number of output gates in step 1 evolution, the number can be varied throughout the evolution. As depicted in Fig. 4, it is here proposed to increase the number of outputs included in the fitness function as the evolution passes on.

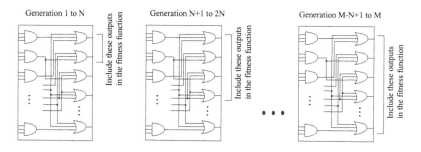

Fig. 4. Variable fitness function throughout the step 1 evolution.

The number of outputs applied by the fitness function is kept constant for N generations, before the number of outputs is doubled. This continues for a total of M generations. The benefit of this approach is that evolution is concentrated on a smaller number of outputs in the beginning of the evolution. In a way, it is a

---
[1] As mentioned earlier only one output bit is "1" for each training vector.

variant of the *increased complexity evolution*. The performance of this approach is reported in the results section.

A variant of this approach is by introducing one *more* step of evolution. After finishing the original step 1 evolution with a fitness measure equal to 16 as described in Section 3.1, the AND-gates and the OR-gates covered by the fitness function F_1 could be fixed, and a new step of evolution could evolve the 16 "floating" OR gates. This is followed by the already described step 2 evolution of the selectors. No experiments have been undertaken by this approach.

These methods could increase the performance, but not necessarily, since the step 2 evolution already removes the bad performing OR gates. Further, this would reduce the generalisation improvement described in Section 3.1.

3.4   Modified Evolution of Selector Bits

Instead of evolving *all* the selector bits in step 2 evolution, it is here proposed a scheme where only a *limited* number of them is evolved. Some of the output lines from each AND-OR unit that were floating in step 1 evolution are now kept permanently connected and is not evolved in step 2 evolution. This enhances the generalisation effect of these gates in the complete system. In this way, the test set performance could be improved since overfitting on the training set could be reduced. The chromosomes for the two systems reported in the results section – consisting of fixing 8 selectors at a time, are illustrated in Fig. 5. The dark areas in the figure indicate the part of the chromosome which is ignored during fitness computation. This is due to these selectors being permanently on.

Fig. 5. Step 2 evolution where some of the selector lines are not evolved.

Earlier experiments based on a fitness measure less than 32 did not result in a test set improvement in average (however, the best runs indead leaded to improvement) [8]. This was when all selector lines were evolved in step 2. Due to the possible overfitting, the number of generations had to be kept small. If not all the selector bits are evolved, this problem is reduced. Thus, the step 2 evolution can be continued for a larger number of generations without the same risk of overfitting on the training data.

3.5   The GA Simulation

Various experiments were undertaken to find appropriate GA parameters. The ones that seemed to give the best results were selected and fixed for all the experiments. This was necessary due to the large number of experiments that would have been required if GA parameters should be able vary through all the experiments. The preliminary experiments indicated that the parameter setting was not a major critical issue.

The simple GA style – given by Goldberg [10], was applied for the evolution with a population size of 50. For each new generation an entirely new population of individuals is generated. Elitism is used, thus, the best individuals from each generation are carried over to the next generation. The (single point) crossover rate is 0.8, thus the cloning rate is 0.2. Roulette wheel selection scheme is applied. The mutation rate – the probability of bit inversion for each bit in the binary chromosome string, is 0.01. For some of the following experiments, other parameters have been used, but these are then mentioned in the text.

The proposed architecture fits into most FPGAs. The evolution is undertaken off-line using software simulation. However, since no feed-back connections are used and the number of gates between the input and output is limited, the real performance should equal the simulation. Any spikes could be removed using registers in the circuit.

For each experiment presented in the Section 4, four different runs of GA were performed. Thus, *each* of the four resulting circuits from step 1 evolution is taken to step 2 evolution and evolved for four runs.

4   Results

This section reports the experiments undertaken to search for an optimal configuration of the prosthetic hand controller. They will be targeted at obtaining the best possible performance for the *test* set.

Table 1. The results of evolving the prosthetic hand controller in different ways.

Type of system	# inp/gate	Step 1 evolution			Step 1+2 evolution		
		Min	Max	Avr	Min	Max	Avr
A: Fitness measure 16 (train)	3	63.7	69.7	65.5	71.33	76.33	73.1
A: Fitness measure 16 (test)	3	50.3	60.7	55.7	44	67	55.1
B: Fitness measure 32 (train)	3	51	57.7	53.4	70	76	72.9
B: Fitness measure 32 (test)	3	40	46.7	44.4	45	54.3	50.1
C: Direct evolution (train)	4	56.7	63.3	59.3	-	-	-
C: Direct evolution (test)	4	32.7	43.7	36.6	-	-	-

Table 1 shows the main initial results – in percentage correct classification [8]. These experiments are based on applying the original fitness function defined in Section 3.2. Several different ways of evolving the controller are included. The training set and test set performances are listed on separate lines in the table. Each gate in the AND-OR unit has three or four inputs. The columns beneath "Step 1 evolution" report the performance after only the *first* step of evolution. That is, each subsystem is evolved separately, and afterwards they become assembled to compute their total performance. The "Step 1+2 evolution" columns show the performance when the *selector units* have been evolved too (step 2 of evolution). In average, there is an improvement in the performance for the latter. Thus, the proposed *increased complexity evolution* give rise to improved performances.

In total, the best way of evolving the controller is the one listed first in the table. The circuit evolved with the best *test set* performance obtained 67% correct classification. The circuit had a 60.7% test set performance after step 1 evolution[2]. Thus, the step 2 evolution provides a substantial increase up to 67%. Other circuits didn't perform that well, but the important issue is that it has been shown that the proposed architecture provides the *potential* for achieving high degree of generalization.

A feed-forward neural network was trained and tested with the same data sets. The network consisted of (two weight layers with) 16 inputs, 40 hidden units and 6 outputs. In the best case, a test set performance of 58.8% correct classification was obtained. The training set performance was 88%. Thus, a higher training set performance but a lower test set performance than for the best EHW circuit. This shows that the EHW architecture holds good generalisation properties.

The experiment B is the same as A except that in B all 32 outputs of each AND-OR unit are used to compute the fitness function in the step 1 evolution. In A, each AND-OR unit also has 32 outputs but only 16 are included in the computation of the fitness function as described in Section 3.1. The performance of A in the table for the step 1 evolution is computed by using *all* the 32 outputs. Thus, over 10% better training set as well as the test set performance (in average) are obtained by having 16 outputs "floating" rather than measuring their fitness during the evolution.

Each subsystem is evolved for 10,000 generations each, whereas the step 2 evolution was applied for 100 generation. These numbers were selected after a number of experiments. The circuits evolved with direct evolution (E) were undertaken for 100,000 generations[3]. The training set performance is impressive when thinking of the simple circuit used. Each motion is controlled by a *single* four input OR gate. However, the test set performance is very much lower than what is achieved by the other approaches.

4.1   Variable Fitness Function in Step 1 Evolution

Table 2 includes the results when the fitness function is changed throughout step 1 evolution. Two and four fitness functions are used which correspond to changing the fitness function one and three times during the evolution, respectively. Except for the number of generations, the GA parameters are the same as for the earlier conducted step 1 evolution experiments. Both the training set

---

[2] Evaluated with all 32 outputs of the subsystems.

[3] This is more than six times 10,000 which were used in the other experiments.

as well as the test set performance are improved – for all variants of the fitness function, compared to applying all output lines for fitness computation throughout the evolution. This is seen by comparing Table 2 to B in Table 1. In the first two lines of Table 2, the evolution is run for the same number (10000) of generations as B in Table 1. However, the performance is not better than that obtained in A.

Table 2. The results of evolving the prosthetic hand controller with a fitness function changing throughout the evolution.

Type of system	Total # generations	Step 1 evolution			Step 1+2 evolution		
		Min	Max	Avr	Min	Max	Avr
Two fitness functions (train)	10k	50.3	58.7	55.2	69	72	70.6
Two fitness functions (test)	10k	49.7	54.7	51.6	49	56.3	53.0
Two fitness functions (train)	20k	51.7	61.7	56.0	70	78.3	73.9
Two fitness functions (test)	20k	45.7	55.3	50.6	47	56.3	52.8
Four fitness functions (train)	10k	57	62.3	60.2	67.7	71.3	69.5
Four fitness functions (test)	10k	50.3	57.3	54.5	47.3	56	53.5
Four fitness functions (train)	40k	55.3	64.7	60.9	69.3	77	71.9
Four fitness functions (test)	40k	50	59	55	49.7	59.7	53.4

The average test set performance is very high for the step 1 evolution when *four* fitness functions are applied. It is about 10% higher than B in Table 1. However, the test set performance is reduced after step 2 evolution. This could be explained by overfitting on the training set. Further, the total number of generations (10k versus 20k and 40k) do not very much influence the performance.

4.2   Modified Evolution of Selector Bits

The first experiment is based on re-running the step 2 evolution of A, but now with fixing 8 selector bits at a time, and evolving for 500 generations. As seen in Fig. 6 the performance is improved – compared to the original scheme, for all runs when selectors 25-32 were fixed. However, when fixing selectors 17-24 it was less. Each column is the average of the four step 2 runs. Thus, the *Run #* on the x axis is the number of the four different circuits evolved in step 1 evolution.

Having selectors 25-32 fixed give an average test set performance of 56.4%. This is slightly better than the best earlier result (55.7%). Thus, in average there is not a large improvement or decrease in performance. This indicate first, that the scheme evolving *all* the selector bits (the original scheme) do not suffer much from overfitting the training data. Second, depending on the random values of the "floating" OR gates, there is a potential of improvement by not evolving all of the selectors.

Evolving for 100 generations provided in average about the same performance as the original scheme. The original scheme, which was evolved only for 100 generations, did not benefit from being evolved for *more* than 100 generations. In another experiments *four* selectors were fixed. However, these experiments gave less performance.

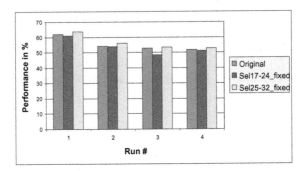

Fig. 6. Results for the test set when fixing 8 selectors permanently on during step 2 evolution (500 generation evolution).

5   Conclusions

In this paper, an EHW architecture for pattern classification – including incremental evolution, has been introduced. Several different fitness functions have been applied. The results indicate that the proposed schemes are able to improve the generalization performance. Thus, this way of evolving a signal processing hardware is promising. Applying other real-world applications is a topic for future work.

References

1. R.N. Scott and P.A. Parker. Myoelectric prostheses: State of the art. *J. Med. Eng. Technol.*, 12:143–151, 1988.
2. S. Fuji. Development of prosthetic hand using adaptable control method for human characteristics. In *Proc. of Fifth International Conference on Intelligent Autonomous Systems.*, pages 360–367, 1998.
3. I. Kajitani and other. An evolvable hardware chip and its application as a multi-function prosthetic hand controller. In *Proc. of 16th National Conference on Artificial Intelligence (AAAI-99)*, 1999.
4. W-P. Lee et al. Learning complex robot behaviours by evolutionary computing with task decomposition. In Andreas Brink and John Demiris, editors, *Learning Robots: Proc. of 6th European Workshop, EWLR-6 Brighton*. Springer, 1997.
5. X. Yao and T. Higuchi. Promises and challenges of evolvable hardware. In T. Higuchi et al., editors, *Evolvable Systems: From Biology to Hardware. First Int. Conf., ICES 96*. Springer-Verlag, 1997. Lecture Notes in Computer Science, vol. 1259.
6. J. Torresen. Scalable evolvable hardware applied to road image recognition. In *Proc. of the 2nd NASA/DoD Workshop on Evolvable Hardware*. Silicon Valley, USA, July 2000.
7. J. Torresen. A divide-and-conquer approach to evolvable hardware. In *Evolvable Systems: From Biology to Hardware. Second Int. Conf., ICES 98*, pages 57–65. Springer-Verlag, 1998. Lecture Notes in Computer Science, vol. 1478.

8. J. Torresen. Two-step incremental evolution of a digital logic gate based prosthetic hand controller. In *Evolvable Systems: From Biology to Hardware. Fourth Int. Conf., ICES'01)*. Springer-Verlag, 2001. Lecture Notes in Computer Science, vol. 2210.

9. M. Yasunaga et al. Genetic algorithm-based design methodology for pattern recognition hardware. In J. Miller et al., editors, *Evolvable Systems: From Biology to Hardware. Third Int. Conf., ICES 2000*. Springer-Verlag, 2000. Lecture Notes in Computer Science, vol. 1801.

10. D. Goldberg. *Genetic Algorithms in search, optimization, and machine learning*. Addison Wesley, 1989.

Efficiently Computable Fitness Functions
for Binary Image Evolution

Róbert Ványi[1,2]

[1] Department of Informatics, University of Szeged
Árpád tér 2., H-6720 Szeged, Hungary
vanyi@inf.u-szeged.hu
[2] Department of Programming Languages, Friedrich-Alexander University
Martensstraße 3., D-91058 Erlangen, Germany

Abstract. There are applications where a binary image is given and
a shape is to be reconstructed from it with some kind of evolutionary
algorithms. A solution for this problem usually highly depends on the
fitness function. On the one hand fitness function influences the convergence speed of the EA. On the other hand, fitness computation is done
many times, therefore the fitness computation itself has to be reasonably
fast. This paper tries to define what "reasonably fast" means, by giving
a definition for the efficiency. A definition alone is however not enough,
therefore several fitness functions and function classes are defined, and
their efficiencies are examined.[1]

1 Introduction

It is very difficult to estimate the number of the possible image processing problems. Some of these problems have fast and easy solutions, however in many
cases one has to use some heuristic method, like evolutionary algorithms [1].
EAs have been used in a wide range of applications from hand tracking [2] or
edge detection [3] to road identification in images [4].

An interesting topic is the reconstruction of a structure, like a Lindenmayer
system [5][6], or an iterated function system [7][8] from a binary image. A solution
of this problem can be used as a structural description [9]. A practical application
is for example the structural description of blood vessels [10].

Just like in other problems solved with EA, the fitness function has to help
the evolutionary process to achieve a greater convergence speed, and to avoid
the local minima as well. The computation of the fitness should also be fast,
because it has to be done many times. Therefore the efficiency of the fitness
functions has to be examined in detail. For this purpose it is useful to use a bit
theoretical approach to get a clear view, and to see the possible bottle-necks.

After giving some common notations, the formal basics of the fitness functions are discussed. This includes the definition of the efficiency and definitions

---

[1] This work is supported by the grants of the Deutscher Akademischer Austauschdienst
(DAAD) and Siemens AG

S. Cagnoni et al. (Eds.): EvoWorkshops 2002, LNCS 2279, pp. 280–291, 2002.
© Springer-Verlag Berlin Heidelberg 2002

of several function classes and functions, together with some properties. In Section 3 some practical computation methods for efficient fitnesses are described and outlined with an algorithm. To predict the behavior of the various fitness functions, some tests are done in Section 4. Finally some conclusions are mentioned in Section 5.

1.1   Digital Images

A digital image is a function $I : \mathbb{Z}_n \times \mathbb{Z}_m \to \mathbb{Z}_k$, where $n, m, z \in \mathbb{Z}^+$. The domain $C_{n,m} = \mathbb{Z}_n \times \mathbb{Z}_m$ will be referred to as a container of size $n \times m$. An element (x, y) of $C_{n,m}$ is called *pixel*. $I(x, y)$ is called the *value or color of the pixel* (x, y). Pixels with value 0 are the *background pixels*. The set of the pixels that are not background pixels is denoted by $|I|$. The positive integer k is called the *color depth* of the image. If $k = 2$, the image is called *bi-level* or *binary image*. In this case $|I|$ is the set of the *foreground pixels*. The set of digital images with color depth k over a container $C_{n,m}$ is denoted by $\mathcal{I}_{n,m}^k$. The set of all images with color depth k is $\mathcal{I}^k$. It can be assumed for any image I that $I(x, y) = 0$, whenever $(x, y) \notin C$. Therefore any two images I and J defined over different containers can be extended on to a common container $C_{I,J}^\top$.

2   Formal Basics of Fitness Functions

Definition 1 (Fitness function). *A fitness function for digital images is a function $\Phi^{k,l} : \mathcal{I}^k \times \mathcal{I}^l \to \mathbb{R}$, where $k, l \in \mathbb{Z}^+$ are positive integers. That is Φ orders a real value to a pair of images with color depths at most k and l respectively.*

One of the images is the destination image, the other image is generated by the evolutionary algorithm. Therefore they are denoted by D and G respectively. If these two images are defined over different containers, they can be extended on to $C_{D,G}^\top$. Thus it can be assumed that D and G are defined over the same container C. When the color depths are unambiguous the index is omitted, and only Φ is used. When it is also clear from the context, the word *"fitness"* can have the meaning of either fitness function (Φ) or fitness value ($\Phi(D, G)$). One possibility to define a fitness function is to compare images pixel by pixel. Therefore it is useful to suppose that the fitness function is in an additive form, which is a sum of the *pixel fitnesses* as follows.

$$\Phi^{k,l}(D, G) = \sum_{(x,y) \in C_{D,G}^\top} \varphi^{k,l}(D, G, x, y)$$

This form can be assumed without loss of generality. One can also speak about multiplicative and marginal forms, when $\Phi(D, G)$ is $\prod_{(x,y) \in C} \varphi_p(D, G, x, y)$, $\min_{(x,y) \in C} \varphi_{min}(D, G, x, y)$ or $\max_{(x,y) \in C} \varphi_{max}(D, G, x, y)$ respectively.

As mentioned in the introduction, the computation of the fitness function has to be reasonably fast. But what does it mean "reasonably fast"? It is difficult to define, but it is sure that if only a few points are in the generated image then only a few pixel comparisons should be made. This way the notion of the efficient fitness can be defined.

Definition 2 (Efficient fitness). *A fitness function is called* efficiently computable *or* efficient *for short when for any given destination image D, and any finite sequence of generated images $G_0, G_1, \ldots, G_{s-1}$ the following holds:*

$$T_\Phi(D, G_0, G_1, \ldots, G_{s-1}) \leq t_{\Phi,p}(D) + t_{\Phi,c} \sum_{i=0}^{s-1} |G_i|,$$

where $T_\Phi(D, G_0, G_1, \ldots, G_{s-1})$ is the shortest time within $\Phi(D, G_i)$ can be computed for all $i \in \{0, 1, \ldots, s-1\}$, $t_{\Phi,p} : \mathcal{I} \to \mathbb{R}$ is a function, usually the preprocessing time, and $t_{\Phi,c}$ is a constant, called the comparison coefficient *of Φ.*

Remark 1. Usually the position of each non-zero pixel of the generated image has to be calculated, therefore the calculation time cannot be better than linear. Thus it is allowed also for the fitness calculation to use linear time. Notice the difference between a fitness function and the computation of a fitness function. The latter can be a program or a procedure, and can be inefficient even if the fitness itself is efficient.

The question is how to construct efficient fitness functions. A good point to start at is to consider fitnesses, where the pixels are compared locally, that is when comparing the images pixel by pixel, only the values of the current pixel are considered.

2.1   Local Fitness Functions

Definition 3 (Local additive fitness). *A fitness function is called* local additive, *when it has the following form:*

$$\Phi(D, G) = \sum_{(x,y) \in C} \varphi(D, G, x, y) = \sum_{(x,y) \in C} f(D(x,y), G(x,y), x, y),$$

that is there exists a comparator *or* error function *such that the fitness value is the sum of the error values over all pixels in the container. Usually an error function is displacement invariant, that is independent from the coordinates. In that case the last two parameters are omitted.*

Remark 2. Similarly local multiplicative and local marginal fitness functions can also be defined, using $\prod$, min and max respectively, instead of $\sum$.

It is easy to see that not all local fitness functions are efficient, since the computation time depends on f. It can be proved however that all local fitness functions having an upper bound for the computation time of f, that is if the computation time of f is $\mathcal{O}(1)$, are efficient. In the case of a fitness function $\Phi^{k,l}$, displacement invariant comparator functions have such an upper bound, namely

$$\max_{0\le i<k,0\le j<l}\{t_f(i,j)\},$$

where $t_f(i,j)$ is the computation time of f with the parameters i and j. This means, local additive fitness functions with displacement invariant error function are effective. However, to prove the efficiency, the fitness functions have to be slightly reformulated.

Theorem 1. *Every local additive fitness function can be given in the following form:*

$$\Phi(D,G) = f'(D) + \sum_{G(x,y)\neq 0} f''(D(x,y),G(x,y),x,y),$$

where $\sum_{G(x,y)\neq 0}$ is a shorthand for $\sum_{(x,y)\in C\wedge G(x,y)\neq 0}$. Furthermore, if the computation time of f is $\mathcal{O}(1)$, then the computation time of f'' is also $\mathcal{O}(1)$.

Proof. Given a local additive fitness function Φ with an error function f.

$$\Phi(D,G) = \sum_{(x,y)\in C} f(D(x,y),G(x,y),x,y) =$$

$$= \sum_{G(x,y)=0} f(D(x,y),G(x,y),x,y) + \sum_{G(x,y)\neq 0} f(D(x,y),G(x,y),x,y)$$

Note, that instead of summing over the zero pixels, the sum over the whole image can be calculated and then the sum over the non-zero pixels can be subtracted. That is

$$\sum_{G(x,y)=0} f(D(x,y),G(x,y),x,y) = \sum_{G(x,y)=0} f(D(x,y),0,x,y) =$$

$$= \sum_{(x,y)\in C} f(D(x,y),0,x,y) - \sum_{G(x,y)\neq 0} f(D(x,y),0,x,y) \qquad (*)$$

Thus

$$\Phi(D,G) = \sum_{(x,y)\in C} f(D(x,y),0,x,y) - \sum_{G(x,y)\neq 0} f(D(x,y),0,x,y) +$$

$$+ \sum_{G(x,y)\neq 0} f(D(x,y),G(x,y),x,y) = \sum_{(x,y)\in C} f(D(x,y),0,x,y) +$$

$$+ \sum_{G(x,y)\neq 0} (f(D(x,y),G(x,y),x,y) - f(D(x,y),0,x,y))$$

Assume that $f'(I) = \sum_{(x,y)\in C} f(I(x,y),0,x,y)$ and $f''(i,j,x,y) = f(i,j,x,y) - f(i,0,x,y)$. In that case

$$\Phi(D,G) = f'(D) + \sum_{G(x,y)\neq 0} f''(D(x,y),G(x,y),x,y).$$

Furthermore if f is $\mathcal{O}(1)$, then the computation time of f'' also has an upper bound. Such a bound can be $2t_f + t_-$, where t_f is an upper bound for the computation time of f, and t_- the upper bound of the time needed for a subtraction.

Remark 3. By using multiplicative operators instead of additive operators, a similar theorem can be found for multiplicative local fitnesses, where

$$\Phi(D,G) = f'(D) \cdot \prod_{G(x,y)\neq 0} f''(D(x,y)G(x,y),x,y),$$

$f'(I) = \prod_{(x,y)\in C} f(I(x,y),0,x,y)$, and $f''(i,j,x,y) = \frac{f(i,j,x,y)}{f(i,0,x,y)}$ (assumed that $f(i,0,x,y) \neq 0$). Unfortunately this cannot be done for marginal fitnesses, since min (max) does not have an inverse operator and therefore equation (*) cannot be applied. However if

$$\min_{G(x,y)=0} (f(D(x,y),G(x,y),x,y)) \geq \min_{(x,y)\in C} (f(D(x,y),G(x,y),x,y))$$

$(\max_{G(x,y)=0}(f(D(x,y),G(x,y),x,y)) \leq \max_{(x,y)\in C}(f(D(x,y),G(x,y),x,y)))$, then the local marginal fitness has the following form:

$$\Phi(D,G) = \min_{G(x,y)\neq 0} f(D(x,y),G(x,y),x,y).$$

(Or max if Φ used max.) This condition shows whether the extrema (or at least one of them) are within the drawn pixels or not. Therefore marginal fitness functions satisfying this condition are called marginal fitness functions with *covered extremum*.

Corollary 1. *Local additive and multiplicative fitness functions and local marginal functions with covered extremum can be written in the form:*

$$\Phi(D,G) = f'(D)\sigma\Omega(f''(D(x,y),G(x,y),x,y)),$$

where σ is multiplication for multiplicative fitness, addition otherwise. For marginal fitnesses $f'(D) = 0$ and $f'' = f$, otherwise as defined previously. Ω is the $\sum, \prod$, min and max operator respectively, applied on elements of C, where $G(x,y) \neq 0$.

These types of fitness functions will be used in the following, with the usual additive notation. For easier reference, two definitions are given.

Definition 4 (Loop operator). *Given an associative binary operator $\omega : M \times M \to M$, with an identity $\varepsilon_\omega \in M$ ($\varepsilon_\omega m = m\varepsilon_\omega = m \; \forall m \in M$). That is $(M,\omega,\varepsilon_\omega)$ is a monoid. A loop operator Ω belonging to ω is defined as follows:*

$$\Omega_X(f) = \begin{cases} \varepsilon_\omega & , \text{ if } X = \emptyset \\ \omega(f(p),\Omega_{X\setminus\{p\}}) \text{ for an arbitrary } p \in X &, \text{ otherwise,} \end{cases}$$

where $X \subseteq P$ is a finite set and $f : P \to M$ is an arbitrary computable function.

Definition 5 (Simple fitness functions). *A fitness function is called* simple, *if it has the following form:*

$$\Phi(D, G) = f'(D)\sigma\Omega_{G(x,y)\neq 0}(f''(D(x,y), G(x,y), x, y)),$$

where σ is a binary operator, Ω is a loop operator belonging to a binary operator ω, and the computation times of σ, ω and f'' are $\mathcal{O}(1)$.

Theorem 2 (Efficiency of simple fitnesses). *Every simple fitness function can be computed efficiently.*

Proof. Since $f'(D)$ is independent from G, it can be precomputed before the evolution process, within a certain time. Assume this time being $t_{\Phi,p}(D)$. The computation time for $f''(D(x,y), G(x,y), x, y)$ also has also an upper bound $t_{f''}$. To compute Ω additional computation has to be done, but it is also bounded by t_ω for each element. Thus defining $t_{\Phi,c}$ as $t_{f''} + t_\omega$ the computation time yields:

$$T_\Phi(D, G_0, G_1, \ldots, G_{s-1}) \leq \sum_{i=0}^{s-1} T_\Phi(D, G_i) \leq t_{\Phi,p}(D) + \sum_{i=0}^{s-1}\sum_{G(x,y)\neq 0}(t_{f''} + t_\Omega) =$$

$$= t_{\Phi,p}(D) + \sum_{i=0}^{s-1} t_{\Phi,c}|G_i| = t_{\Phi,p}(D) + t_{\Phi,c}\sum_{i=0}^{s-1}|G_i|$$

Corollary 2. *Local additive and multiplicative fitness functions and marginal fitness functions with covered extremum, where the computation time of the error function is $\mathcal{O}(1)$, can be computed efficiently.*

2.2   Simple and General Error Functions

A usual error function is the quadratic error, where $f(i, j) = (i - j)^2$. This error function can be modified by distinguishing the different errors, and this way we also get the most general form of an error function for (binary) images: $f(i, j) = W_{i,j}$ where the $W_{i,j}$ values are predefined constants.

2.3   Multi-level Fitness

The problem with quadratic error is that it does not say anything about the structure of the destination image. If $i = 0$ and $j = 1$ we know that we have drawn a pixel to a wrong place. But how wrong? To take such a distance into consideration the fitness of a pixel can be modified in the following way: $\varphi(D, G, x, y) = h(D, x, y)$, if $D(x, y) = 0$ and $G(x, y) = 1$. Otherwise the fitness value is a predefined constant. $h(D, x, y)$ is the distance of (x, y) and the set of 1 pixels in D, which is usually the Euclidean distance, but other distance functions can also be used. For this fitness we cannot give a comparator function, since the fitness φ of a pixel (x, y) depends not only on its own values, thus it is not a local fitness function. However, it can be seen that

$$\sum_{(x,y)\in C}\varphi(D, G, x, y) = \sum_{(x,y)\in C} f(f_D(D, x, y), D(x, y), G(x, y), x, y),$$

where $f_D(D,x,y) = h(D,x,y)$, and the computation time of f is $\mathcal{O}(1)$. The only problematic part is the computation of h. Notice that h does not depend on G. Therefore it can be precomputed. The idea is to modify D and f as follows:

$$D' : C \to \mathbb{R} \times \mathbb{R}, \ D'(x,y) := [f_D(D,x,y), D(x,y)]$$
$$f'(v,g,x,y) := f(v[1], v[2], g, x, y).$$

Thus

$$\sum_{(x,y)\in C} \varphi(D,G,x,y) = \sum_{(x,y)\in C} f'(D'(x,y), G(x,y), x, y),$$

which means this fitness function can be given in a local form, by preprocessing D. Also the computation time of f' is $\mathcal{O}(1)$. This means the multi-level and any similar fitness can be computed efficiently, as stated by Lemma 1, assumed that D can be processed. This preprocessing is nothing else than writing the appropriate distances into D, similarly to the *distance maps* used in [11].

Remark 4. According to Definition 2. x and y can be arbitrary large. But it is impossible to preprocess an infinite image. Therefore D *can be preprocessed*, and converted to D' only if there exists a function D_o, and two integers x_b and y_b, such that $(x \geq x_b \wedge y \geq y_b) \Rightarrow D'(x,y) = D_o(x,y)$, and the calculation time of D_o is $\mathcal{O}(1)$. From now on it is always assumed that D can be preprocessed.

Lemma 1. *Any fitness in the form*

$$\Phi(D,G) = \sum_{(x,y)\in C} \varphi(D,G,x,y) = \sum_{(x,y)\in C} f(f_D(D,x,y), D(x,y), G(x,y), x, y),$$

can be computed efficiently, when the calculation time of f is $\mathcal{O}(1)$.

2.4   Distance Based Fitness

Using the idea from the previous fitness function a more sophisticated version can also be given, where the fitness is a function of the distance. That is $\varphi(D,G,x,y) = W(h(D,x,y))$ if $D(x,y) = 0$ and $G(x,y) = 1$.

Lemma 2. *The distance based fitness can given in the form*

$$\sum_{(x,y)\in C} \varphi(D,G,x,y) = \sum_{(x,y)\in C} f(W(h(D,x,y)), D(x,y), G(x,y), x, y),$$

Corollary 3. *The distance based fitness can be computed efficiently.*

2.5   Semi-local Fitness Functions

In the previous section it was seen that some non-local fitness functions can be converted into local functions. In this section a more general form of non-local fitness functions will be given, which can be converted into local functions.

Definition 6 (Semi-local fitness function). *A fitness function is called* local *(additive) in G, when it has the following form:*

$$\Phi(D,G) = \sum_{(x,y)\in C} \varphi(D,G,x,y) = \sum_{(x,y)\in C} f(D,G(x,y),x,y).$$

Similarly fitness functions local in D can also be defined. A fitness function is called semi-local, *when it is local in either D, or G.*

Remark 5. As previously, semi-local multiplicative and marginal fitness functions can be defined. In this paper only semi-local additive fitnesses are used.

Theorem 3. *Any fitness function $\Phi^{k,2}$ local in G can be converted into a form*

$$\Phi(D,G) = f'(D) + \sum_{G(x,y)\neq 0} D'(x,y).$$

where $D'(x,y) = f(D,1,x,y) - f(D,0,x,y)$.

Proof. Modify D using for example the Gödel-coding as follows: $D^*(x,y) := [D]$, that is code the whole image into each pixel. This means the fitness function is now in local form, so it can be modified according to Theorem 1 as follows:

$$\Phi(D,G) = \sum_{(x,y)\in C} f(D,G(x,y),x,y) = \sum_{(x,y)\in C} f(D^*(x,y),G(x,y),x,y) =$$

$$= f^*(D^*) + \sum_{G(x,y)\neq 0} f''(D^*(x,y),1,x,y),$$

where

$$f^*(D^*) = \sum_{(x,y)\in C} f(D^*(x,y),0,x,y) = \sum_{(x,y)\in C} f(D,0,x,y) =: f'(D),$$

and $f''(D^*(x,y),1,x,y) = f''(D,1,x,y) = f(D,1,x,y) - f(D,0,x,y)$. Define D' as follows: $D'(x,y) := f(D,1,x,y) - f(D,0,x,y)$. Hence:

$$\Phi(D,G) = f'(D) + \sum_{G(x,y)\neq 0} D'(x,y)$$

Corollary 4 (Efficiency of semi-local fitnesses). *Given a fitness function $\Phi^{k,2}$. If Φ is local in G then Φ can be computed efficiently.*

2.6   Hausdorff Distance

Given two point sets D and G. The Hausdorff distance is the maximum of the two half-distances $h_G(D)$ and $h_D(G)$, where $h_G(D) = \max_{d\in D} h_p(d,G)$. $h_p(d,G)$ is the distance of a point d and a point set G, defined as usual.

Unfortunately the Hausdorff distance cannot be computed efficiently, since $h_G(D)$ depends on G and cannot be precomputed, though it is semi-local in D. Whereas $h_D(G)$ is semi-local in G therefore it can be computed efficiently. One can try to use a fake Hausdorff distance, where the min and max computations are exchanged in $h_G(D)$. This fitness can be computed efficiently, but the use of this as fitness function is not recommended (see results).

3   Fast Computation of Efficient Fitnesses in Practice

In the previous sections several fitness functions were presented that are efficient. It is however important to be able to write efficient computer programs for fitness computation. Let us consider the local fitness functions. If we write a loop over all pixels, it will not be efficient. However Theorem 1 can be applied. Thus we have to use a loop over the non-zero pixels in G and compute f''.

For distance based fitnesses the destination image can be modified to contain distances instead of zeroes. In this way using Lemma 1 only a loop is needed over the non-zero pixels in G, and for each of them $W(D(x, y))$ has to be computed. The cleverest idea is to write $W(h(D, x, y))$ into D, so in the loop only some addition has to be done.

Theorem 3 gives the most general result, and thus the best way for the implementation. First D has to be modified using the formula for D' then only the pixel values in D' have to be summed, where $G(x, y) \neq 0$. The outline of this method is given by Algorithm 1. At the beginning of the evolution process, the procedure **preprocess** must be called. Later for each generated image the fitness is calculated by procedure **calculate_fitness**. This can be (should be) merged with the drawing of the generated image.

---

procedure preprocess
$f':=0$
for all (x, y) pixels of D image **do**
$\quad f1 := f(D, 1, x, y);\ f0 := f(D, 0, x, y)$
$\quad D'(x, y):=f1 - f0;\ f' := f' + f0$
end for

procedure calculate_fitness(G:Image)
fitness:$=f'$
for all (x, y) foreground pixels of G **do**
$\quad$ fitness:=fitness+$D'(x, y)$
end for

Algorithm 1. The Outline of the Fast Fitness Computation Algorithm

---

4   Convergence of Fitness Functions

Examining the efficiency of different fitness functions makes no sense, if the quadratic-error is enough for use with evolutionary algorithms. Therefore it is important to examine the convergence properties too. During this paper several fitness functions were presented. To predict their behavior is very hard, and can be approximated by doing many tests in different environments. However to get the idea how the fitness landscapes may look like, a simple test environment is introduced. A simple destination image was created and it was compared with a "generated" image that is with a displaced, rotated and scaled version of the original. This means the solutions can be represented by a four dimensional real vector. To visualize the fitness three test were made for each fitness function. In the first test the optimal solution was taken and then the parameter vector was mutated several times. Then the distance from the optimum in the four dimensional space and the fitness value were calculated. This was done with 1 upto 100 mutations, each 250 times.

As it can be seen from the results in Figure 1, using quadratic error the fitness suddenly drops after leaving the optimal solution, and then remains in approximately the same level. With distance based functions this drop is not so sudden, but also noticeable. As going further the fitness does not stay in the same level, but gets lower. The ideal case is the Hausdorff distance, where the fitness gets lower as going further without drops or jumps.

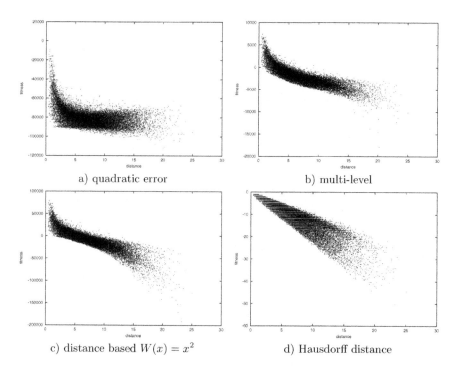

a) quadratic error   b) multi-level

c) distance based $W(x) = x^2$ d) Hausdorff distance

Fig. 1. Results of the first test

The second two tests tried to visualize the fitness landscape. In the first landscape test the image was only displaced, in the second one the image was scaled and rotated. For each possible transformations the fitness were computed, and plotted onto the z axis. The results can be seen in Figure 2. The x and y axes present the displacement in the x and y direction in the first image, and the rotation angle, scaling rate in the second one.

5 Conclusion

In this paper several fitness function classes were introduced, and were examined with respect to the efficiency. It was effectively proved that in many cases local fitness functions and also functions local in G can be computed efficiently. Some tests were also made to estimate the convergence properties. As it can be predicted (and was seen already in experiments, not mentioned in this paper)

quadratic
error

multi-
level
fitness

distance-
based
fitness
with
$W(x)=x^2$

fake
Hausdorff
distance

Hausdorff
distance

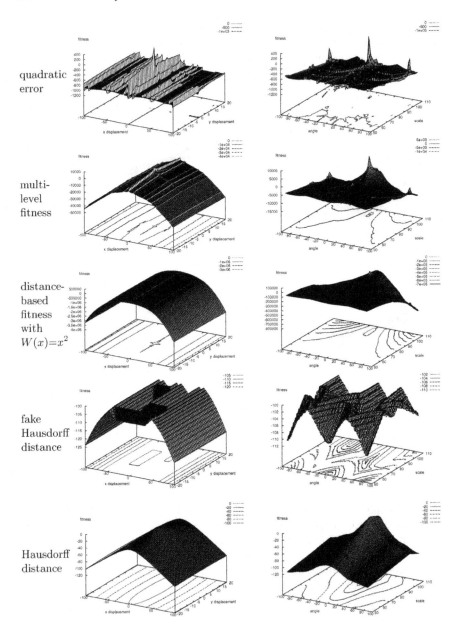

Fig. 2. Fitness landscapes

the distance based functions are much better then the simplest quadratic error, without the need of extra computation time. Using the results of this paper other, more complex fitness functions can also be defined, *without risking any loss of computation speed.*

References

1. Riccardo Poli. Genetic programming for feature detection and image segmentation. In Terence C. Fogarty, editor, *Evolutionary Computing*, number 1143, pages 110–125. Springer-Verlag, University of Sussex, UK, 1-2 1996.
2. Tony Heap and Ferdinando Samaria. Real-time hand tracking and gesture recognition using smart snakes. In *Proceedings of Interface to Human and Virtual Worlds*, Montpellier, France, June 1995.
3. Christopher Harris and Bernard Buxton. Evolving edge detectors with genetic programming. In John R. Koza, David E. Goldberg, David B. Fogel, and Rick L. Riolo, editors, *Genetic Programming 1996: Proceedings of the First Annual Conference*, pages 309–315, Stanford University, CA, USA, 28–31 1996. MIT Press.
4. Julian Eugene Boggess. A genetic algorithm approach to identifying roads in satellite images. In *Proceedings of the Ninth Florida Artificial Intelligence Research Symposium, FLAIRS-96*, pages 142–146, Eckerd Coll., St. Petersburg, FL, USA, 20-22 May 1996.
5. Christian Jacob. Genetic L-system programming. In *Proceedings of the Parallel Problem Solving from Nature, International Conference on Evolutionary Computation, PPSN III*, volume 866 of *Lecture Notes of Computer Science*, pages 334–343. Springer-Verlag, Berlin, 1994.
6. John R. Koza. Discovery of rewrite rules in Lindenmayer systems and state transition rules in cellular automata via genetic programming. In *Symposium on Pattern Formation (SPF-93)*, Claremont, California, USA, 1993.
7. Pierre Collet, Evelyne Lutton, Frédéric Raynal, and Marc Schoenauer. Polar IFS + parisian genetic programming = efficient IFS inverse problem solving. *Genetic Programming and Evolvable Machines*, 1(4):339–362, October 2000.
8. John C. Hart. Fractal image compression and the inverse problem of recurrent iterated function systems. *IEEE Computer Graphics and Applications*, 16(4), 1996.
9. Róbert Ványi and Gabriella Kókai. Giving structural descriptions of tree-like objects from binary images using genetic programming. In Lee Spector, Erik D. Goodman, Annie Wu, William B. Langdon, Hans-Michael Voigt, Mitsuo Gen, Sandip Sen, Marco Dorigo, Shahram Pezeshk, Max H. Garzon, and Edmund Burke, editors, *Proceedings of the Genetic and Evolutionary Computation Conference (GECCO-2001)*, pages 163–172, San Francisco, California, USA, 7-11 July 2001. Morgan Kaufmann.
10. Róbert Ványi, Gabriella Kókai, Zoltán Tóth, and Tünde Pető. Grammatical retina description with enhanced methods. In Riccardo Poli, Wolfgang Banzhaf, William B. Langdon, Julian F. Miller, Peter Nordin, and Terence C. Fogarty, editors, *Genetic Programming, Proceedings of EuroGP'2000*, volume 1802 of *LNCS*, pages 193–208, Edinburgh, 15-16 April 2000. Springer-Verlag.
11. Evelyne Lutton, Jacques Levy-Vehel, Guillaume Cretin, Philippe Glevarec, and Cidric Roll. Mixed IFS: Resolution of the inverse problem using genetic programming. Technical Report No. 2631, INRIA, August 1995.

Evolutionary Based Autocalibration from the Fundamental Matrix

Anthony Whitehead[1,*] and Gerhard Roth[2]

[1]School of Computer Science, Carleton University, Ottawa, Canada
awhitehe@scs.carleton.ca

[2]National Research Council of Canada, Ottawa, Canada K1J 6H3
Gerhard.Roth@nrc.ca

Abstract. We describe a new method of achieving autocalibration that uses a stochastic optimization approach taken from the field of evolutionary computing and we perform a number of experiments on standardized data sets that show the effectiveness of the approach. The basic assumption of this method is that the internal (intrinsic) camera parameters remain constant throughout the image sequence, i.e. they are taken from the same camera without varying the focal length. We show that for the autocalibration of focal length and aspect ratio, the evolutionary method achieves comparable results without the implementation complexity of other methods. Autocalibrating from the fundamental matrix is simply transformed into a global minimization problem utilizing a cost function based on the properties of the fundamental matrix and the essential matrix.

1 Introduction

Advances in the field of projective vision make it possible to compute various quantities from an uncalibrated image sequence: in particular the fundamental matrix between image pairs [1, 2]. Autocalibration has become popular due to these recent advances because of the desire to create 3D reconstructions from a sequence of uncalibrated images without having to rely on a formal calibration process. The standard model for an uncalibrated camera has five unknown intrinsic parameters found in a 3x3 calibration matrix K. These parameters are the focal length, aspect ratio, skew and the center of projection x and y (the principal point). The accurate estimation of these 5 parameters is the fundamental goal of autocalibration.

Autocalibration algorithms can be divided into two basic classes. In class A algorithms, we compute the calibration matrix K from the fundamental matrix (the recovered epipolar geometry) [4, 5, 6, 7, 8] and class B algorithms compute K from a projective reconstruction [9, 10, 11] of the scene. Since the projectively reconstructed frames must all be warped to a consistent relative base, Class B algorithms are computationally difficult in comparison to simply finding the fundamental matrix between image pairs. It is claimed that Class B autocalibration algorithms are superior to Class A algorithms because the Class A algorithms do not enforce the constraint that the

---

* Partially funded by Nortel Networks Scholarship.

S. Cagnoni et al. (Eds.): EvoWorkshops 2002, LNCS 2279, pp. 292–303, 2002.

plane at infinity be the same over the entire image sequence [1]. It is precisely this constraint that makes Class B algorithms computationally difficult and we show that Class A algorithms combined with the use of evolutionary systems are as accurate as their Class B counterparts.

Another concern with Class A algorithms is the existence of extra degenerate motions, these being pure rotations, pure translations, affine viewing and spherical camera motions [1, 12]. However, there exist many practical situations that do not contain these degenerate motions where autocalibration is necessary. For example, there are many photographs and video clips in existence for which there is no knowledge of the camera. In order to reconstruct some of these image sequences, autocalibration is the only means.

Autocalibration has been criticized [13] in the past because many different possible calibrations will always provide a 3D reconstruction with almost perfect Euclidean structure. In essence, the only thing we can really measure, the skews and aspect ratios, are very close to what they should be because of manufacturing accuracy. Because of this, the corresponding reconstruction will always look good i.e. the different right angles look square and the different length-ratios look correct. Commonly, the "look" of a reconstruction is used as a ground truth element, but it is clearly a weak one, and any algorithm using such a comparison as a measure of goodness is highly suspect. Because of the manufacturing accuracies, we attempt to autocalibrate only the focal length and the aspect ratio and make assumptions about the remaining parameters.

The constraining equations for the two autocalibration methods presented in this paper are based on the fundamental matrix, and are non-linear. In what follows, we will show that it is possible to reformulate the process of autocalibration as the minimization of a cost function of the calibration parameters. This type of reformulation has not been achieved for all autocalibration algorithms, specifically the class B algorithms which are thought to be superior. For example, the modulus constraint is a non-linear relationship between the camera calibration parameters and the projective camera matrices that have been used as the basis of a class B autocalibration algorithm [10]. The application of the modulus constraint produces a set of polynomial equations for every pair of images, and a system of polynomial equations for the entire image sequence. The solution of such a polynomial system is very difficult to compute, but one possibility is to find all the permutations of exact solutions in closed form and then to combine the results [8]. This is rather cumbersome, and another way to solve such a polynomial system is to use a continuation method [17]. Unfortunately, continuation methods only work well for a small number of equations, and are not suitable for the large polynomial systems that are generated by long image sequences.

In this paper, we examine two autocalibration algorithms that use fundamental matrices and an evolutionary approach to estimating the parameters; one based on Kruppa's equation [6, 4, 8], and the second based on the idea of finding the calibration matrix which optimally converts a fundamental matrix to an essential matrix [7]. In both cases the problem can be formulated as the minimization of a cost function that we describe in sections 2 and 3. The correct camera calibration is the global minimum of this cost function over the space of possible camera parameters. In the past, the claim has been that such minimization approaches to autocalibration are sensitive to the initial starting point of the gradient descent algorithm, but when computing only one parameter, the starting point is irrelevant because we can solve the associated 1D optimization problem using standard numerical approaches [14]. When there is more than one parameter, such as focal length and aspect ratio, we use a simple stochastic approach [15] from the field of evolutionary computing to overcome this problem. We show experimentally that for this type of cost function this stochastic method reliably finds the global minimum. As well, a number of experiments are performed on image sequences with known camera calibration, some of which have been described in the autocalibration literature and utilize class B algorithms. We show that the stochastic approach achieves results that are as good as the class B algorithms. The next section describes the two autocalibration methods, and the theory behind them. The third section describes the experiments, and the fourth presents the conclusions and future avenues of research to improve accuracy.

2 Autocalibration from the Fundamental Matrix

The goal of autocalibration is to compute the camera calibration matrix K. The standard linear camera calibration matrix (K), used to convert from image coordinates in pixels to world coordinates on the camera-sensing element in millimeters, has the following entries [1, 2]:

$$
K = \begin{pmatrix} fk_u & -fk_u \cot(\theta) & u_0 \\ 0 & fk_v/\sin(\theta) & v_0 \\ 0 & 0 & 1 \end{pmatrix}
\tag{1}
$$

Here f is the focal length in millimeters, and k_u and k_v are the number of pixels per millimeter for the camera. If we let α_u and α_v be fk_u and fk_v respectively by multiplying the focal length (f) in mm and the mm/pixel (k), we have the focal length in pixels. The ratio α_u / α_v is the aspect ratio and is often (but not always) one because of manufacturing, and θ is the skew angle. The skew angle θ is almost always 90 degrees, again because of manufacturing. This leaves us with four free intrinsic camera parameters α_u, α_v, u_0 and v_0. The calibration matrix K can therefore be rewritten in a much simpler form as:

$$
K = \begin{pmatrix} \alpha_u & 0 & u_0 \\ 0 & \alpha_v & v_0 \\ 0 & 0 & 1 \end{pmatrix}
\tag{2}
$$

The fundamental matrix F is a 3x3 matrix of rank 2 that defines the epipolar geometry between two images [2]. Given two corresponding points m_1 and m_2 from images I_1 and I_2, the epipolar constraint specifies:

$$
m_2 F m_1 = 0
\tag{3}
$$

The fundamental matrix can be computed from a set of 2D correspondences between the two images [16]. If we know the epipolar geometry and thus the Fundamental Matrix, it is possible to compute the intrinsic camera parameters.

2.1 Autocalibration via Equal Essential Eigenvalues

The essential matrix can be considered as the calibrated version for the fundamental matrix. Given the camera calibration matrix K and the fundamental matrix F, then the essential matrix E is related by the following equation:

$$E = K^T F K \tag{4}$$

Since F is a 3x3 matrix of rank two with the condition that there are exactly two non-zero eigenvalues, E is also of rank two. E however, has an added constraint that the two non-zero eigenvalues must be equal [2]. It is this constraint that is used to create the autocalibration algorithm [7]. The idea is to find the calibration matrix K that makes the two eigenvalues of E equal, or in the case of estimation, as close as possible. Given two non-zero eigenvalues of E, σ_1 and σ_2 where $\sigma_1 > \sigma_2$, then in the ideal case $(\sigma_1 - \sigma_2)$ should be zero. Consider the difference $(\sigma_1 - \sigma_2) / \sigma_1$, which can be written as:

$$1 - (\sigma_2/\sigma_1) \tag{5}$$

If the eigenvalues of E are equal, (5) computes to zero; as they differ, equation (5) approaches one. Clearly, (5) becomes the cost function to be minimized.

As we are dealing with a sequence of N images, we can have at most N-1 adjacent image pairs and therefore we have N-1 different fundamental matrices F_i (i=1..N-1). Based on our assumption that the same camera with invariant intrinsic parameters is used, our goal is to find K by minimizing the cumulative values of (5) for all the fundamental matrices F_i in the sequence. Assume F_i is the fundamental matrix relating image I_K and I_{K+1}. To autocalibrate over the N image sequence, we must find the K that minimizes:

$$\sum_{i-1}^{N-1} \omega_i (1 - \sigma 2 / \sigma 1) \tag{6}$$

Where ω_i is a weight factor, between zero and one, which defines the confidence of the computed fundamental matrix F_i. ω_i is defined in more detail in the next section.

2.2 Autocalibration via Kruppa's Equations

Another way to perform autocalibration from the fundamental matrix is to use Kruppa's equations [1, 2]. To understand these equations we must first define the absolute conic. In Euclidean space the absolute conic lies on the plane at infinity, and has the equation:

$$x^2 + y^2 + z^2 = 0 . \tag{7}$$

The absolute conic contains only complex points that satisfy $x^T x = 0$. If we consider a standard camera projection matrix P = K[R|-Rt]. Where R is the rotational motion of between camera positions and t is the translation component of the camera motion, thus a 3D point x on the absolute conic projects to a 2D point:

$$u = P(x) = KRx. \tag{8}$$

Thus, $x = R^T K^{-1} u$, and since $x^T x = 0$, this implies:

$$u^T K^{-1} R R^T K^{-1} u = u^T K^{-T} K^{-1} u = 0 . \tag{9}$$

This clearly shows that any 2D point u is on the image of the absolute conic if and only if it lies on the conic represented by the matrix $K^{-T} K^{-1}$. From projective geometry, KK^T is the dual absolute conic, and is labeled as C. If we can find C, then we can directly compute the camera parameters K by Cholesky factorization.

Kruppa's equations relate the fundamental matrix to the terms of the dual absolute conic. The first form of these equations required the computation of not just the fundamental matrix, but also of the two camera epipoles, which are known to be unstable [2]. Recently, a new way of relating the fundamental matrix and the dual absolute conic was described which does not require the computation of the camera epipoles [4]. Consider the singular value decomposition of a fundamental matrix F to be UDV^T. We let the column vectors of U and V be u_1, u_2, u_3 and v_1, v_2, v_3 respectively. This gives the new form of Kruppa's equation to be:

$$\frac{v_2^T C v_2}{r^2 u_1^T C u_1} = \frac{-v_2^T 2 C v_1}{s r u_1^T C u_1} = \frac{v_1^T C v_1}{s^2 u_2^T C u_2} \tag{10}$$

To autocalibrate we must find the C which makes these three ratios equal, or in the case of estimation, as close to equal as possible. We let $ratio_1$ be equal to:

$$\frac{v_2^T C v_2}{r^2 u_1^T C u_1} - \frac{-v_2^T 2 C v_1}{s r u_1^T C u_1} \tag{11}$$

And define $ratio_2$ and $ratio_3$ similarly as the other two possible permutations in the ratios. Autocalibration can then be achieved by finding the C (KK^T) that minimizes the sum of the ratios squared. Given the same image sequence that produced equation 6, the Kruppa ratios over n images minimizes:

$$\sum_{i-1}^{N-1} \omega_i (ratio_1^2 + ratio_2^2 + ratio_3^2) \tag{12}$$

Again, ω_i is a weight factor, between zero and one, which defines the confidence of the computed fundamental matrix F_i.

2.3 Evolutionary Idea

Since the two autocalibration methods based on the fundamental matrix have an associated cost function we can use a gradient descent algorithm to find the solution. The caveat here is that there are often many local minima in the cost function, so the solution that is found depends on the starting point. However, we note that the calibration parameters can all be bounded; i.e. the center of projection rarely varies from the image center, the aspect ratio is generally 1 and the skew is 90 degrees. Thus we are attempting to find the global minimum for a set of real-valued, bounded optimization parameters. This problem has been dealt with in the field of evolutionary computing. We use an approach called Dynamic Hill Climbing (DHC), that combines genetic algorithms, hill climbing and conjugate gradient methods to be an optimization algorithm that is very successful in solving such real valued optimization problems [15].

The idea is to repeatedly perform gradient descent in the search space and to restart the gradient descent in an area of the search space that is as far removed as possible from previous solutions. We call such a method Statistically Distributed Random Starting (SDRS), and in this way we cover the search space as effectively as possible as seen in Fig 1.

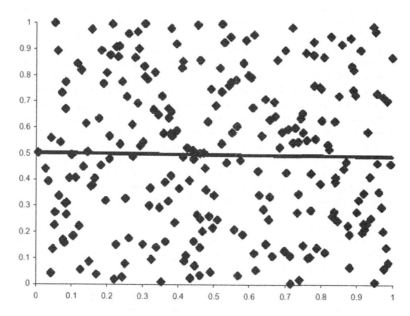

Fig. 1. : Scatter plot of 2D search space generated by SDRS. 250 points with a trend line indicating an even disbursement of start points.

The pseudo-code for SDRS:

```
SDRS()
For each optimization parameter in the search space
    Find the largest region that has not had a start point
    Compute a random point in this region
    Set this point to the start point for this dimension
Endfor
Return N-dimensional startPoint
```

SRDS allows for the most complete coverage of the search space with a user specified number of runs. This allows the DHC algorithms to successfully find the global minimum throughout the search space.

The pseudo-code for estimating K:

```
ESTIMATE_K()
For n times
    StartPoint = SRDS()
    Perform the DHC gradient descent from StartPoint.
    IF Cost function (Equal Eigenvalues OR Kruppa) is minimal
        Save this K.
    ELSE
        Discard this K
    Endfor
Return K
```

The algorithm ESTIMATE_K returns the calibration parameters in the matrix K that produced the minimum value from the cost function. Evaluating the cost function for the two different autocalibration methods is very efficient. A single gradient descent of the cost function uses the Powell optimization algorithm, which is in turn based on repeated applications of the one dimensional Brent method [14]. The equal eigenvalues approach requires only the computation of the eigenvalues of a three by three matrix, and for the Kruppa approach the computation of three ratios. In both cases the weights ω_i are set in proportion to the number of matching 2D feature points that support a given fundamental matrix. The larger the number of points that support the epipolar geometry characterized by F, the more confidence we have in that fundamental matrix, and therefore the greater the weight. We next show experimentally that the global minimum is found very reliably by this approach.

3   Experimental Results

For many autocalibration algorithms the evaluation of performance consists of a simple visual inspection of the resulting 3D reconstruction. This is not adequate because it has been shown that the quality of the final reconstruction is visually acceptable for a wide variety of calibration parameters [13]. In order to compare the capabilities of the evolutionary method, we performed a variety of experiments that compared against the results from the literature that used "look" as a goodness criteria. This allowed us to compare the evolutionary method against other algorithms, and specifically we show that it has comparable results to the more complicated Class B algorithms. A secondary measure for experimentation is comparison against ground truth, i.e. the intrinsic parameters are already known a-priori. Finally, we take several sequences taken from the same uncalibrated camera and show that the evolutionary computing based algorithm is consistent and repeatable.

The first set of experiments described in Table 1 show how the autocalibration process works when we are calibrating only the focal length. Table 1 shows the results for a number of different test sequences that have been processed in previous autocalibration papers [6, 8, 10, 19]. In particular, the castle sequence is used as a test case for comparison of the class B approach that requires a projective reconstruction [10]. We see that our autocalibration results are comparable to those of other algorithms.

Table 1: Results of autocalibration for focal length vs other algorithms. Focal length is in pixels. Correspondences are computed automatically.

Name	# of Images	Stated Focal	Computed focal len (Eq.Eigen)	% error vs. Stated	Computed focal len (Kruppa)	% error vs. Stated
Castle	27	1100	1156.50	5	1197.7	8
Valbone	9	682	605.5	11	685.71	0.5
Nekt	6	700	798.58	14	872.44	24.6
etluueshiba	5	837	857.25	2.4	1233.85	47.4

It is important to note that the stated focal lengths are those computed in the literature, and the assessment of goodness was how the reconstruction looked. In the table we compare how close our autocalibration results are to the previously published results, which we assume to be reasonably correct but cannot confirm. In the last example shown in the table [19], the error with the Kruppa autocalibration is quite large, possibly because the motion is close to being a pure translation which is known to be a degeneracy motion for the Kruppa algorithm [1,12]. It is also a good indicator of how the Equal Eigenvalues method performs well against these degenerate motions.

Finally, because the ground truth is not really known, and the methods for computing F in the literature are not available, it is possible that the stated focal lengths are incorrect.

In the set of experiments outlined in Table 2, the 2D feature points were selected by hand as part of a photogrammetric model building process. From these manually selected correspondences we compute the fundamental matrix between all image pairs in the sequence. In this experiment we know the intrinsic parameters of the camera a-priori from the projects of the photogrammetric package [20]. We therefore assume that all the intrinsic parameters are set a-priori, except for the focal length which we autocalibrate.

Table 2: Results of autocalibration for focal length for photogrammetric sequences. Focal length is in mm., and reprojection error is in pixels. Correspondences selected by hand.

Name	# of Images	True focal	Eigen focal	% error	Kruppa focal	% error	Correct reproj.	Eigen reproj.	Kruppa Reproj.
Curve	4	6.97	4.71	32.4	7.49	1.13	7	2.23	1.44
Cylinder	3	28	26.35	5.9	31.70	13.21	0.96	2.07	2.60
Plant	6	24.20	22.55	6.8	24.39	0.78	0.80	1.49	1.04
Statue	7	5.11	3.67	28.2	5.29	3.5	3.93	9.61	1.95

Table 2 shows the autocalibrated focal length in millimeters versus the true focal length, along with the percentage error for both autocalibration methods. Since we have the associated 3D reconstructions for the corresponding 2D features we can compute more sophisticated performance measures. For a given autocalibrated focal length we compute the reprojection error for all the corresponding feature points. The reprojection errors are the pixel differences between the projection of the 3D feature points into 2D and the original corresponding 2D features. We compute the median of the reprojection errors using the correct focal length, the focal length found by the eigenvalue method, and the focal length found by Kruppa's method. The median of the reprojection errors is a good indicator of the quality of the reconstruction for a given focal length. We see that the median reprojection error increases for the autocalibrated focal lengths, but only slightly. This implies that the error in the autocalibrated focal lengths would not have a significant impact in terms of reconstruction quality and this independently verifies the claims of Bougnoux [13].

In the next experiment we attempt to autocalibrate both aspect ratio and focal length using the two methods. We are again using as input a series of photogrammetric projects for which we know the 2D feature correspondences as well as the ground truth of the intrinsic camera parameters.

Table 3: Results of autocalibration for focal length and aspect ratio for photogram-metric sequences. The equal eigenvalue method is used and focal length is in mm.

Name	True aspect	Eigen Aspect	Variance	% error	True Focal	Eigen focal	Variance	% error
Curve	1.0	1.08	0.003	8	6.97	3.46	0.062	50
Cylinder	1.0	0.98	0.002	2	28	26.72	0.52	4.5
Plant	1.0	0.98	0.012	2	24.2	22.96	0.39	5.1
Dam	0.81	0.972	0.0001	20	30.75	38.52	0.089	9.8

While the results as shown in Tables 3 and 4 are reasonable, the errors when autocalibrating two camera parameters are sometimes higher than autocalibrating just one parameter. The error again compounds when we attempt to autocalibrate all parameters. In particular, the percentage error in focal length increases slightly. One possible explanation is that the gradient descent algorithm is stuck in a local minima, to verify this the results shown in these two tables were computed by averaging over one hundred separate runs of the optimization algorithm. The variance as shown in the table for the autocalibrated aspect ratio and focal length is very small over these runs and indicates that it is highly likely that the stochastic optimization algorithm is finding the correct global minimum.

Table 4: Results of autocalibration for focal length and aspect ratio for photogrammetric sequences. The Kruppa autocalibration method is used.

Name	True aspect	Kruppa Aspect	Variance	% error	True Focal (mm)	Kruppa focal (mm)	Variance	% error
Curve	1.0	0.997	0.011	1.3	6.97	7.56	0.21	8.4
Cylinder	1.0	1.03	0.0001	3	28	32.91	0.0001	17.5
Plant	1.0	0.92	0.003	8	24.2	26.33	0.12	8.8
Dam	0.81	0.997	0.0001	19.75	30.75	38.43	0.0001	24.9

The final experiment, as shown in Table 5, has as input three image sequences that were taken with the same camera with invariant intrinsic parameters.

Table 5: Results for autocalibration of focal length for three sequences

Name	# of Images	Eigen focal	Kruppa Focal
Chapel	12	27.82	31.31
Climber	13	27.91	33.88
Workshop	8	26.19	38.09

Test cases chapel and workshop are almost pure translation while the climber sequence has a motion with significant translation and rotation. We autocalibrate only the focal lengths, which should be equal for all three sequences. The variance of the computed focal length for the eigenvalue method is 0.96 mm and for Kruppa approach is 3.42mm. It is not surprising that the autocalibration results differ, since certain motions are degenerate with regards to the Kruppa based autocalibration [1]. What these results clearly show is that for a given camera, and substantially different sequences, the evolutionary algorithms (especially the equal eigenvalues method) are convergent.

In summary, the experiments show that the evolutionary approach is as good as any complicated Class B algorithm, e.g. the castle sequence in Table 1. Computationally the fundamental matrix based approaches are very efficient since a single evaluation of the cost functions does not take long. The time taken for autocalibration is in the order of seconds for all the image sequences on a 400 MHz Pentium II processor. It also becomes clear that the equal eigenvalues method is superior to the Kruppas method for degeneracy cases. There are cases, however, where the Kruppas method is clearly outperforming the equal eigenvectors method. Further investigation is necessary to determine whether or not a heuristic can be developed to choose one algorithm over the other by pre-determining the camera motion.

4   Conclusions

In theory the autocalibration methods that use fundamental matrices should not perform as well as those that use the camera projection matrices of a projective reconstruction [1, 12, 2]. However, we show that for non-degenerate motions both methods perform equally well when we are calibrating only the focal length, or the focal length and aspect ratio. Similarly in [11], the principle point was not computed accurately using the class B algorithm and was also subsequently assumed.

The equal eigenvalues approach is very simple and works just as well as any Class B method we compared against. While it is theoretically equivalent to the Kruppa approach, it performs better numerically in situations where we are close to a degenerate motion, such as pure translation. The usual class B approach to autocalibration requires the solution of a set of polynomial equations but this is not computationally feasible for long image sequences. With our evolutionary computing based approach we can process long image sequences, which is an advantage to these algorithms. The argument against the optimization-based methods has been that they are sensitive to the starting point of the optimization process [3, 5]. We have shown that the SDRS method helps to find the global minimum of the cost function reliably. In our experiments we have shown that the error in the autocalibration of the focal length is usually in the range of 5% to 15%. This is adequate for applications in which the final results are used for visualization purposes, such as model building but clearly not for applications that require exact depth information.

What may not be an obvious next step is to move forward and decrease the error is to utilize the two (and more as they become available) autocalibration routines (equal eigenvalues and Krupps equations) in yet another evolutionary step. In essence this means that we want to minimize the difference between two calibration matrices K_{KRUPPA} and K_{EIGEN}. This can be measured in a variety of ways, but clearly the

Frobineus norm measure does this exact difference assessment for us. As well, results may become more stable by performing the SDRS algorithm in a windowed manner to ensure better coverage of the search space.

Acknowledgements

Thanks to Photomodeler corporation [20] for providing the image sequences for the photogrammetric projects.

References

[1] O. Faugeras and Q.-T. Luong, The Geometry of Multiple Images. The MIT Press, 2001.

[2] R. Hartley and A. Zisserman, Multiple view geometry in computer vision. Cambridge University Press, 2000..[3] A. Fusiello, Uncalibrated Euclidean reconstruction: a review," Image and Vision Computing, vol. 18, pp. 555-563, 2000.

[3] A. Whitehead and G. Roth, The Projective Vision Toolkit, in Proceedings Modelling and Simulation, (Pittsburgh, Pennsylvania), May 2000.

[4] R. Hartley, "Kruppa's equations deri ved from the fundamental matrix," IEEE Trans. On Pattern Analysis and Machine Intelligence, vol. 19, pp. 133-135, February 1997.

[5] Q.-T. Luong and O.D.Faugeras, "Self-calibration of a moving camera from point correspondences and fundamental matrices," International Journal of Computer Vision, vol. 22, no. 3, pp. 261-289, 1997.

[6] L. Lourakis and R. Deriche, "Camera self-calibration using the svd of the fundamental matrix," Tech. Rep. 3748, INRIA, Aug. 1999.

[7] P. Mendonca and R. Cipolla, "A simple technique for self-calibration," in Proceedings of IEEE Conference on Computer Vision and Pattern Recognition, (Fort Collins, Colorado), pp. 112-116, June 1999.

[8] C. Zeller and O. Faugeras, "Camera self-calibration from video sequences: the kruppa equations revisited," Tech. Rep. 2793, INRIA, Feb. 1996.

[9] M. Pollefeys, R. Koch, and L. V. Gool, "Self calibration and metric reconstruction in spite of varying and unknown internal camera parameters," in International Conference on Computer Vision, pp.90-96, 1998.

[10] M. Pollefeys, Self-calibration and metric 3d reconstruction from uncalibrated image sequences. PhD thesis, Catholic University Leuven, 1999.

[11] M. Pollefeys, R. Koch, and L. V. Gool, "Self-calibration and metric reconstruction in spite of varying and unknown intrinsic camera parameters," International Journal of Computer Vision, vol. 32, no. 1, pp. 7-25, 1999.

[12] P. Sturm, A case against kruppa's equations for camera self-calibration," IEEE Trans. On Pattern Analysis and Machine Intelligence, vol. 22, pp. 1199-1204, Oct. 2000.

[13] S. Bougnoux, "From projecti ve to Euclidean space under any practical situation, a criticism of self-calibration," in Proc. 6th Int. Conf. on Computer Vision, (Bombay, India), pp. 790-796, 1998.

[14] W. H. Press and B. P. Flannery, Numerical recipes in C. Cambridge university press, 1988.

[15] M. Maza and D. Yuret, "Dynamic hill climbing," AI Expert, pp. 26{31, 1994.

[16] R. Hartley, "In defense of the 8 point algorithm," IEEE Transactions on Pattern Analysis and Machine Intelligence, vol. 19, no. 6, 1997.

[17] A. Morgan, Solving polynomial systems using continuation for science and engineering. Prentice Hall, Englewoord Clifis, 1987

[18] G. Roth and A. Whitehead "Using projective vision to find camera positions in an image sequence," in Vision Interface 2000, (Montreal, Canada), pp. 87-94, May 2000.

[19] T. Ueshiba and F. Tomita, "A factorization method for projective and Euclidea reconstruction," in ECCV'98, 5th European Conference on Computer Vision, (Freiburg, Germany), pp. 290-310, Springer Verlag, June 1998.

[20] Photomodeler by EOS Systems Inc. http://www.photomodeler.com.

Medical Image Registration
Using Parallel Genetic Algorithms

Yong Fan[1,2], Tianzi Jiang[1], and David J. Evans[2]

[1] National Laboratory of Pattern Recognition, Institute of Automation,
Chinese Academy of Sciences, Beijing 100080, China
{yfan, jiangtz}@nlpr.ia.ac.cn
[2] Department of Computing and Mathematics, Nottingham Trent University,
Nottingham, NG1 4BU, UK
{yong.fan, dj.evans}@ntu.ac.uk

Abstract. Registration of medical image data of different modalities
and multiple times is an important component of medical image analysis. A variety of robust and accurate voxel-based approaches have been
proposed, and mathematically almost all of them are associated with
optimization problems that are highly non-linear and non-convex. This
article presents a parallel genetic strategy to attack mutual information
based registration. The experimental results show robust registration
with high speedup achieved. Furthermore, this method is readily applicable for other voxel-based registration methods.

1 Introduction

The recent remarkable progression of information technology makes many medical imaging methods available today. Anatomical structures can now be investigated precisely by means of computer-assisted x-ray tomography (CT) and by
magnetic-resonance imaging (MRI). Functional information about the brain can
be obtained with single-photon-emission computed tomograhpy (SPECT) and
with positron-emission tomography (PET). These different modalities present
different, complementary information about the anatomy of the brain and its
function. It might be useful to fuse them for the purposes of diagnosis or therapy
planning. So, registration that establishes the correspondence of spatial information in medical images is fundamental to medical image interpretation and
analysis.

Medical image registration can be divided into feature based methods and
voxel intensity based methods. In the feature based methods, the correspondence
of pre-segmented features or markers in data sets is first established which should
be sufficient to establish the transformation between images, then a certain transformation is defined based on the above features. The main advantages of such
methods are high accuracy and high computational efficiency when accurate correspondence is available. However, the process of establishing correspondence is
more complicated and difficult since segmentation is very difficult in most cases
and no segmentation and measurement is perfectly accurate. Although artificial

S. Cagnoni et al. (Eds.): EvoWorkshops 2002, LNCS 2279, pp. 304–314, 2002.
© Springer-Verlag Berlin Heidelberg 2002

markers can provide easy correspondence, it is not welcomed by patients due to its invasiveness [2,3].

Recently, the widely used methods are voxel intensity based methods that use the gray-level information alone without any requirement to segmentation of features. In such methods, registration is achieved with the transformation that maximizes similarity measurement of images. The popular and successful similarity measurement is mutual information derived from the information-theoretic approach [4,6]. This measurement is generally usable for registration of different modalities and more accurate than other measurements, such as squared intensity differences, the variance of intensity ratios algorithm, and the partitioned intensity uniformity algorithm, etc. [7,9,8]. Mathematically, such registration can be modelled as a global optimization problem. However, the optimization problem is difficult to solve, since it is non-linear, non-convex, and computationally expensive. Most of the existing local optimization methods including gradient descent, simplex methods and so on are sensitive to local minima and therefore will not find the global solution. Global optimization methods like simulated annealing or evolutionary algorithms often require a large number of iterations to yield a better solution, but are time consuming.

In this paper, we present a parallel global optimization algorithm to attack this problem. By adopting a local optimization method, i.e., the simplex method in the parallel genetic algorithm achieves not only the global solution but also fast convergence. This parallel strategy has been applied successfully to register 3D MR/CT brain images and the results are encouraging.

The rest of this paper is organized as follows. Section 2 is devoted to a detailed description of a registration method. In Section 3, the parallel genetic algorithm and its implementation are described in detail. Section 4 contains the experimental results and Section 5 the conclusion.

2   Registration Using Mutual Information

Currently several approaches based on voxel similarity measurements are available for the registration of medical images. The most popular similarity measurement is mutual information (MI) that is derived from the information theory approach. For two images, mutual information is a measure of how well one image explains the other, or vice versa. The computation of the mutual information of two images is directly based on the joint probability distribution of the voxel intensities without any assumption about the two images, so it can be used in both intermodality and intramodality registration.

By denoting the two image data sets that are to be registered as reference image I and floating image J, and associated voxel intensity information as u and v, respectively, mathematically, the mutual information based registration problem can be modelled as an optimization problem, i.e., find a transformation T between reference image I and floating image J that gives the maximal mutual information:

$$T^* = arg\max_T MI(u(x), v(T(x)))$$

(1)

where x is viewed as a random variable over coordinate locations, and $MI(\cdot, \cdot)$ is the mutual information.

2.1   Mutual Information

The concept of mutual information comes from communication theory and is directly related to the joint entropy of two information systems or two random variables.

For a random variable, entropy is a statistic measurement of its randomness

$$H(x) = - \int p(x) \ln p(x) dx \qquad (2)$$

where $p(x)$ is the probability distribution of the random variable x.

For two random variables, we can use joint entropy to describe their dependence or their similarity. Given two random variables x and y, their joint entropy is defined as:

$$H(x, y) = - \int p(x, y) \ln p(x, y) dx dy \qquad (3)$$

where $p(x, y)$ is the joint probability distribution of x and y.

Mutual information is then defined in terms of entropy as below [4,6]

$$MI(x, y) = H(x) + H(y) - H(x, y). \qquad (4)$$

By treating the reference image and the float image as two information systems which contain m different signals $\{r_1, r_2, \ldots, r_m\}$ and n different signals $\{f_1, f_2, \ldots, f_n\}$ respectively, the information contained in the two images can be interpreted by the above mutual information concept. The more similar the images are, the more mutual information is contained in them. Such effect can be illustrated in Fig 1. (In fact, a partition of the intensities:$\{I_0, I_1, \ldots, I_m\}$ is referred to as an information system in which the intensities are treated as signals.)

Since mutual information was first introduced in medical image registration[4,6], there are two methods available to compute the joint entropy: the 2D histogram method and the Parzen Window method. Due to the latter being more time consuming, here we use the 2D histogram method to estimate the joint entropy.

2.2   Estimation of Entropies

Mutual information is directly based on the entropies described above that are defined in terms of the probability distributions associated with the random variables. One efficient way to estimate the distributions is to calculate histograms of the voxel intensities. The joint probability distribution according to the transformation T is defined as below

$$p_{ij} = \text{Card}(\{v(x) \in J_i \quad \text{and} \quad u(T(x)) \in I_j\})/N, \qquad (5)$$

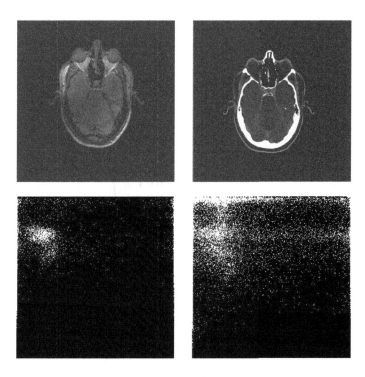

Fig. 1. Illustration of the joint entropy. From top to bottom, They are MR image , CT image, joint histogram of registered images and joint histogram of misregistered images, respectively.

where J_i and I_j are certain partitions of the float image J and reference image I and N is the number of partitions. The distributions of p_i and p_j according to J and I can also be defined similarly.

Here, we use a center based histogram algorithm to compute the distributions. Given a partition number of the intensities, a fuzzy clustering is first implemented on the intensities of the image data to get the centers that determines the partition $\{I_1, I_2, \ldots, I_n\}$. According to the partitions, the marginal probabilities and joint probability are computed. By this method, we can get more accurate distributions, especially when a small partition number is used.

2.3 Spatial Transformation

The spatial transformation is the key factor that determines the search space and computational complexity. Here we focus on the nine parameter affine transformation which consists of translation, rotation and scaling. The transformation can be represented in matrix form as [1]

$$
\begin{pmatrix} x' \\ y' \\ z' \\ 1 \end{pmatrix} = translation * yaw * roll * pitch * rescaling * \begin{pmatrix} x \\ y \\ z \\ 1 \end{pmatrix}
\tag{6}
$$

where *translation*, *yaw*, *roll*, and *rescaling* are defined respectively below

$$
translation = \begin{pmatrix} 1 & 0 & 0 & x_t \\ 0 & 1 & 0 & y_t \\ 0 & 0 & 1 & z_t \\ 0 & 0 & 0 & 1 \end{pmatrix},
$$

$$
yaw = \begin{pmatrix} \cos(\phi) & \sin(\phi) & 0 & 0 \\ -\sin(\phi) & \cos(\phi) & 0 & 0 \\ 0 & 0 & 1 & 0 \\ 0 & 0 & 0 & 1 \end{pmatrix},
$$

$$
roll = \begin{pmatrix} \cos(\omega) & 0 & -\sin(\omega) & 0 \\ 0 & 1 & 0 & 0 \\ \sin(\omega) & 0 & \cos(\omega) & 0 \\ 0 & 0 & 0 & 1 \end{pmatrix},
$$

$$
pitch = \begin{pmatrix} 1 & 0 & 0 & 0 \\ 0 & \cos(\theta) & \sin(\theta) & 0 \\ 0 & -\sin(\theta) & \cos(\theta) & 0 \\ 0 & 0 & 0 & 1 \end{pmatrix},
$$

$$
rescaling = \begin{pmatrix} mx & 0 & 0 & 0 \\ 0 & my & 0 & 0 \\ 0 & 0 & mz & 0 \\ 0 & 0 & 0 & 1 \end{pmatrix}.
$$

For brain images, the x-axis is often defined as the axis that passes from left to right, the y-axis as the axis from front to back, and the z-axis as the axis from bottom to top.

Because of the discrete nature of the images, the transformed voxels generally do not coincide with the reference image coordinates. Tri-linear interpolation is used in our algorithm as a compromise between efficiency and accuracy, although the nearest interpolation is faster and sinc interpolation is more accurate [8,11,10].

3   Parallel Genetic Algorithm

As stated above, mutual information based registration can be modelled as an optimization problem. Since the cost function is non-linear and non-convex and has multiple local maxima, the local optimization methods are not suitable for

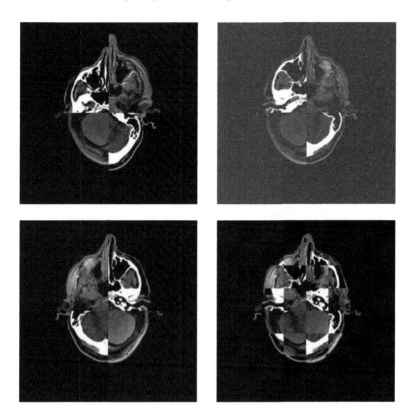

Fig. 2. Fused slices of CT and MRI images. Top left is the unregistered result, others are the registered results

such a problem. Classical global optimization methods like genetic algorithms or simulated annealing can achieve a better, often the optimal solution, but require more iterations. So, they are also not the most favourite methods due to the expensive computation requirement to evaluate the cost function (3D transformation and interpolation are involved). Here we present a parallel genetic algorithm for such a problem.

The parallel genetic algorithms have been demonstrated to be successful for optimization, especially for speed purposes. There are several methods to parallelize the serial genetic algorithm, such as global genetic algorithms, migration genetic algorithms and diffusion genetic algorithms [14]. Due to the hardware availability and its robustness, we focus on the migration genetic algorithms. The basic form for the parallel genetic algorithm can be described as below:

STEP 1 Define a suitable representation and the genetic operators, then generate randomly a population of candidate solutions and partition it into several sub-populations; decide a migration strategy for sharing individuals between the sub-populations.

STEP 2 Each sub-population executes the step 3 and 4.

STEP 3 Perform self-evolution based on the chosen genetic operators: selection, crossover, mutation, local hill-climbing.

STEP 4 Send the best individuals to the neighboring sub-populations based on the migration strategy, receive their best ones and replace the worst ones of the subpopulation.

STEP 5 Determine whether the stopping criteria are satisfied. If satisfied, stop the iteration; otherwise go to step 2.

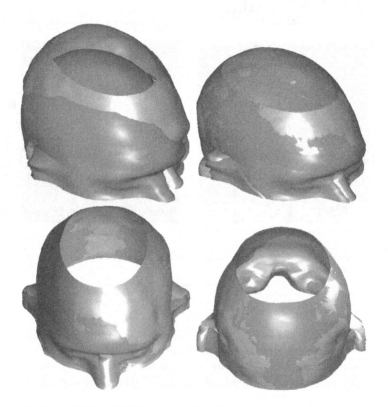

Fig. 3. 3D views of CT and MRI images. Top left is the unregistered result and others are registered results from different views.

Thus, the above basic algorithm with different migration strategies and different strategies of sub-populations' self-evolution produces different instantiations for the parallel genetic algorithm. For the strategy of subpopulation self-evolution, we can apply the successful and popular accepted genetic algorithm. As to the migration strategy, the most popular models are the island model and the stepping stone model [13]. In the island model, best individuals are allowed to be sent to any other subpopulation, i.e. all sub-populations are neighbors. In a sense, the island model represents a better model of nature at the cost of the high communication overhead produced in it. In order to reduce the communication overhead, the stepping stone model defines a close chain neighborhood structure in which any subpopulation is merely able to share individuals with its two neighbors. As a compromise between the island model and the stepping stone model, the random migration strategy randomly determine who and how many sub-populations will migrate their best individuals to others. This strategy not only reduces the communication but also makes the parallel genetic algorithm behavior more close to nature.

It is well known that, although the genetic algorithm is a relatively robust global optimization algorithm, it requires more iterations to satisfy the convergence criteria. As a local optimization method, the simplex downhill method is quite fast [12]. So, here we incorporate a local hill-climbing strategy in our parallel genetic algorithm. In every evolution, we compute the fitness of each chromosome and view the fitness as a distribution function. Then we compute the entropy of the distribution function. When the entropy does not change again, we decide that the genetic algorithm has reached a region of attraction of a good minimum and begin to implement local hill-climbing. The local hill-climbing method is the simplex downhill method in multidimensions in which we put the best individual as the initial starting point.

4   Experimental Results

In order to evaluate the performance of the parallel genetic algorithm, the above approach was implemented and tested on representative 3D MRI/CT registration problems. In the experiments, the subpopulation evolution strategy uses the steady state genetic algorithm with a real-valued representation which works directly on the parameter vector to be optimized. The uniform crossover and the gaussian mutation are used in the subpopulation evolution strategy. The selection is implemented by means of a tournament scheme and the mutual information value is directly used as fitness function. The parallel genetic algorithm was programmed in C++ and implemented on a PC cluster in MPICH. The configuration for our parallel genetic algorithm is as below

No. of Subpopulation	16	No. of Generation	100
Size of Subpopulation	30	Migration Interval	5
Probability of Crossover	0.9	Migration Rate	randomly determined
Probability of Mutation	0.05		

In our experiments, the images were provided as part of the project, "Evaluation of Retrospective Image Registration", National Institutes of Health, Project Number 1 R01 NS33926-01, Principal Investigator, J. Michael Fitzpatrick, Vander-bilt University, Nashville, TN.

The CT data consists of 29 slices of 512×512 pixels whose size is 0.653595mm square and the slice spacing is 4mm. The MRI data consists of 26 slices of

256×256 pixels whose size is 1.25mm square and the slice spacing is 4mm. The CT data is used as reference image while the MRI data is served as float image. Even using the parallel algorithm, the computation task is still quite heavy for full 3D data registration. So in our implementation, the mutual information is computed by using a low-resolution approach [5] in which the float image is sub-sampled by a factor of $f_x = 4, f_y = 4, f_z = 1$ and the image intensity partition numbers for both images are fixed as 64. The search space is set as:

$$x_t \in (-30, 30), \quad y_t \in (-30, 30), \quad z_t \in (-30, 30),$$
$$\phi \in (-\pi/6, \pi/6), \quad \omega \in (-\pi/18, \pi/18), \theta \in (-\pi/18, \pi/18),$$
$$mx \in (0.95, 1.05), \; my \in (0.95, 1.05), \quad mz \in (0.95, 1.05).$$

The 2D fused slices of the experimental results and 3D views are presented in Fig. 2 and Fig. 3 which are used as visual assessments for the registration accuracy.

In order to assess the robustness of the registration, we compared the experimental results with the results obtained from the full data. The registration errors are presented in Table 1.

Table 1. Registration errors compared with the full data registration. The average errors are computed for ten runs.

parameters	average errors	parameters	average errors	parameters	average errors
x_t	0.202	ϕ	0.004	mx	0.011
y_t	0.230	ω	0.002	my	0.008
z_t	0.196	θ	0.002	mz	0.007

The Speed up analysis of the parallel genetic algorithm is very difficult to define because it is completely different from the sequential version. Here we evaluated the speed up by fixing the total number of individuals and the fitness function value which is shown in Fig. 4. The total number of the individuals is 480 and the stop criterion is that the fitness function value is over a certain value, here we used 2.25. So, for the implementation of 1 subpopulation, there are 480 individuals and no migration.

5   Conclusion

We have presented a parallel genetic strategy for medical image registration that produces robust results with quasi-linear (or super-linear) speedup according to the subpopulation number. The encouraging results show that registration using the parallel genetic algorithm is a robust method. Furthermore, this method is readily applicable for other voxel intensity based registration.

Here, we only presented fused registered results for visual assessments for the registration accuracy. The quantitative accuracy assessments for Vanderbilt data are underway. We have evaluated affine transformation based registration. Later we shall apply our parallel genetic algorithm on non-rigid registration.

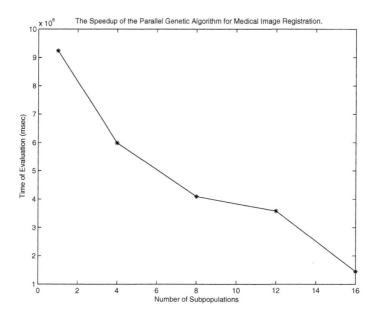

Fig. 4. Speedup of the algorithm. In this experiment, a certain mutual information value is fixed for the implementations with the same configuration and different subpopulation numbers.

Acknowledgements

This work was partially supported by Hundred Talents Programs of the Chinese Academy of Sciences, the Natural Science Foundation of China, Grant No. 60172056 and 697908001, State Commission of Science and Technology of China, Grant No. G1998030503, and EPSRC Visiting Fellowships, Grant No. GR/N10813.

References

1. Isaac N. Bankman, *Handbook of Medical Imaging*, Academic Press, 2000.
2. Joseph V. Hajnal, Derek L.G. Hill, and David J. Hawkes, *Medical Image Registration*, CRC Press, 2001.
3. J.B.A. Maintz and M. A. Viergever, " A Survey of Medical Image Registration", *Medical Image Analysis*, Vol. 2, No. 1, pp.1-36, 1998.
4. F. Maes, A. Collignon, D. Vandermeulen, G. Marchal, and P. Suetens, " Multimodality Image Registration by Maximization of Mutual Information", *IEEE Tran. Medical Imaging*, Vol. 16, pp.187-198, 1997.
5. F. Maes, D. Vandermeulen, P. Suetens, "Comparative evaluation of multiresolution optimization strategies for multimodality image registration by maximization of mutural information", *Medical Image Analysis*, Vol. 3, No. 4, pp.373-386, 1999

6. W.M. Wells, III, P.Viola, H. Atsumi, S. Nakajima, and R. Kikinis, "Multi-modal Volume Registration by Maximization of Mutual Information", *Medical Image Analysis*, Vol.1, pp.35-51, 1996.

7. R.P. Woods, S.R. Cherry, and J.C. Mazziotta, "Rapid Automated Algorithm for Aligning and Reslicing PET Images", *J. Comput. Assist. Tomogr.*, Vol.16, pp.620-633, 1992

8. J.V. Hajnal, N. Saeed, E.J. Soar, A. Oatridge, I.R. Young, and G.M. Bydder, "A Registration and Interpolation Procedure for Subvoxel Matching of Serially Acquired MR Images", *J. Comput. Assist. Tomogr.*, Vol. 19, pp. 289-296, 1995.

9. R.R. Woods, J.C. Maziotta, and S.R. Cherry, "MRI-PET Registration with Automated Algorithm", *J. Comput. Assist. Tomogr.*, Vol.17, pp.536-546, 1993

10. Josien P. W. Pluim, J. B. Antoine Maintz, Max A. Viergever, "Interpolation Artefacts in Mutual Information-Based Image Registration", *Computer Vision and Image Understanding*, Vol.77, No.2, pp. 211-232, 2000

11. Neil A. Thacker, Alan Jackson, David Moriarty, Elizabeth Vokurka, "Improved quality of re-sliced MR images using re-normalized sinc interpolation", *Journal of Magnetic Resonance Imaging*, Vol.10, No.4, pp.582-588, 1999.

12. Press, W. H.,et al., Numerical Recipes in C: The Art of Scientific Computing Second Edition, Cambridge University Press, Cambridge, 1992

13. B. Wilkinson and M. Allen,*Parallel Programming: Techniques and Applications using networked workstations and Parallel Computers,* Prentice Hall, New Jersey, 1999.

14. Albert Y. H. Zomaya, *Parallel and Distributed Computing Handbook*, McGraw-Hill, New York, 1996.

Disruption Management for an Airline – Rescheduling of Aircraft

Michael Løve[1], Kim Riis Sørensen[2], Jesper Larsen[3], and Jens Clausen[3]

[1] FDB, Logistic Development, Albertslund, Denmark
[2] Carmen Systems Denmark, Copenhagen, Denmark
[3] Informatics and Mathematical Modelling, Technical University of Denmark,
Lyngby, Denmark

Abstract. The Aircraft Recovery Problem (ARP) involves decisions concerning aircraft to flight assignments in situations where unforeseen events have disrupted the existing flight schedule, e.g. bad weather causing flight delays. The aircraft recovery problem aims to recover these flight schedules through a series of reassignments of aircraft to flights, delaying of flights and cancellations of flights. This article demonstrates an effective method to solve ARP. A heuristic is implemented, which is able to generate feasible revised flight schedules of a good quality in less than 10 seconds. This article is a product of the DESCARTES project, a project funded by the European Union between the Technical University of Denmark, British Airways and Carmen (see [1]).

1 Introduction

The Aircraft Recovery Problem arises when unforeseen events have disrupted an existing flight schedule, e.g. bad weather causing flights to be delayed. The first priority for the airline is then to restore the flight schedule as much as possible, i.e. minimize the number of cancellations and the total delay.

ARP has been given other names by different researchers and its precise definition varies accordingly. Hence, it is important to define ARP as it is understood here: Given an original flight schedule and one or more disruptions, the Aircraft Recovery Problem consists of *delaying* flights, *cancelling* flights and *swapping* aircraft to flight assignments in order to create a feasible and more preferable revised flight schedule.

The *flight schedule* includes all flights flown within a certain period of time by a given fleet including the original departure and arrival times, the expected flight durations and the *tail assignments*. Tail assignments refer to specific aircraft being assigned to specific flights. The term *swap* denotes that two flights, designated to be undertaken by two specific aircraft, are interchanged between these aircraft.

1.1 Decision Costs

The decision costs are of paramount importance in ARP. In mathematical modelling, costs reflect how *preferable* each decision possibility is. Preferable here

S. Cagnoni et al. (Eds.): EvoWorkshops 2002, LNCS 2279, pp. 315–324, 2002.

means a better objective function value, thus the ability to quantify the quality of a solution is essential.

The most common approach to quantifying decision costs is to estimate the *real* costs associated with each decision. If the actual costs of these decision possibilities are calculated, they must include factors such as ill-will from customers, costs of customers missing down-line flights, crew planning issues, etc. These factors are important to include, yet most of them are difficult, if not impossible to estimate.

It may therefore be futile to base an objective function on **real** *operating costs when solving ARP. Instead, we have have focussed on quantifying the basic operating principles that flight controllers adhere to.*

In general, the objectives are to minimize the number of cancellations, the total delay and to make as few swaps as possible. However, there is a clear trade-off between these 3 objectives and the quality of a solution to ARP is ultimately a matter of preference. By focusing on the basic operating principles, different solution strategies are easily applied that can accommodate these preferences. One example is the trade-off between the number of swaps and the total delay: Depending on the situation on hand, the acceptable number of swaps varies. A strategy can thus be applied that attempts to recover the flight schedule by making a minimum number of swaps while accepting a number of delays.

2   Previous Work on ARP

Many authors have worked on problems similar to ARP. Teodorović and Gubernić [2] introduce a method of solving ARP that focuses on minimizing the total passenger delay. Later Teodorović together with Stojković [3] introduce a different model that aims to minimize the number of cancellations and the total passenger delay. In [4] Jarrah et al. introduce two separate models that minimize delays and cancellations respectively. The models thus are not able to consider the trade-off between cancelling and delaying. Yan and Yang formulate a model in [5] that aims to minimize the period of time in which the flight schedule is disrupted. In [6] Cao and Kanafani introduces the first model which seemingly can consider delays and cancellations simultaneously while solving problems of a realistic size. However, as it is demonstrated in [7], this model contains several errors. Lastly, Argüello et al. demonstrate a heuristic approach to solving ARP in [8,9]. The method solves realistic examples of ARP effectively.

3   Motivation for Using Heuristics to Solve ARP

Aircraft recovery is an activity that occurs every day in most airlines. Presently it is done manually with the assistance of various tools such as computer based graphical interfaces that allow flight controller easy access to information on the present situation.

An algorithm to solve the ARP has to be able to handle problems of a realistic size, e.g. 50 airports, 100 aircraft and 500 flights. Secondly, an algorithm preferably has to produce a result in less than 3 minutes. And finally, an algorithm must be flexible. This means that the algorithm has to accommodate for new or modified restrictions frequently.

At the same time ARP is a very complex problem: The cost of a certain decision depends on other decisions made earlier, e.g. the ready-time of an aircraft at a given airport depends on whether the aircraft was delayed or reassigned earlier that working day. This dependency along with a large solution space and integral decision variables makes ARP a very difficult problem to solve. Given all these factors, heuristics seem a clear choice.

4 Basic Heuristic Design

4.1 Definition of Network

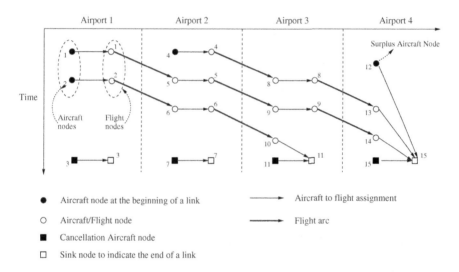

Fig. 1. *Underlying network to use with a ARP-heuristic.*

Before the ARP-heuristic can be designed, a network is defined to base the heuristic on. This network is shown in Figure 1. The vertical axis depicts the time of day. At each airport, the nodes on the left are called aircraft nodes, because they represent aircraft. These nodes are placed at the point in time when the aircraft is ready to depart. The nodes immediately to the right of the aircraft nodes are flight nodes and they represent scheduled departures of flights. They are also placed at the point in time when the flight is scheduled to depart.

The arcs connecting the aircraft and flight nodes represents that the particular aircraft is assigned to carry out the flight to which it is connected. The heuristic modifies the flight schedule by altering these assignments through *swaps*. A swap is illustrated in Figure 2.

Before Swapping

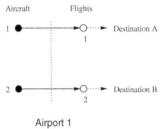

After Swapping

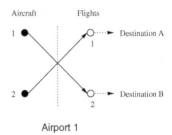

Fig. 2. *A swap.*

The network in Figure 1 is similar to the network introduced in [6], however, there are significant differences. The term *flight link* is used to describe a sequence of flights performed by one aircraft. In Figure 1 an example of a flight link is the sequence $2 \to 2 \to 6 \to 6 \to 10 \to 11$. A time horizon spanning the remaining part of the working day is necessary if such lay over restrictions have to be considered.

To cater for the lay over restrictions, each link terminates in a sink node, thus indicating where the aircraft will lay over, e.g. aircraft 10 in Figure 1 ends its working day at airport 3. This makes it possible to keep track of how many aircraft end at each particular airport.

Another feature in the underlying network is the cancellation aircraft node, e.g. nodes 3 and 7 in Figure 1. By default, all cancellation aircraft nodes are assigned to the sink node. However, the cancellation aircraft can be assigned to all flights, thus cancelling them. If for example cancellation aircraft 3 were swapped with aircraft 2, then aircraft 2 would remain at airport 1 and flights 2 and 6 would be cancelled.

Finally, there are surplus aircraft. By default, these are connected by a forward arc to the sink node. Consequently, if the surplus aircraft is not used, it remains at the airport. Alternately, it can be assigned to a flight, thus keeping it from getting cancelled or delayed.

4.2   Basic Parameters and Decision Variables

Following is a definition of the variables, parameters and sets used in all the heuristics that have been implemented to solve ARP.

A = set of nodes representing aircraft.

a = index for aircraft nodes.

F = set of nodes representing flights.

f = index for flight nodes.

F_a = subset of F consisting of candidate flights considered for aircraft a. If aircraft a is delayed beyond the time horizon, F_a is set to empty. In Figure 1, F_a could reasonably consist of flights $\{4, 5, 6, 7\}$ for aircraft $a = 4$.

F_c = dynamic set containing flights f that have been cancelled.

r_f = the revenue of flight f.

d_{af} = the delay incurred if aircraft a is assigned to flight f. d_{af} is calculated as needed and takes all relevant previous assignments into consideration. d_{af} is measured in minutes.

α_f = delay cost multiplier associated with each flight f.

β_f = cancellation cost multiplier associated with cancelling flight f.

The decision variable is x_{af}. If this binary variable is 1 aircraft a is assigned to flight f.

To illustrate how the decision variable works, refer to Figure 1. Here $x_{1,1} = 1$ because aircraft 1 is assigned to flight 1. Conversely, $x_{1,8} = 0$ because aircraft 1 is not assigned to flight 8.

4.3   Objective Function

Given the network from Figure 1 and the basic parameters and decision variables, the objective function can be defined.

$$Objective = \sum_{a \in A} \sum_{f \in F} r_f \cdot x_{af} \tag{1}$$

$$- \sum_{a \in A} \sum_{f \in F \setminus F_c} \alpha_f \cdot DF \cdot r_f \cdot d_{af} \cdot x_{af} \tag{2}$$

$$- \sum_{a \in A} \sum_{f \in F_c} \beta_f \cdot r_f \cdot x_{af} \tag{3}$$

where (1) is the total revenue and (2) the total cost of delays. The constant DF is the percentage of the revenue r_f which is subtracted per minute delay of flight f. Finally, (3) is the cost associated with cancellations.

In both (2) and (3) the revenue r_f is used directly to measure the cost because it seems a natural way to prioritize the flights.

As a true revenue r_f can not be calculated until weeks after the flight has been flown – only then is the necessary data available. A revenue r_f calculated before the actual flight can only be based on predicted numbers. Disruptions to the flight schedule may render these forecasts obsolete, which is why α and β are relevant. They can quickly be assigned values which encourages a heuristic to find solutions that prioritize the flights according to the actual situation on hand.

4.4 Choice of Solution Neighbourhood

Given the network shown in Figure 1, a neighbourhood can be defined. In short, the neighbourhood to an ARP solution consists of all those solutions that can be reached by making 1 feasible swap between 2 different aircraft. By default, an aircraft can be assigned to any flight departing from the airport where the aircraft is located. In terms of complexity this corresponds to $O(n^2)$ where n is the number of flights. Notice that the neighbourhood size can be halved by utilizing that only $\frac{n^2}{2}$ *different* neighbours exist.

4.5 Evaluating Solutions

The time it takes to explore a neighbourhood also depends on how effectively the objective value is calculated for each solution. A reasonable way of doing this would be to track each aircraft through its link and calculate the contribution to the objective value. Cancellations would also be included this way, because cancellation aircraft are treated as a normal aircraft in this respect. In other words, each flight would be considered once, causing this method to result in a complexity of $O(n)$, where n is the number of flights. It is possible, however, to reduce this calculation significantly by only recalculating the cost of those links between which the reassignments occurred. The overall complexity of evaluating all solutions in a neighbourhood is $O(n^3)$ where n is the number of flights. For practical purposes, however, the complexity is significantly smaller.

4.6 Creating Problem Instances

To test the heuristics, which have been implemented to solve ARP, 25 problem instances were created using a problem generator outlined (further details can be found in [10]). Simply put, a link is created for each aircraft by repeatedly selecting a destination airport at random until the time horizon is exceeded. This limits the length of a link to a maximum of 5 flights. The resulting problem instances are listed in table 1.

It is important to notice that because airports are randomly selected, flights will not be concentrated around a few airports like they would be in a hub-and-spoke system – instead they will be more evenly distributed. Likewise, aircraft will typically not travel back and forth between the same two airports. These 2 factors mean that the generated flight schedules do not resemble real flight schedules from a practical point of view. However, it seems reasonable to assume that the complexity remains unchanged.

5   Choice of Heuristics

Several heuristics were implemented with the overall aim of finding the heuristic, which could produce the best results in 3 minutes or less. Initially, various versions of a heuristic were implemented that had the possibility of escaping local

Table 1. Dimensions of the problem instances.

Instance No.	Number of Airports	Number of Aircraft	Number of Flights
1	10	20	80
2	10	30	125
3	10	40	166
4	10	50	206
5	20	20	78
6	20	40	158
7	20	60	231
8	20	80	306
9	20	100	382
10	30	30	111
11	30	60	221
12	30	90	326
13	30	120	440
14	30	150	559
15	40	40	148
16	40	80	290
17	40	120	432
18	40	160	588
19	40	200	741
20	50	50	195
21	50	90	333
22	50	110	404
23	50	150	562
24	50	200	753

optima, namely the Iterated Local Search (ILS) with a Variable Neighbourhood Search (VNS) incorporated (see [11,12]). A simple Steepest Ascent Local Search (SALS) heuristic was also implemented along with a repeated SALS (RSALS), which functions exactly like SALS but is repeated for different initial solutions. Surprisingly, the SALS algorithms were most effective.

5.1 Steepest Ascent Local Search Heuristic

The ILS algorithms were able to find reasonably good solutions quickly – typically within the first 10 seconds. Significantly better solutions were not found even in 24-hour test runs and with a wide range of different parameter settings. This prompted a number of questions concerning the search space structure and whether or not other types of heuristics would be more effective.

A SALS algorithm quickly finds a local optimum. However, once it has reached such an optimum, it is trapped. This is not a problem if the local optimum is near the global optimum and this turned out to be the case as will be explained later (for a more in-depth analysis see [10,7]).

5.2 Implementing the SALS Heuristic

The simplified program structure of the SALS algorithm is shown in Figure 3. There are 3 main elements and these are:

InitialSolution: The initial solution is the original flight schedule including, aircraft ready-times, scheduled departures, original tail assignments and all

```
program Steepest Ascent Local Search (SALS)
begin
     x_af = InitialSolution()
     repeat
          x'_af := LocalSearch(x_af)
          x_af := AcceptanceCriterion(x_af, x'_af)
     until a better solution cannot be found
end.
```

Fig. 3. *Outline of a SALS procedure*

delays/cancellations. All these data are represented in the manner illustrated in Figure 1.

LocalSearch: The local search procedure is initiated by a solution x_{af} in the form of a flight schedule. A best improvement strategy is chosen so that all of the neighbours to x_{af} are evaluated before the best solution x'_{af} among the neighbours to x_{af} is used as a starting point for the next iteration.

AcceptanceCriterion: This procedure determines if the latest local search yielded an improved solution. If so, it allows the algorithm to continue.

5.3 Solution Quality of SALS

In table 2 the results achieved by the SALS algorithm are listed. For the sake of comparison, the column *Best Result* is introduced. This column refers to best solution ever found to each problem instance and it should be noted that they are not necessarily optimal. The best results have been found using the ILS heuristic in 24-hour test runs with different parameter settings and in test runs where RSALS was repeated 2000 times using different initial solutions. We believe that the solution values are very close to the optimumal values.

In an operational environment the running time of the algorithm is highly critical. Therefore we have focused on simple and fast methods rather than more computing-intensive methods like Tabu Search, Simulated Annealing etc.

SALS is clearly very fast. In table 2, it takes 2.11 seconds on average to find a local optimum using SALS and never more than 13 seconds. The best solutions found to the problem instances listed in table 1 improved the revenue in the original flight schedule by 59.0% on average given the parameters chosen. Hence, SALS improves the revenue approximately 57% on average. The ILS heuristics produced slightly better results than SALS in terms of revenue. However, the speed of SALS makes it very attractive.

The prototype has also been tested on data from British Airways. 10 flight schedules were extracted from data obtained from British Airways. Each flight schedule consists of all the BA short-haul flights that should have been flown on a particular day in November 2000. Realistic delays (between 30 and 240 minutes) are introduced on approx. 20% of all aircraft. The average size of the instances is 79 active aircraft, 44 airports, 339 flights and 16.7 disrupted lines of work. In the

Table 2. Overview of the SALS heuristic results

Instance No.	Number of Flights	Best Result	SALS	Gap (in %)	Time (in secs.)
1	80	264051	255213	3.35	0.09
2	125	402593	375699	6.68	0.12
3	166	614664	594911	3.21	0.50
4	206	690592	675224	2.23	0.90
5	78	201030	196291	2.36	0.03
6	158	508363	507043	0.26	0.26
7	231	814305	803711	1.30	0.63
8	306	1096463	1086028	0.95	1.29
9	382	1292384	1268061	1.88	3.16
10	111	376766	369218	2.00	0.02
11	221	782862	776675	0.79	0.34
12	326	1110793	1100121	0.96	1.12
13	440	1569761	1538055	2.02	2.62
14	559	1763103	1683127	4.54	6.77
15	148	437393	417642	4.52	0.07
16	290	963745	958332	0.56	0.57
17	432	1521328	1506960	0.94	1.70
18	588	1954775	1917987	1.88	5.19
19	741	2407012	2336639	2.92	12.15
20	195	612507	602827	1.58	0.15
21	333	1127078	1104572	2.00	0.63
22	404	1473042	1449719	1.58	0.85
23	562	1773247	1721653	2.91	4.05
24	753	2595359	2547788	1.83	9.42

Average gap between best solution and SALS: 2.22   (secs.) 2.11

preliminary tests made so far British Airways have been very positive regarding the results obtained. Furthermore solutions are obtained using an average of 6.07 seconds. This fast execution time makes it possible for the controllers at British Airways to run the algorithm several times in order to produce structually different solutions. More intensive testing and an experimental integration of the algorithm with key people within British Airways operations control have been planned.

6   Conclusion

We have demonstrated that the Aircraft Recovery Problem (ARP) can be solved effectively by a relatively simple SALS-heuristic using a network representation. On average, less than 10 seconds are required to find a feasible revised flight schedule that includes all planned flights on a given day. Our method allows the necessary flexibility that flight controllers need in terms of creating different types of revised flight schedules. The effect of applying various strategies has also been shown: The trade-off between delays, cancellations and swaps can be adjusted and our SALS-heuristic can then generate feasible revised flight schedules that accommodate these trade-offs.

The results give a strong indication that ARP is a relatively simple problem to solve. The future challenge lies in incorporating crew and passenger considerations.

Acknowledgements

The authors thank Alex Ross and Nicki Davis from British Airways who supplied us with airline industry insights and evaluated and commented our results. We also thank Allan Larsen from IMM, DTU for commenting and assisting.

References

1. N. Davis, A. Larsen, and S. Tiourine. Descartes – decision support for integrated crew and aircraft recovery. Presentation at the AGIFORS Airline Operations Study Group Meeting in Jamaica, May 2001.
2. D. Teodorović and S. Guberinić. Airline Optimization. *European Journal of Operational Research*, 15:178–182, 1984.
3. D. Teodorović and M. Stojković. Model for Operational Daily Airline Scheduling. *Transportation Planning and Technology*, 14:273–285, 1990.
4. A. I. Z. Jarrah, G. Yu, N. Krishnamurthy, and A. Rakshit. A Decision Support Framework for Airline Flight Cancellations and Delays. *Transportation Science*, 27:266–280, 1993.
5. S. Yan and D.-H. Yang. A Decision Support Framework for Handling Schedule Perturbations. *Transportation Research*, 30:405–419, 1996.
6. J.-M. Cao and A. Kanafani. Real-Time Decision Support for Integration of Airline Flight Cancellations and Delays Part I & II. *Transportation Planning and Technology*, 20:183–217, 1997.
7. M. Løve and K. R. Sørensen. Disruption management in the airline industry. Master's thesis, Informatics and Mathematical Modelling (IMM). Technical University of Denmark, March 2001.
8. M. F. Argüello, J. F. Bard, and G. Yu. A grasp for aircraft routing in response to groundings and delays. *Journal of Combinatorial Optimization*, 5:211–228, 1997.
9. M. F. Argüello, J. F. Bard, and G. Yu. Models and Methods for Managing Airline Irregular Operations. In Gang Yu, editor, *Operations Research in the Airline Industry*. Kluwer Academic Publishers, Boston, 1998.
10. M. Løve, K. R. Sørensen, J. Larsen, and J. Clausen. Using heuristics to solve the dedicated aircraft recovery problem. Technical report, Informatics and Mathematical Modelling, Technical University of Denmark, 2001.
11. T. Stützle. Iterated Local Search for the Quadratic Assignment Problem. Technical report, Technische Hochschule Darmstadt, 1999.
12. N. Mladenović and P. Hansen. Variable Neighborhood Search. *Computers & Operations Research*, 24:1097–1100, 1997.

Ant Colony Optimization
with the Relative Pheromone Evaluation Method

Daniel Merkle[1] and Martin Middendorf[2]

[1] Institute for Applied Computer Science and Formal Description Methods
University of Karlsruhe, Germany
merkle@aifb.uni-karlsruhe.de
[2] Computer Science Group
Catholic University of Eichstätt, Germany
martin.middendorf@ku-eichstaett.de

Abstract. In this paper the *relative pheromone evaluation method* for Ant Colony Optimization is investigated. We compare this method to the standard pheromone method and the summation method. Moreover we propose a new variant of the relative pheromone evaluation method. Experiments performed for various instances of the single machine scheduling problems with earliness costs and multiple due dates show the potential of the relative pheromone evaluation method.

1 Introduction

The Ant Colony Optimization (ACO) metaheuristic [1,2] has been applied to several combinatorial optimization problems like the Traveling Salesperson problem, vehicle routing problems, scheduling problems, and other problems (for an introduction and overview see [3]). Many of these problems are permutation problems where a permutation of n items has to be found that minimizes the value of some cost function.

An ACO algorithm for permutation problems consists of several iterations where in every iteration each of m ants constructs one permutation. One standard approach is that every ant constructs a solution by deciding iteratively which job is at the next place starting with the first place of the schedule. For the selection of an item the ant uses pheromone information which stems from former ants that have found good solutions. In addition an ant may also use heuristic information for its decisions. The pheromone information, denoted by τ_{ij}, and the heuristic information, denoted by η_{ij}, are indicators of how good it seems to have item j at place i of the permutation.

Usually the probabilistic decisions of the ants are made according to a local decision rule where the probabilities of the possible outcomes are correlated to the relative amount of pheromone on the corresponding edges in the decision graph. For the considered permutation problems the ant chooses the next item for place i, $i \in [1 : n]$ from the set S of items that have not been placed so far according to a probability distribution that is based on the pheromone values

S. Cagnoni et al. (Eds.): EvoWorkshops 2002, LNCS 2279, pp. 325–333, 2002.
© Springer-Verlag Berlin Heidelberg 2002

τ_{ij}, $j \in \mathcal{S}$. In [4] an extended local decision rule with some lookahead at the pheromone values of following edges in the decision graph was used.

In [5] it was shown that it can be advantageous to take also other pheromone values than those corresponding to the edges of the possible outcomes of the actual decision into account. A particular method for such a more global evaluation of the pheromone information called *summation evaluation* was proposed for solving permutation problems where good solutions satisfy a similarity property, namely, for every two good permutations the places of every item differs only slightly. For the summation method the probability distribution to decide which item to put on place i is based on the values $\sum_{l=1}^{i} \tau_{lj}$, $j \in \mathcal{S}$, i.e., on pheromone values $\tau_{1j}, \ldots, \tau_{ij}$. It was shown in [5] that the summation method gives better results than the standard method for the Single Machine Weighted Total Tardiness problem. Meanwhile other authors have shown that the summation method gives very good results also for the Job Shop problem [6] and the Flow-Shop problem [7]. For the Resource-Constrained Project Scheduling problem the summation method performed better than the standard method but a combination of both methods was best [8]. Tests for Constraint Satisfaction problems have shown that the summation method performed only slightly better than the standard method, but the instances for this problem usually do not satisfy the desired similarity property that was mentioned above [9].

Another global pheromone evaluation method called *relative pheromone evaluation* was proposed recently by the authors in [10]. It was suggested that this method could be useful for permutation problems where the desired similarity property of the good solutions for using the summation method does not hold. But so far the relative evaluation method has not been tested on any non-trivial problem (The focus of [10] was to obtain insights into the behavior of ant algorithms by studying their behaviour on certain toy problems that can easily be solved optimal in linear time).

In this paper we study *relative pheromone evaluation* on a non-trivial problem and propose to combine *relative pheromone evaluation* with a new pheromone update strategy. It is shown that both methods perform significantly better than the standard method (and the summation method) on certain types of permutation problems. In particular, experimental results are presented for single machine scheduling problems with earliness costs and multiple due dates which clearly show the potential of the method.

The paper is organized as follows. The definition of the single machine scheduling problem with earliness costs and multiple due dates is given in Section 2. In Section 3 we describe the ACO algorithm with standard local and *summation evaluation*. The relative pheromone evaluation method is described in Section 4. The choice of the parameter values of the algorithms that are used in the test runs and the test instances are described in Section 5. Results are reported in Section 6 and a conclusion is given in Section 7.

2   The Single Machine Total Earliness with Multiple Due Dates Problem

The optimization problem that we use as our test problem is the Single Machine Total Earliness with Multiple Due Dates problem (SMTEMDP). This problem is to find a one machine schedule without idle times that minimizes the sum of the earliness values of the jobs for a given set of jobs where each job has several due dates. The earliness of every job is computed with respect to the next due date that is following the completion time of a job. The SMTEMDP arises when each customer comes only at certain dates to the producer to take away the goods that have been produced for her. Hence, every good that has been produced has to be stored by the producer until the next time when the corresponding customer comes. The problem is then to find a production schedule that minimize the sum of the storing times.

Formally, SMTEMDP is for n given jobs, where job j, $1 \leq j \leq n$ has a processing time p_j, due dates $d_{j_1} > \ldots > d_{j_{n_j}}$ to find a non-preemptive one machine schedule without idle times that minimizes $E = \sum_{j=1}^{n} \{\min_{k=1,\ldots,n_j}\{\max\{0, dj_k - C_j\}\}$ where C_j is the completion time of job j. E is called the total earliness of the schedule. We assume that the last due date is a dummy due date that equals the sum of the processing times, i.e., $d_{j_{n_j}} = \sum_{j=1}^{n} p_j$.

3   The ACO Algorithm

In this section we explain how an ACO algorithm with standard or *summation evaluation* works. As an example problem we use SMTEMDP. In ACO algorithms each of m ants in a generation constructs one solution for the problem. For the SMTEMDP the ants select the jobs in the order in which they will appear in the schedule. For the selection of a job an ant uses a probabilistic decision rule that is based on pheromone information and heuristic information. The heuristic information, denoted by η_{ij}, and the pheromone information, denoted by τ_{ij}, are indicators of how good it seems to have job j at place i of the schedule. The heuristic value is generated by some problem dependent heuristic whereas the pheromone information stems from former ants that have found good solutions. The next job is chosen according to the probability distribution over the set of unscheduled jobs S determined by one of the following methods. In the following α and β are parameters that determine the relative influence of the pheromone information and the heuristic.

Standard local method ([1,2]):

$$p_{ij} = \frac{\tau_{ij}^{\alpha} \cdot \eta_{ij}^{\beta}}{\sum_{h \in S} \tau_{ih}^{\alpha} \cdot \eta_{ih}^{\beta}} \tag{1}$$

Summation method ([5]):

$$p_{ij} = \frac{(\sum_{k=1}^{i} \tau_{kj})^{\alpha} \cdot \eta_{ij}^{\beta}}{\sum_{h \in S} (\sum_{k=1}^{i} \tau_{kh})^{\alpha} \cdot \eta_{ih}^{\beta}} \qquad (2)$$

It should be noted that it is possible for both methods to give the pheromone values and the heuristic values a different influence using exponents as weights. The heuristic values η_{ij} are computed according to

$$\eta_{ij} = \frac{1}{\min_{k=1,\ldots,n_j} \{\max\{1, dj_k - (t + p_j)\}\}}$$

where t is the finishing time of the last scheduled job. The idea is that an ant should prefer jobs that will cause small earliness costs when scheduled next.

After all m ants of the generation have constructed a solution the ant that found the best solution in that generation is allowed to update the pheromone matrix. But before that some of the old pheromone is evaporated according to

$$\tau_{ij} = (1 - \rho) \cdot \tau_{ij}$$

where the parameter ρ determines the strength of the evaporation. The reason for this is that old pheromone should not have a too strong influence on the future. Then, for every job j in the schedule of the best solution found in the generation some amount of pheromone is added to element (ij) of the pheromone matrix where i is the place of job j in the schedule. The amount of pheromone added is $1/E$ where E is the total earliness of the schedule, i.e.,

$$\tau_{ij} = \tau_{ij} + \frac{1}{E} \qquad (3)$$

Note, that it is also possible for ACO algorithms to add just some constant amount of pheromone for the update instead of a solution depended amount like $1/E$. The algorithm stops when a certain number of iterations has been done.

4 Relative Pheromone Evaluation

Relative pheromone evaluation was proposed in [10] and motivated by the following observation: Assume that an item j has not been placed by an ant on a place $> i$ although a large part of the pheromone in column j lies on the first $i - 1$ elements $\tau_{1j}, \ldots, \tau_{(i-1)j}$ in the column. In this case all values τ_{kj} for $k \geq i$ are small. Hence it is unlikely that the ant will place item j on place i when some other item has high pheromone value for place i. This is true even when τ_{ij} is the highest of all pheromone values $\tau_{ij}, \ldots, \tau_{nj}$ on the remaining places. Since for many optimization problems it might still be important on which place $l \geq i$ item j is, even when these places are not the most favorite ones. For such problems it was proposed to use relative pheromone evaluation, i.e., not to use directly the pheromone values in each row but to normalize them with the relative amount of pheromone in the rest of their columns.

Relative method:

$$p_{ij} = \frac{\tau_{ij}^* \cdot \eta_{ij}}{\sum_{h \in S} \tau_{ih}^* \cdot \eta_{ih}} \quad \text{where} \quad \tau_{ij}^* := \left(\frac{\sum_{k=1}^{i} \tau_{kj}}{\sum_{k=1}^{n} \tau_{kj}} \right)^{\gamma} \cdot \tau_{ij} \tag{4}$$

Parameter γ determines the strength of the normalization operation. A interesting observation can be made for the relative method when using the standard pheromone update according to formula 3 where ants add the same amount of pheromone to every pheromone value τ_{ij}. The effect is that small pheromone values receive a stronger relative increase than large pheromone values. This effect can also be found for the standard method and is intended to give small pheromone values a realistic change to become large when they are part of a good solution. But in contrast to the standard method, for relative evaluation two elements of the pheromone matrix with an equal amount of pheromone might profit to a different extend from the same amount of update. The smaller the sum of the pheromone values in the remaining places in the column, i.e., $\tau_{ij} + \ldots + \tau_{nj}$, the larger is the relative amount of update for τ_{ij}.

To study the influence of this effect we investigate the following modification of *relative pheromone evaluation* which has a modified pheromone update not showing this effect. Since the evaluation and the update are different from the standard method we call this variant *relative pheromone utilization* method. For this method the amount of update is adapted to the relative amount of pheromone in the rest of the column. The larger the relative amount of pheromone in the rest of the column for an element the stronger is the update. Assume that the update in some column j is done for element τ_{ij}. Then, for elements τ_{lj} with $l < i$ it should make no difference in which row $i > l$ the update was. Therefore, after adding some amount of pheromone to τ_{ij}, all pheromone values τ_{lj} with $l \geq i$ are multiplied with the same factor $\omega \geq 1$ such that the total amount of update on all elements τ_{lj} with $l \geq i$ is $1/E$. Formally, the update is done in two steps as follows.

Pheromone update of relative pheromone utilization method: Set

$$\tau_{lj} = \begin{cases} \tau_{lj} + \frac{1}{E} \cdot \frac{\sum_{l=i}^{n} \tau_{lj}}{\sum_{l=1}^{n} \tau_{lj}} & \text{for } l = i \\ \tau_{lj} & \text{else} \end{cases} \tag{5}$$

and then for

$$\omega = \frac{\sum_{l=1}^{n} \tau_{lj} + \frac{1}{E}\left(1 - \frac{\sum_{l=i}^{n} \tau_{lj}}{\sum_{l=1}^{n} \tau_{lj}}\right)}{\sum_{l=1}^{n} \tau_{lj}}$$

set

$$\tau_{lj} = \begin{cases} \omega \cdot \tau_{lj} & \text{for } l \geq i \\ \tau_{lj} & \text{else} \end{cases} \tag{6}$$

5   Test Instances and Parameters

We tested the different variants of the ACO algorithm on instances of the SMTEMDP of size 100 jobs. These instances were generated as follows: for each job $j \in [1, 100]$ an integer processing time p_j is taken randomly from the interval $[1, 10]$, and integer due dates $d_{j_1}, \ldots, d_{j_{n_{j-1}}}$ are taken randomly from the interval $[1 : \sum_{j=1}^{n} p_j]$. Recall that each job j has a dummy due date $d_{j_{n_{j-1}}} = \sum_{j=1}^{n} p_j$, $j \in [1 : n]$. Four instances with number of due dates 2, 3, 5, and 10 (plus the dummy due date) were generated. We used parameters $\alpha = 1$, $\gamma = 1$, $\rho = 0.01$ and the number of ants in every generation was $m = 10$. Test were done without heuristic ($\beta = 0$) and with heuristic ($\beta = 1$). For each of the four test instances and each parameter combination 10 runs were performed. Every run was stopped after 15000 generations.

6   Experimental Results

In order to compare the power of the pheromone evaluation (respectively utilization) methods we performed tests without any heuristic (i.e. $\beta = 0$) and also with heuristic (i.e. $\beta = 1$). The results for different numbers of due dates are shown in tables 1 and 2.

Table 1. Comparison of four different pheromone evaluation methods for different number of due dates: Results for SMTEMDP when no heuristic is used ($\beta = 0$); average total earliyness (best values are bold) and in brackets average number of iteration when best solution was found; Relative* denotes relative pheromone utilization

	$\beta = 0.0$			
# Due dates	Standard	Summation	Relative	Relative*
2	3782.5 (5025)	4281.0 (8314)	**1710.0** (6994)	3097.7 (5300)
3	2441.1 (5068)	3496.4 (8433)	**1182.7** (7001)	2087.2 (5368)
5	1793.2 (5082)	2420.8 (8266)	**862.4** (7363)	1547.4 (5675)
10	964.2 (5537)	1317.0 (8527)	**512.9** (7453)	781.2 (5731)

In all cases the results of *relative pheromone evaluation* are the best. They clearly outperforme the standard method with an improvement between 46,8% and 54,8% when no heuristic is used ($\beta = 0$) and between 38,4% and 41,9% when the heuristic is used ($\beta = 1$). The average number of iterations that were used until the best result was found was not much larger for *relative pheromone evaluation* compared to the standard method (at most 44,9% for $\beta = 0$ and at most 24,1% for $\beta = 1$).

In all cases of tables 1 and 2 *relative pheromone utilization* (i.e., *relative pheromone evaluation* with the modified pheromone update) was the second best method. The average number of iterations used to find its best results was in all cases lower than for the *relative pheromone evaluation*. It is interesting that this

Table 2. Comparison of four different pheromone evaluation methods for different number of due dates: Results for SMTEMDP when heuristic is used ($\beta = 1$); average total earlyness (best values are bold) and in brackets average number of iteration when best solution was found; Relative* denotes relative pheromone utilization

# Due dates	$\beta = 1.0$			
	Standard	Summation	Relative	Relative*
2	755.1 (3261)	534.1 (5062)	**460.2** (3292)	498.7 (3260)
3	555.0 (2783)	550.5 (5933)	**341.7** (3453)	389.3 (3277)
5	353.6 (3207)	271.6 (6579)	**211.3** (3979)	254.4 (3730)
10	236.9 (3864)	199.2 (7622)	**137.7** (4401)	158.4 (3833)

method could not beat the *relative pheromone evaluation*. This indicates that it might be possible to improve the standard method by using different strengths of pheromone update for the decisions corresponding to a solution. So far all authors use the same amount of pheromone for every decision that corresponds to a solution. The only exception is [4] where the amount of pheromone sinks for later decisions).

Without heuristic (i.e. for $\beta = 1$) *summation evaluation* was clearly the worst method (with respect to quality of solution and number of iterations used). This is no surprise since the problem structure does not fit to this method. It seems the heuristic forces summation evaluation to concentrate the search for each job to places around a few due dates. So that it could beat the standard method when using the heuristic.

Figures 1 and 2 show the average solution quality over time for the case of 10 due dates (plus the dummy due date) and $\beta = 0$ respectively $\beta = 1$. Only at an initial phase all three methods perform similar. After that *summation*

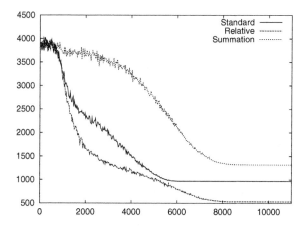

Fig. 1. Average solution quality for 10 due dates plus the dummy due date for $\beta = 0$

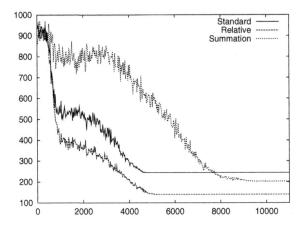

Fig. 2. Average solution quality for 10 due dates plus the dummy due date for $\beta = 1$

evaluation performes nearly always worst. Only for $\beta = 1$ it becomes better than the standard method was after these method has converged. This confirms the expectation that *summation evaluation* is not good for such scheduling problems with many due dates. *Relative pheromone evaluation* is the best method over the whole run after the initial phase for $\beta = 0$ and $\beta = 1$.

7 Conclusion

We have shown that the relative pheromone evaluation method can be used successfully for solving permutation problems with ACO algorithms. In particular the performance of an ACO algorithm for the Single Machine Total Earliness with Multiple Due Dates problem (SMTEMDP) was studied. The results show clearly that *relative pheromone evaluation* outperforms the standard method (and what is expected for this type of problem also the summation evaluation method).

References

1. M. Dorigo. *Optimization, Learning and Natural Algorithms* (in Italian). PhD thesis, Dipartimento di Elettronica , Politecnico di Milano, Italy, 1992. pp. 140.
2. M. Dorigo, V. Maniezzo, and A. Colorni. The ant system: Optimization by a colony of cooperating agents. *IEEE Trans. Systems, Man, and Cybernetics – Part B*, 26:29–41, 1996.
3. M. Dorigo and G. Di Caro. The ant colony optimization meta-heuristic. In D. Corne, M. Dorigo, and F. Glover, editors, *New Ideas in Optimization*, pages 11–32. McGraw-Hill, 1999.
4. R. Michels and M. Middendorf. An ant algorithm for the shortest common supersequence problem. In D. Corne, M. Dorigo, and F. Glover, editors, *New Ideas in Optimization*, pages 51–61. McGraw-Hill, 1999.

5. D. Merkle and M. Middendorf. An ant algorithm with a new pheromone evaluation rule for total tardiness problems. In *Proceedings of the EvoWorkshops 2000*, number 1803 in Lecture Notes in Computer Science, pages 287–296. Springer Verlag, 2000.
6. T. Teich, M. Fischer, A. Vogel, and J. Fischer. A new ant colony algorithm for the job shop scheduling problem. In *Proceedings of the Genetic and Evolutionary Computation Conference (GECCO-2001)*, page 803, 2001.
7. C. Rajendran and H. Ziegler. Ant-colony algorithms for flowshop scheduling. *European Journal of Operational Research*, 2001. submitted.
8. D. Merkle, M. Middendorf, and H. Schmeck. Ant colony optimization for resource-constrained project scheduling. *IEEE Transactions on Evolutionary Computation*, 2000. to appear.
9. M. Dorigo A. Roli, C. Blum. Aco for maximal constraint satisfaction problems. In *Proceedings 4th Metaheuristics International Conference*, pages 187–191, 2001.
10. D. Merkle and M. Middendorf. On the behaviour of ant algorithms: Studies on simple problems. In *Proceedings 4th Metaheuristics International Conference*, pages 573–577, 2001.

Improving Street Based Routing
Using Building Block Mutations

Neil Urquhart, Peter Ross, Ben Paechter, and Kenneth Chisholm

School Of Computing
Napier University
10 Colinton Road
Edinburgh, Scotland
Eh10 5DT
n.urquhart@napier.ac.uk

Abstract. Street based routing (SBR) is a real-world inspired routing problem that builds routes within an urban area for mail deliveries. The authors have previously attempted to solve this problem using an Evolutionary Algorithm (EA). In this paper the authors examine a heuristic mutation based on concept of building blocks. In this case a building block is defined as a group of genes, which when placed together within a genotype result in a useful feature within the phenotype. After evaluation on three test data sets our experiments conclude that the explicit use of heuristic building blocks makes a significant improvement to the SBR algorithms results.

1. Introduction

1.1 Introduction to Street Based Routing

The authors previously introduced the concept of Street Based Routing (SBR) in [2]. A great deal prior research into routing concentrated on solving the Traveling Salesman Problem (TSP) [9][13][14], the arc routing inspired problems such as Chinese Postman Problem, the Rural Postman Problem [3][5][6] and garbage collection [1]. More complex problems involving capacity constraints and multiple routes include [15][16]. An example of a real-world problem may be found in [4].

If the pattern of deliveries to households within an urban area is examined, it will become immediately apparent that significant adjacent clusters of delivery points occur as they are grouped into streets. It follows that all delivery points within a street will normally be serviced in sequence, before moving on to the next street. Some considerable advantage may be drawn from grouping delivery points into street sections and solving the problem by ordering street sections rather than individual delivery points. For practical purposes a street section is defined as all the delivery points on one side of a street between two junctions. Thus most named streets within a town are divided into several street sections.

Within each street section, one of three pre-defined delivery patterns may be applied to obtain the order in which the delivery points within the section are to be

S. Cagnoni et al. (Eds.): EvoWorkshops 2002, LNCS 2279, pp. 334–341, 2002.

serviced. Within each street section the possible patterns of delivery have been identified, as:

1. Deliver to all the households on one side of the street, then cross over and deliver to the opposite side, ending up at the start point.
2. Traverse the street from end to end, delivering to both sides, crossing over as required, finishing at the opposite end of the street from the starting point.
3. Deliver to all of the households on one side, then deliver to the opposite side at a later stage in the route.

1.2 The Building Block EA

The details of the Evolutionary Algorithm used to implement the SBR system were discussed by the authors in [2]. The algorithm used employed a steady state population of size 100 was utilized incorporating elitism and tournament selection and replacement. Each individual genotype consists of a list of street sections that are incorporated into a route based on the SBR delivery patterns outlined in section 1.1. incorporated in the route. Having built a route based on the above methods, the fitness value is the length of the route under consideration. Within each generational cycle 50 child individuals are created, each child may be created by copying a randomly selected genotype from the main population or by recombining two parent genotypes selected, using a tournament of size 2, from the main population. The probability of a child being created by recombination is 0.3.

The mutation operator randomly selects a sub-string of genes from within the chromosome and moves this sub-string to another point selected at random within the gene. The recombination operator is based on uniform crossover, but makes use of a global precedence operator; in this case it is the total length of the route. When two parents have been selected, the first gene from one parent is copied to the child. Subsequent genes are copied from either parent, based on selection of the gene that results in the smallest increase in walk length. In some cases the choice will be restricted by the requirement for the child chromosome to contain a valid permutation.

1.3 The Building Block Concept

Previous research into building blocks has been undertaken by Gero et al. in [7] and [8]. Gero et al. divide the population into two groups, those whose fitness is above a threshold and those whose fitness is beneath this threshold. They then attempt to distinguish which genetic feature or features are prevalent in the high fitness group and absent in the low fitness group. This feature may be isolated as a specific string of genetic material. Subsequently two sets of genotypes are created randomly one with and one without the string in question. If the average fitness of the chromosomes with the string is greater than the fitness of those without then the string is deemed to be useful. Useful strings identified in this manner are encapsulated into a single gene, preventing the modification of the string by mutation or recombination.

The methods used by Gero et al. were not felt to be entirely suitable to the SBR problem for two reasons. Firstly the encapsulation may force convergence on a local optimum, a sequence of streets is 'locked' in an encapsulated gene, cannot later have another street inserted within it. Secondly the string analysis techniques used to discover the genes are computationally expensive.

In authors of [11] investigate a genetic encoding for the Royal Road functions. The encoding described uses a biologically inspired solution incorporating extra genetic material within the encoding, to locate building blocks and assign them an identifier. The use of building blocks within Genetic Programming (GP) is examined in [12], the authors are able to make use of the subroutines evolved as part of the GP solution and useful subroutines are regarded as building blocks.

The representation used within SBR is a permutation of integer numbers each number referencing a specific street section. An SBR genotype might be something like this:

| 34 | 56 | 22 | 34 | 67 | 3 | 5 | 28 | ... |

A building block as proposed here consists of two genes e.g.:

| 22 | 34 |

Within this paper we initially only examine contiguous blocks of length two, and later of length three and four. It may be possible to look for larger blocks, but that would rapidly become computationally expensive. It is proposed that locating such building blocks and then duplicating them into other members of the population may improve the performance of the EA under consideration.

2. Locating Building Blocks

2.1 Frequency Based Building Blocks

The authors have examined two methods of discovering building blocks; the first method notes the frequency with which genes are placed next to one another within the population. It is then used to identify building blocks as follows: assuming there exists a genome that consists of the genes A, B, C, D, E and F arranged in a permutation. At the end of each generation the chromosome may be analysed to record the neighbour of each gene in a matrix. Suppose the Genome 'ABCDEF' is analysed this would result in the pairs AB, BC, CD, DE, and EF. These would be recorded in the matrix as per table 1. If one subsequently added the genome 'BCEFAD' the matrix would appear as per table 2. After the GA has run for many generational cycles the matrix would end up resembling that shown in table 3.

Table 1. An ajacency matrix for the genome ABCDEF.

	A	B	C	D	E	F
A		1				
B			1			
C				1		
D					1	
E						1
F						

Table 2. The matrix shown in table 1 after the genome BCEFAD has been added.

	A	B	C	D	E	F
A		1		1		
B			2			
C		1		1	1	
D					1	
E						2
F	1					

By scanning each row and finding the column with the maximum value we can establish which pairs of genes that have occurred most often within the population. Based on table 3 these pairs would be (note that we might read these pairs from right to left as well as left to right as ABCD will result in a different route to ACBD):

Table 3. An example of a matrix after many generations of the GA have been executed.

AC, BD, CB, DC, ED, FA

	A	B	C	D	E	F
A		15	22	10	13	15
B	13		22	24	6	15
C	23	43		13	29	14
D	33	23	34		12	24
E	19	33	43	62		21
F	62	33	51	17	37	

The hypothesis is that by 'encouraging' the use of these building blocks will result in the EA creating 'better' quality individuals. These blocks may be termed 'frequency' based building blocks.

Using one of our data sets, which we will call "The Grange" data set, we repeat the process outlined in above to attempt to identify building blocks after 250, 500 and 750 evaluations (within this EA we create 50 'children' per generation, thus incurring 50 evaluations per generation). Fig 1 outlines the percentage of building blocks that appear in the 'best' individual for each generation.

Using building blocks identified at generation 100 80 percent of the pairs identified are still present in the final solution at generation 2614. Obtaining building blocks in generation 100 appears to be more accurate that those obtained in generation 50. Only 65 percent of those building blocks identified in generation 50 were identified at generation 100, hence the difference in the results obtained. The experiment was repeated for blocks identified at generation 150, 89 percent of the building blocks obtained in generation 100 were still present in generation 150 leading to similar results for 100 and 150.

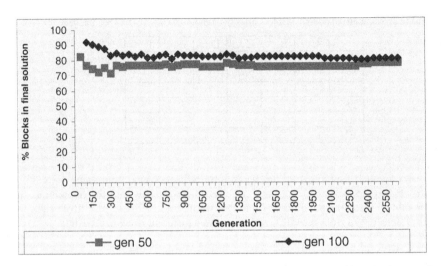

Fig. 1. Percentage of building blocks discovered in generations 50 and 100 and subsequently present in the best individual of each generation during the remainder of the run.

2.2 Local Fitness Based Building Blocks

The second method of discovering building blocks is based on "local fitness". Because the fitness value of each individual is simply the length of the route, it is possible to calculate during the fitness function what distance each gene contributes to the overall fitness.

This contribution may be termed the local fitness and is calculated as the distance from the previous street section to the current street section plus the distance traveled to the next street section. It will be recalled from our discussion in Section 1 that the SBR representation assigns a gene for each side of the street section, and if they are adjacent in the chromosome then it treats them as a single delivery operation covering both sides of the street. When this occurs the local fitness value covers both genes.

After a set number of evaluations the local fitness may be calculated for all the genes in the population. For each street section a building block may be constructed by finding the instance of that gene with the lowest local fitness. The building blocks then consist of the gene plus its adjacent neighbors. Unlike the frequency based building blocks which always have just two genes the local fitness based building blocks have 3 or 4 genes per block. Four gene building blocks occur where the lowest local fitness is allocated to an instance of two adjacent building blocks.

3. Building Block Mutation

Whether using the frequency or local-fitness based methods of discovering building blocks, the same technique may be utilised to make use of them. In each case a list of building blocks is constructed using the appropriate method. A new operator "heuristic building-block mutation" is utilised. Building block mutation, selects an individual from the population, then randomly selects a building block from the current list. The first gene of the building block is then located in the chromosome and the building block is then inserted into the chromosome starting from that point. If the inserted genes are duplicated within the chromosome, the originals are deleted to ensure the modified chromosome represents a legal solution to the problem. Finally the list of building blocks is cleared and new building blocks added in each generation.

4 Experimental Results

4.1 Experimental Technique

Both methods were tested over three different problems entitled "Edin", "Grange" and "Fairmilehead". Each data set represents an area within the City Of Edinburgh with 417, 609 and 1592 delivery points respectively. The "Edin" data set was previously used in [2]. Each run was carried out 20 times. Three versions of the algorithm were tested on each dataset, a version incorporating the frequency heuristic, a version incorporating the local fitness heuristic and as a control the standard unmodified algorithm. In each case the pseudo random number generator was seeded with the run number (i.e. an integer in the range 1 to 20). Seeded runs provide a better base for comparison rather than runs based on a purely random seed.

4.2 Results

In total 60 runs of the GA were carried out in accordance with the procedures described in 4.1 . The average route lengths for each data set may be seen in table 4. In each case the local fitness based heuristic gives a consistently shorter route length than the standard SBR-EA that does not incorporate building block mutations.

The average improvement achieved by using the local fitness based heuristic are 6.4 percent for the Edin data set, 3.2 percent for the Grange data set and 4.9 percentage for the Fairmilehead data set. Because these improvements are relatively small, a Students T-test has been used to give an indication as to the probabilities that this improvement is statistically significant. The t-test results are shown in table 5.

Table 4. Average route length for each data set (Averaged over 20 runs of each algorithm).

	Standard	Frequency based building blocks	Local Fitness based building blocks
Edin	1626.068	1543.214	1522.608
Grange	9442.2279	9242.3148	9136.2614
FMH	24777.953	23711.716	23574.759

Table 5. T-Test values comparing the results obtained using the building block heuristics with those obtained using the standard SBR algorithm.

	Edin Data	Grange Data	Fairmilehead Data
Frequency	0.999	0.979	1
Local fitness	0.999	0.999	0.999

The t-test results shown in table 5 represent the confidence factor that decrease in route length is due to the results obtained with the building blocks being part of a different distribution. For instance there is a probability of 0.979 that the results obtained with the 20 runs on the Grange data set with the building block heuristic improved are due to a specific change, rather than just being part of the normal distribution of results obtained when running the SBR-EA.

The best results were obtained on the Edin and Grange data sets. These data sets are based on a street topology that is largely random. The Grange data set is based on a street topology that is largely based on a grid scheme. The reason for the success on the data sets based random topology may be that the building blocks represent route sections of route covering streets that are difficult for the EA to place within the route.

5. Conclusions and Future Work

A small but significant improvement in the SBR-GA performance has been noted with the addition of the building block mutation operator. Of the two methods of obtaining building blocks the local fitness based method gives the best results. One of the major differences between the two methods is that the frequency method defines building blocks based not only on the data contained in the current population but also on previous generations. If the GA encounters a local optimum, and genetic diversity is reduced, the existence of similar chromosomes may influence the statistics used define the building blocks.

Unlike previous work [8] and [9] where string analysis is used to define the building blocks, the local fitness method highlights 3 or 4 gene building blocks using a simple modification to the fitness function. The additional computational complexity involved in this operation is minimal.

The concept of defining building blocks and storing them in a separate data structure may be used with other GA representations. There may also be other heuristics that can be used to define building blocks.

Future research will concentrate on monitoring the increase in fitness noted by including a building block within a chromosome. Each building block may be given a 'score' value. Those blocks that produce the greatest improvement in fitness will have their score value increased accordingly. Building blocks that acquire a high score will more likely to be used. The highest scoring building blocks will be retained from one generation to the next. To prevent convergence on a local optimum blocks will have a finite life, requiring their deletion after a number of generations. The SBR application uses a permutation-based representation; future research may include the use of the operators described in this paper on other permutation-based problems. It is anticipated that the length of the permutation will have an effect on performance.

References

1. Bousonville, T. Local Search and Evolutionary Computation for Arc Routing in Garbage Collection. *Proceedings of the Genetic and Evolutionary Computation Conference 2001. Eds L Spector, E Goodman, A Wu, W B Langdon, H M Voigt, M Gen, S Sen, M Dorigo, S Pezeshk, M Garzon E Burke*. Morgan Kaufman Publishers 2001.
2. Urquhart N, Paechter B, Chisholm K. Street-based Routing Using an Evolutionary Algorithm. *Proceedings of EvoWorkshops 2001, Como, Italy Eds, E.J.W. Boers et al.* Springer-Verlag 2001.
3. Lacomme P, Prins C, Ramdane-Cherif W. A Genetic Algorithm for the capacitated Arc routing problem and its extensions. *Proceedings of EvoWorkshops 2001, Como, Italy Eds, E.J.W. Boers et al.* Springer-Verlag 2001.
4. Hart E, Ross P, Nelson J. Scheduling Chicken Catching – An Investigation Into The Success Of A Genetic Algorithm On A Real World Scheduling Problem. *Annals Of Operations Research 92 Baltzer Science Publishers 1999.*
5. Balaji R and Jeyakesavan V A 3/2-Approximation Algorithm for the Mixed Postman Problem. *SIAM Journal on Discrete Mathematics Vol 12 No 4, 1999*
6. Kang, M and Han, C. Solving the rural postman problem using a genetic algorithm with a graph transformation. *Proceedings of the 1998 ACM symposium on Applied Computing.* ACM Press New York 1998.
7. Gero, J S and Krazakov, V. Evolving design genes in space layout problems. *Artificial Intelligence in Engineering 12(3) pp 193-176.*
8. Gero, J. S., Kazakov, V. and Schnier, T. Genetic engineering and design problems. *in D. Dasgupta and Z. Michalewicz (eds), Evolutionary Algorithms in Engineering Applications, pp.47-68.* Springer-Verlag, Berlin 1997
9. Freisleben B. and Merz P. New Genetic Local Search Operators for the Traveling Salesman Problem. Parallel Problem Solving from Nature - *PPSN IV Eds: Hans-Michael Voigt, Werner Ebeling Ingo Rechenberg, Hans-Paul Schwefel* Springer-Verlag 1996..
10. Michalewicz Z. Genetic Algorithms + Data Structures = Evolution Programs (Third, Revised and Extended Edition). Springer-Verlag 1996.
11. Wu A S , Lindsay R K. A comparison of the fixed and floating building block representation in the genetic algorithm. *Evolutionary Computation Vol4, No 2 pp 169-193.* MIT Press 1996.
12. Rosca J. Towards automatic discovery of building blocks in genetic programming. *Working notes for the AAAI Symposium on Genetic Programming pp 78-85.* 1995.
13. Tamaki H, Kita H, Shimizu N, Maekawa K, Nishikawa Y. A Comparison Study of Genetic Codings for the Travelling Salesman Problem. *Proceedings of the First IEEE Conference on Evolutionary Computionary Computation* 1994.
14. Bui T, Moon B. A new Genetic Approach for the Traveling Salesman Problem. *Proceedings of the First IEEE Conference on Evolutionary Computation* 1994.
15. Thangiah S, Vinayagamoorthy R, Gubbi A. Vehicle Routing with Time Deadlines using Genetic and Local Algorithms. *Proceedings of the Fifth International Conference on Genetic Algorithms* Forrest S Ed. Morgan Kaufmann 1993.
16. Blanton, J.L. Jr. and Wainwright, R.L. Multiple Vehicle Routing with Time and Capacity Constraints using Genetic Algorithms.
    Proceedings of the Fifth International Conference on Genetic Algorithms Forrest S Ed. Morgan Kaufmann, 1993.

Author Index

Lecture Notes in Computer Science

For information about Vols. 1–2221
please contact your bookseller or Springer-Verlag